MW00838056

An Introduction to the Environmental Physics of Soil, Water and Watersheds

This introductory textbook describes the nature of the earth's environment and its physical processes so as to highlight environmental concerns arising from human use and misuse of soil and water resources. The book provides the basic knowledge required to understand how land and water resources can be used sustainably through action at the scale of watersheds or catchments. Whilst the book provides an understanding of physical and some chemical environmental processes, no prior knowledge of the physical sciences or calculus is assumed.

Topics considered include the development of soil at the earth's surface and the environmentally significant issues that arise from its interactions with water, salt and other contaminants. This is followed by a description of how water moves over the earth's surface, enters beneath it, and moves within the soil. The factors governing the exchange of water and energy between the earth's surface and the atmosphere are considered and the main causes of soil and water degradation, both on land and in rivers, are discussed in the context of what can be done to mitigate such environmental effects.

The author provides a thorough introduction to the basic issues regarding the sustainable, productive use of land resources that is vital in maintaining healthy rivers and good groundwater qualities. The book is intended to inform and develop the capacity to take a quantitative approach to studying these growing environmental concerns involving land and water quality. The straightforward writing style, lack of prerequisite knowledge and copious illustrations make this textbook suitable for introductory university courses, as well as being a useful primer for research and management staff in environmental and resource-management organisations. Each chapter ends with a set of student exercises for which solutions are available on the Internet at http://publishing.cambridge.org/resources/0521536790.

Following an initial career designing fighter jet planes in Australia, CALVIN ROSE made a major career change to spend ten years teaching physics and researching in Makerere College, the University of East Africa. Here he completed a Ph.D. degree with London University and wrote the innovative text *Agricultural Physics* (1966) that was widely used for many years. He subsequently took up a position with the Environmental Biology group of the CSIRO, Australia, where he embarked on research evaluating sustainable pastoral and agricultural land uses in tropical regions. In 1973 he was appointed inaugural Dean and Professor of Environmental Sciences at Griffith University, Queensland, where he is now an Emeritus Professor. Through his international research and teaching, Professor Rose has promoted the role of physical processes in seeking solutions to invariably complex environmental and sustainability issues, and has helped to establish the sub-discipline of environmental physics. He is a Fellow of the Australian Academy of Technological Sciences and Engineering and the Australian Institute of Agricultural Sciences and Technology, a life member of the Australian Society of Soil Science and a board member of the International Soil Conservation Organisation.

An Introduction to the Environmental Physics of Soil, Water and Watersheds

Calvin W Rose
Faculty of Environmental
Sciences, Griffith University

CAMBRIDGE
UNIVERSITY PRESS

CAMBRIDGE UNIVERSITY PRESS
Cambridge, New York, Melbourne, Madrid, Cape Town, Singapore,
São Paulo, Delhi, Dubai, Tokyo, Mexico City

Cambridge University Press
The Edinburgh Building, Cambridge CB2 8RU, UK

Published in the United States of America by Cambridge University Press, New York

www.cambridge.org
Information on this title: www.cambridge.org/9780521536790

First published 2004
Reprinted with corrections 2006

A catalogue record for this publication is available from the British Library

Library of Congress Cataloguing in Publication Data
Rose, C. W.
An introduction to the environmental physics of soil, water, and watersheds/
by Calvin W. Rose.
    p. cm.
Includes bibliographical references.
ISBN 0 521 82994 1 – ISBN 0 521 53679 0 (paperback)
1. Soil physics. 2. Soils – Environmental aspects. 3. Soil moisture.
4. Groundwater – Environmental aspects. I. Title.
S592.3.R67 2004
631.4'3 – dc22   2003055754

ISBN 978-0-521-82994-6 Hardback
ISBN 978-0-521-53679-0 Paperback

Contents

Preface	*page* xi

1 **Environmental systems of rock, soil and earth
energy exchanges**   1

1.1	The planet earth	1
1.2	The rock cycle	2
1.3	Soils, the interface of earth environments	5
1.4	Processes of soil formation	8
1.5	Energy- and water-exchange systems at the earth/atmosphere interface	18
1.6	How things are measured – units and conversions	29
Main symbols for Chapter 1		42
Exercises		42
References and bibliography		44

2 **Soil and soil strength**   46

2.1	Introduction	46
2.2	Introductory mechanics	48
2.3	Physical characteristics of soil	51
2.4	Strength and behaviour of sediments	59
2.5	Stress, strength and strain	63
2.6	Soil strength	67
2.7	Effects of water content on soil strength	72
2.8	The consistency of soil	76
2.9	Some environmental implications of the mechanical characteristics of soil	79
Main symbols for Chapter 2		80
Exercises		80
References and bibliography		83

3	**The behaviour of liquids**	**84**
3.1	Introduction	84
3.2	The environmental significance of liquids	85
3.3	Fluid pressure and buoyancy	86
3.4	Liquids in motion	90
3.5	Energy and fluid flow	99
3.6	Flow around submerged solids	107
3.7	Oceans and waves	110
3.8	An introduction to ocean-wave–coastline interactions	116
	Main symbols for Chapter 3	120
	Exercises	121
	References and bibliography	124
4	**Soil, water and watersheds**	**125**
4.1	Introduction	125
4.2	Precipitation and runoff measurement	131
4.3	Soil water content and its profile storage	135
4.4	The water budget for a watershed using the principle of mass conservation of water	137
4.5	Water-balance accounting	141
	Main symbols for Chapter 4	149
	Exercises	149
	References and bibliography	152
5	**Evapotranspiration and exchange of energy at the earth's surface**	**153**
5.1	An introduction to vegetation-based ecosystems	153
5.2	Atmospheric humidity	157
5.3	Evaporation from an open-water surface	161
5.4	Measurement of evapotranspiration using the principle of conservation of energy	162
5.5	Determination of evapotranspiration from vegetated surfaces using standard meteorological data	175
5.6	Non-radiative sensible heat exchange between the land surface and the lower atmosphere	182
5.7	Ground heat flux and soil temperature	185
	Main symbols for Chapter 5	189
	Exercises	191
	References and bibliography	195

6	**Infiltration at the field scale**	197
6.1	Introduction: from rainfall to the sea	197
6.2	Infiltration into small areas of soil	199
6.3	Spatial variability in infiltration rate	207
6.4	A case study of infiltration rate at the plot scale with infiltration-excess overland flow	213
6.5	A general model of spatially variable infiltration at the field-plot scale	216
	Main symbols for Chapter 6	222
	Exercises	222
	References and bibliography	225

7	**Overland flow on watersheds**	226
7.1	Introduction	226
7.2	Runoff-model components	228
7.3	A small-scale runoff model neglecting hydrological lag	229
7.4	Overland flow of water	234
7.5	The dynamics of overland flow	242
7.6	A dynamic runoff model recognising hydraulic lag	249
	Main symbols for Chapter 7	254
	Exercises	255
	References and bibliography	257

8	**Erosion and deposition by water**	259
8.1	Introduction	259
8.2	An overview of soil-erosion and deposition processes	261
8.3	Deposition characteristics of sediment	266
8.4	Soil erosion as a rate process	270
8.5	Erosion theory and soil-erodibility characteristics of sediment	273
8.6	Effective averaging over a soil-erosion event	278
	Main symbols for Chapter 8	285
	Exercises	286
	References and bibliography	287

9   Watersheds and rivers   289

9.1   Introduction   290
9.2   Hydrological considerations at watershed and
       catchment scale   292
9.3   Sediment transport in watersheds   300
9.4   The transport of nutrients and other chemicals
       in watersheds   307
9.5   Rivers   311
9.6   River health   316
Main symbols for Chapter 9   320
Exercises   320
References and bibliography   322

**10   Movement of water through the
       groundwater zone**   324

10.1   Introduction   324
10.2   Groundwater at equilibrium   327
10.3   Movement of groundwater   330
10.4   Groundwater in natural subsurface formations   340
Main symbols for Chapter 10   342
Exercises   343
References and bibliography   348

**11   Movement of water through the
       unsaturated zone**   350

11.1   Introduction   350
11.2   The capillary fringe – a common start to the
       unsaturated zone   352
11.3   Water and solids in equilibrium   356
11.4   Movement of water in the unsaturated zone   364
Main symbols for Chapter 11   376
Exercises   377
References and bibliography   378

12   Salinity and contaminant transport   380

12.1   Introduction   381
12.2   Dryland salinity processes   383
12.3   Irrigation-induced salinity processes   388
12.4   Movement of contaminants in groundwater   391
12.5   Contaminant transport in the unsaturated zone   397

12.6   Solute and contaminant transport in
           agricultural contexts   404
Main symbols for Chapter 12   414
Exercises   415
References and bibliography   417
Appendix   419
Answers to all exercises   421
Index   434

Preface

This book is intended to provide the basic physical knowledge required to understand the processes involved in the sustainable use of the earth's land and water resources. Description of the physical science of soil and water processes is carried through to application at the watershed scale. Consideration of processes at this scale is necessary since this is the scale at which land-management decisions begin to be made and at which activities with environmental and water-quality implications occur.

The book is introductory in the sense that no prior knowledge of physics or calculus is assumed. Arithmetic and elementary algebra are used. No experience of computer-spreadsheet use is required of the reader, though the utility of such aids to calculation is illustrated on a few occasions. Though elementary in this sense, in some issues consideration is given to ideas at the frontier of research and understanding.

How theory can be applied to field data is illustrated using many examples. Exercises that further illustrate the application of theory are given at the end of each chapter. Answers to all exercises are given in the book, and fully worked solutions are available on the world-wide web at http://publishing.cambridge.org/resources/0521536790.

The methodologies of environmental physics and engineering dominantly employed in this book can provide only one link in the chain of activity necessary to deal effectively with typically complex environmental issues. Other links or areas of knowledge required to approach the environmental issues addressed are at least hinted at. This book illustrates how a physical viewpoint plays a vital role in the increasingly urgent task of understanding, ameliorating, avoiding or transforming the undesirable consequences of human action which can threaten the quality and perhaps the sustainability of future life on our planet.

It is a great pleasure to acknowledge the inspiration of students and colleagues from all over the world with whom I have worked. Gratitude is particularly due to colleagues in Australia, South-east Asia and the USA. Within Australia, teaching and research with colleagues in the Faculty of Environmental Sciences of Griffith University and joint research with colleagues in the Commonwealth Scientific and Industrial

Research Organisation and in the Queensland Department of Primary Industries have been particularly valuable in expanding the experience on which this book has drawn. Whilst persons' names are far too many to mention, I would like to give particular thanks to Mr Cyril Ciesiolka, Drs Bofu Yu and Kep Coughlan, Professors Bill Hogarth and J.-Yves Parlange and Cambridge University Press referees.

1

Environmental systems of rock, soil and earth energy exchanges

1.1 The planet earth

If we were in a space craft looking back at earth we would see something like the view shown in the picture above. That view has affected how we regard our planet earth, our lifeboat in the ocean of space. How thin the layer of atmosphere appears to be compared with the earth's diameter might surprise us, as might just how much of the earth's surface is covered by oceans and how perfectly spherical the solid earth seems to be, despite our awareness of the majestic mountains and deep canyons that are so impressive to the earth's land dweller.

Yet the earth's land surface is not quite as solid as we imagine. Despite being a rocky planet 12 700 km in diameter, the occasional emergence of molten magma from active volcanoes betrays the presence beneath the earth's surface of enormously hot material. The impression that the land surface is a strong and rather rigid framework is correct, but only for the outer shell of the earth, the 'lithosphere', which is a bit like the thin hard crust on bread, or the skin on an apple.

Geological research has shown that the lithosphere behaves like a collection of enormous coherently moving 'plates', or a slowly distorting three-dimensional jigsaw puzzle, with the intermeshing pieces carrying continents and parts of adjacent oceans. These continental-sized plates some 100 km thick are in continuous, imperceptibly slow, but inexorable motion, powered by thermally driven motion of the earth's 'mantle' beneath the lithosphere. Thus continents are very slowly drifting and altering their positions with respect to each other. As these plates try to move past or over one other, friction prevents easy sliding and, when this frictional resistance is suddenly overcome, an earthquake results. The San Andreas fault through San Francisco is a well-known example of such a transverse fracture zone.

As the edges of plates push into each other, mountain ranges can form from the folding and buckling of the plates. For example, the high Himalayan mountains are still being formed by this buckling process as the plate that carries India and Australia pushes up into the plate carrying the Asian and European land masses.

When a plate with a continent at its margin moves over an oceanic crust, the continental plate tends to ride up over the oceanic crust, which is then pushed down beneath the continent or is 'subducted'. A deep trench in the ocean floor develops offshore over the subduction zone, where heating augmented by friction and compression can melt rock at depth, and this can emerge as a chain of volcanoes.

The movement of these plates is very slow, only centimetres per year, and thus the time scales in continental drift are enormous, allowing living things time to adjust and evolve to cope with the slow climatic changes due to drift and other causes. The drifting apart of what we know today as separate continents is believed to have commenced some 600 million years ago, with the commencement of life forms on earth much earlier still.

1.2 The rock cycle

The lithosphere consists dominantly of rock-like material, which provides the major focus of the science of geology. Study of the earth's crust and its rock materials has its own fascination, and the resulting information and insights have many applications, such as in prospecting for minerals and in identifying possible sites for the extraction of geothermal energy.

In this text a major interest in rock material is that it provides the original raw material from which soils are formed. Furthermore, as will be developed in a number of later chapters in this book, the nature and properties of soils are intimately involved in environmental concerns about the sustainability of a productive soil resource in the face of land-management practices that degrade some important characteristics of the soil. Also the nature and properties of soils are greatly affected by the properties of the rock material from which the soil is derived. Thus, prior to considering the processes involved in the formation of soils, we will very briefly review why it is that rocks of very different characteristics can be found, sometimes exposed on the soil surface, but often buried beneath the layer of soil formed by breakdown of the rock material.

Most rock in the earth's crust is 'igneous' in origin, which means that its origin is solidified molten magma. The volatile components of such magma have also supplied many of the components of the atmosphere, except for the vital component, oxygen, which is released by the 'photosynthetic' process exhibited by many bacteria and in plants and trees. By releasing oxygen when absorbing carbon dioxide in the presence of sunlight, this process not only constitutes part of life itself, but also provides the basic requirement for other life forms, including human life.

When they are exposed to solar radiation and to the motion of water and wind, which is ultimately driven by the same source of energy, the igneous rocks undergo processes of physical disintegration and chemical decomposition referred to as 'weathering'. The products of weathering may just fall away from the parent rock under gravity, a process called 'mass wasting'. Weathered rock material also can be transported away by wind or flowing water, and, when it settles out or becomes deposited, this can form extensive layered deposits. These deposits may become buried over time, by other deposits, by lakes that may form, or by the ocean if the deposit undergoes subsidence or the sea level rises. The increase in pressure on sedimentary deposits following such burial leads to compaction and hardening, a process termed 'lithification'. The end product is called 'sedimentary rock'. This type of rock is one of the components of what is called the cycle of rock transformation, illustrated in Fig. 1.1.

Whenever you see an exposed cliff or road cutting with distinguishably different layers of rock it is likely that this is sedimentary rock. Though the original layers would have been almost horizontal sheets, this need no longer be so, due to uplifting of the crust by the impact of tectonic plates, or other deforming processes. Exposure of the sedimentary

Figure 1.1 Illustrating the cycle of rock transformation. Descriptions of material within solid boundaries represent different types of rock material. Process descriptions are enclosed by dashed boundaries. Sources of energy are enclosed by both solid and dashed boundaries. (Adapted from Strahler and Strahler (1973).)

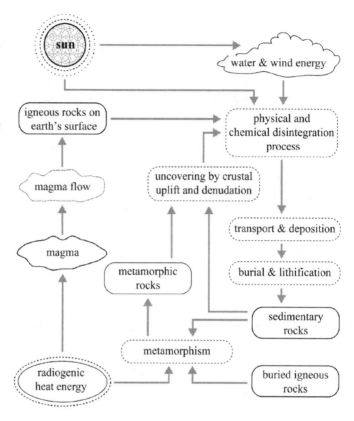

rock at the earth's surface may have been caused by crustal uplift or by denudation of surrounding more erodible material (Fig. 1.1).

Figure 1.2 shows an example from Bryce Canyon in southern Utah, where sedimentary material formed under a freshwater lake has now become exposed and dramatically eroded in the currently rather arid climate of that state.

Buried igneous or sedimentary rock can be subject to enormous pressures and to 'radiogenic' heat, the earth's internal energy source whose origin is the radioactive decay of minerals such as uranium. At extreme pressures and temperatures the prior rock material can be kneaded like dough and become so transformed that the original characteristics of the rock are almost lost and a new type of rock material is formed. This third major class of rock is referred to as 'metamorphic rock' (Fig. 1.1). Like sedimentary rock, metamorphic rock can be greatly raised in height by general uplift or buckling of the lithospheric plates. This uplift, perhaps accompanied by natural erosion processes, can be the reason for the surface exposure of the metamorphic rock. Alternatively, it may be

Figure 1.2 Natural erosion at Bryce Canyon, Utah, has exposed layers of sedimentary deposits that have somewhat different colours. Dramatic erosion, which is characteristic of arid regions, has left standing flat segments of the previously deposited almost-horizontal layers of sediment.

erosion processes alone that lead to the exposure of metamorphic rock over geological time (Fig. 1.1).

As implied by Fig. 1.1, all types of rock are initially derived from igneous rocks. However, outcrops of igneous rocks currently occupy less than 30% of the earth's land surface, with sedimentary rocks being dominant, of which sandstone, shales and limestones are the most important types. Of the outcropping igneous rocks, the two dominant types are granite and basalt.

1.3 Soils, the interface of earth environments

The disintegration of rock appears to have been involved in very early steps in the formation of life on earth. There is evidence that a very early life form on this planet was associated with the emergence of 'stromatilites', rock-like structures that incorporate blue–green algae (or, more properly, cyanobacteria). These bacteria are capable of photosynthesis, the process whereby atmospheric carbon dioxide is incorporated into more complex building materials in the presence of sunlight. Importantly, photosynthesis released oxygen into the previously oxygen-free atmosphere of the earth. Fossil records suggest that, over three billion (3×10^9) years before the present, stromatilites were common on

Figure 1.3 Soils have developed and continue to change at the interface of the major earth environments. (After McTainsh and Boughton (1993).)

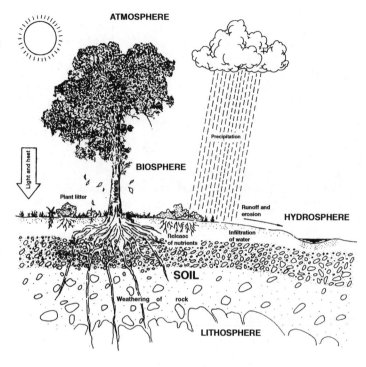

the earth's surface, thus contributing to a slow, but vital, build up in oxygen in the previously oxygen-free atmosphere.

Certainly the weathering of rocks is what provides the substrate from which soils can form, thus providing that vital thin mantle of mineral and organic material that blankets most of the earth's land surface.

Soils have developed at the outer edge of the lithosphere where it interfaces with the other major earth environments: the atmosphere, the biosphere and the hydrosphere (Fig. 1.3).

The formation of soil is intimately involved in the flows of energy and matter through the biosphere (Strahler and Strahler, 1973). The soil plays a vital role in the use and reuse, or 'cycling', of the range of chemical elements necessary to sustain the growth of vegetation. Both carbon and nitrogen, whilst very minor constituents of the earth as a whole, are major essential components of living matter. The soil plays a vital role in concentrating these two elements essential for plant growth in the 'biosphere', or life-bearing layer of the earth (Fig. 1.3).

Both products of rock weathering and the breakdown of vegetative litter are sources of the nutrients required for vegetation to grow. Even rainfall adds a very weak solution of salts whipped up by wind from the sea and the nitrate ion (NO_3^-) from lightning discharges in the atmosphere.

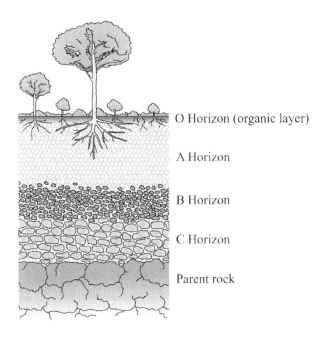

Figure 1.4 Possible characteristics of a developed soil profile.

O Horizon (organic layer)

A Horizon

B Horizon

C Horizon

Parent rock

Whilst Fig. 1.3 suggests that soil has formed *in situ* from rock weathering at the same site, this is not always so. 'Alluvial material', transported by flowing water, or 'aeolian material', transported by wind, are also important sources of sediment on which soil can form. Irrespective of whether soil is formed *in situ* from parent rock or on transported material, the form of the 'soil profile', called the 'soil morphology', provides some clues regarding its origin and the sequence of soil-formation processes which the science of 'pedology' seeks to interpret.

A road or rail cutting through a hill can expose such a soil profile, and the characteristics of such a profile are described by the nature of the sequential horizons (Fig. 1.4).

Factors of soil formation

In trying to understand the enormous variety in soil material on the earth's land surface, there is a long history of seeking some association between soil characteristics and what might be factors affecting them. The five factors traditionally investigated are lithospheric material, topography, biospheric factors, climate and time. There is no doubt about the importance of such factors, but the search for associations between these factors and soil type or characteristics, though sometimes fruitful in restricted geographical regions, has in general been rather frustrating at a global level. The reason for this frustration seems to be that the actual processes believed to be involved in soil formation are but loosely correlated with these five traditionally investigated factors. Also there is the strong likelihood that interaction between such factors is as important as the factors themselves.

Nevertheless, certain useful generalisations have emerged from this style of approach. For example, sedimentary rocks in general can be more readily broken down than igneous rocks, resulting in more rapid soil formation on the former type of rock. Also quite rapid changes in soil and vegetation type can be seen to accompany, and presumably result from, a corresponding change in rock material, such as from volcanic to sedimentary. However, relationships are not always so obvious or simple.

Transported sediment can arise from glacial action, from mass movement on steep slopes, or from erosion by water and wind, and soil characteristics are strongly affected by the type of transporting mechanisms involved. Topography is both an outcome of translocational mechanisms and a contributor to them, for example affecting the location of net erosion and net deposition in erosion by water. However, vegetation and other biotic characteristics affected by climate also strongly affect sediment transport. Just as there are very variable outcomes for soils developed *in situ* from igneous or metamorphic rocks, the variation being only partly explainable in terms of the traditionally defined five factors of soil formation or their interactions, so there is a corresponding variability in outcomes for soils formed on transported sediment.

The fifth listed soil-forming factor, time, is certainly important in itself in all change processes, but the other four factors also change with time. The great climatic and biotic change between the present and the last glacial period, some 15 000 years ago, is one obvious example.

In conclusion, whilst consideration of the five traditional factors of soil formation has been of some value, the complexity involved in soil formation and the strong and largely unknown interactions between such factors have led many investigators to consider alternative approaches to understanding soil formation. The main alternative involves seeking to ascertain the actual processes involved in the formation of soil material. A very brief outline of this style of approach is given in the next section.

1.4 Processes of soil formation

Figure 1.1 shows rock materials as emerging from their formation under conditions of high temperature and pressure to the surface of the lithosphere where both temperature and pressure are low, and where they become exposed to oxygen, liquid water and biotic influences. This emergence is rapid in volcanic eruptions, but in general can be very slow.

As rock material of any kind becomes exposed to the interfacial surface environment, a process of adjustment and change takes place. Processes of weathering commence with physical and chemical disintegration of the rock (Fig. 1.1).

The nature and characteristics of that original rock material significantly affect the rate at which weathering takes place and also the degree and types of disintegration, with consequences for the size distribution of ultimate particles in the soil which subsequently forms. This size distribution in turn affects soil characteristics such as its ability to transmit water through it and its ability to resist erosion by water or wind. The environmental significance of such soil characteristics will be discussed in later chapters of this book.

Weathering and clay formation

Sometimes the term 'weathering' is used in the restricted sense of referring solely to physical, chemical and biotically induced breakdown processes. Whilst some breakdown products, such as quartz, are rather stable and unreactive, many are not, and new or 'secondary minerals' are commonly formed. The term weathering is also used in a more inclusive way to describe this whole sequence of processes from physical and chemical disintegration of the primary minerals to the formation of quite new secondary minerals.

A simple picture of a soil profile is given in Fig. 1.5, indicating the locations of some of the dominant processes continuously at work during soil formation. Figures 1.3 and 1.5 both picture the case in which soil development has continued for a sufficient time and in a suitable situation for typical profile characteristics to be developed. In Fig. 1.5 rock weathering has evidently proceeded for a long time, and is currently occurring dominantly at the base of the soil profile.

Tree roots are able to take up chemicals released during weathering, transporting them up to the tree's leaves, which ultimately fall, returning to the earth's surface. Leaves, twigs and branches that accumulate on the soil surface provide a substrate or mulch very suitable for breakdown by biological activity, provided that some moisture is available. This biologically driven breakdown of plant material releases some of the chemicals previously taken up by the root systems of the vegetation. These nutrients released at the top of the soil profile can be carried into it by infiltrating water. At a rate depending on the solubility of the nutrient and the rate of entry of water, the nutrients will be moved down the soil profile in 'translocation' (Fig. 1.5), whereupon they become available for uptake again by the roots of growing vegetation.

This whole sequence of processes of nutrient uptake by vegetation, return of nutrients to the soil surface and subsequently their uptake again

Figure 1.5 Some of the major processes involved in soil formation.

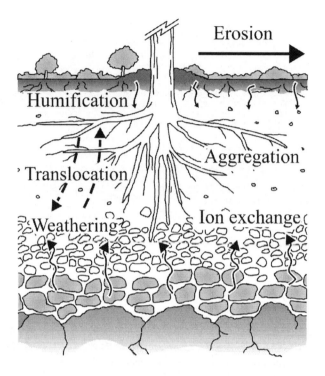

by the root system is referred to as 'nutrient cycling'. Texts such as Brady and Weil (1999) discuss in detail the nutrients required for plant growth and the cycles of these various plant-growth chemicals. Continuous harvesting and removal of vegetation disrupts this natural cycling of essential plant nutrients, thus leading to a depletion of the pool of plant-available nutrients. For the harvesting activity to be sustainable in the long term, this depletion must be recognised and some form of management activity that re-establishes the nutrient cycle must be instituted. Research in agriculture and forestry is directly involved in investigating options in the manner in which this can be done.

The secondary minerals formed by the alteration through time in the breakdown products of primary rock minerals differ very substantially in chemistry and structure from the original rock material. The formation of secondary minerals involves water as well as the products of weathering. The ability of products of weathering to move within the soil profile depends on the stability in water of these breakdown products. Ions such as sodium and calcium are stable and highly mobile, whereas products involving silicon and aluminium are much less free to move. Secondary minerals formed include clays, hydrated oxides of iron and aluminium, and salts.

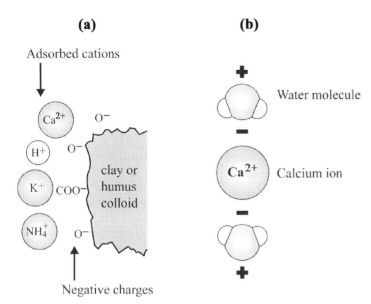

(a)

Adsorbed cations

Ca^{2+} O^-

H^+ O^-

K^+ COO$^-$ clay or humus colloid

NH_4^+ O^-

Negative charges

(b)

Water molecule

Ca^{2+} Calcium ion

Figure 1.6 (a) Illustrating cations attracted to net negative charges on humus and many clays. (b) The calcium cation surrounded by water molecules oriented to the radial electric field surrounding the cation.

Members of the family of secondary minerals called clays are particularly important, both in terms of their common significant amount and in terms of the substantially altered chemical and physical characteristics they impart to soil. Clay minerals are very finely divided, the fundamental particles of clay typically being smaller than two micrometres (2 μm or 2×10^{-6} m), and so are classed as 'colloids'. Whilst being relatively small, colloids are still large compared with the molecular scale. Clay particles commonly have a sheet-like structure, have an enormous surface area relative to their mass, and commonly (though not invariably) bear a negative surface charge. This surface charge often arises from replacement in the clay sheet of one ion by another with a different magnitude of charge, or different 'valency'. The resulting negative charge gives clays the capacity to attract soluble positively charged ions ('cations') such as calcium, an important plant nutrient (Fig. 1.6(a)).

Of the great variety in types of clay, many are plate-like in shape, and so are called 'layer-lattice clays'. These clays consist of sheets of oxygen (O^{2-}) and hydroxyl (OH^-) ions, with silicon and aluminium being common metallic ions. Layer-lattice clays are commonly distinguished by the ratio of the numbers of silica and alumina layers, this ratio being 1 : 1 in kaolinitic clays and 2 : 1 in other clays. Some of the 2 : 1-lattice clays expand on wetting and shrink on drying, making it difficult for tree establishment to occur and providing problems for the foundations of buildings. The small sheet-like structures in clays are lubricated by water molecules when the clays are moist, providing the typical plastic character of soils high in clay content.

The 2 : 1-lattice clays are generally indicative of lower intensities of weathering and leaching, higher intensities of these processes being associated with 1 : 1-lattice clays such as kaolinite. Even further or more intensive leaching and weathering leads to material dominated by the oxides of iron and aluminium.

If the electrical charge on clay is negative, this charge is neutralised by attracted cations, anions being repelled. Because of the 'polar' nature of water, ions are surrounded by a cloud of water molecules, which are preferentially oriented to the radial electric field surrounding the ion, as illustrated in Fig. 1.6(b) for the calcium cation. (Water is a polar liquid because there is a small distance between the centre of effect of its two positive hydrogen atoms and its oxygen atom.)

There is a cloud of various cations located in any water surrounding negative clay particles, as shown in Fig. 1.6(a). Each 'hydrated' cation is so small that it is in continuous thermal agitation, known as 'Brownian motion', which is caused by inequality at any instant in the bombardment by surrounding molecules. Thus any given cation in the cloud can readily be replaced by others, a process called 'cation exchange'. This is an important process in the formation of clays and other rock-decomposition products. It is also involved in plant nutrition. For example, plant roots can release hydrogen ions and in exchange take up important plant-nutrient cations adsorbed on clay colloids.

Figure 1.6(a) shows the cloud of positively charged cations of various 'valences' (or numbers of units of charge) surrounding the surface of a clay particle. Since this cloud of cations forms a diffuse layer surrounding the negatively charged surface, this cloud is referred to as a 'double layer' of charge. If monovalent cations such as H^+ are replaced by divalent cations, such as Ca^{2+}, then the diffuse layer becomes compressed. Since like electrical charges repel one another, the cation sheath surrounding clay particles tends to prevent the clay particles from bumping into each other during the erratic Brownian motion they continuously experience, so that there is a tendency for clay that is in a 'suspended' state to remain in this state. However, some conditions enhance the likelihood of clay particles coming into close proximity to each other. Under such conditions the electrostatic repulsive force between particles is overcome by a quite different type of force, which attracts clay particles that have been able to come into close proximity to each other. This 'short-range' attractive force arises from the attraction an atom of one clay particle has for an atom of another clay particle very close to it. For historical reasons the name van der Waals forces is given to these short-range attractive forces.

The balance between this short-range attractive force and the longer-range ionic repulsive forces between clay particles is tilted more strongly in favour of attraction at higher electrolyte concentrations, and, in particular, the higher the valency of the cation in the electrolyte. Divalent calcium (Ca^{2+}) is the most common cation in soil. However, if sodium (Na^+) is the dominant cation, or if it becomes so, perhaps due to irrigation with somewhat saline water, then clay particles can repel each other, with noticeable consequences for the soil as a whole. The result of net repulsion between clay particles leads to a 'dispersed' or 'deflocculated' condition in which the tendency of clay particles to attract one another to yield a 'flocculated' state is destroyed. Whilst these processes can be most readily observed in pure clay suspensions, the consequences of such processes for soil in the field are obvious in terms of resulting behaviour. For example, a fall of rain (with its very low electrolyte level) on saline clay soils can lead to clay dispersion, which blocks the soil pore space, thus seriously impeding infiltration of water. On drying, soil in this dispersed state forms a very hard and dense deposit, which is almost impossible for roots to penetrate and disperses on rewetting. The serious implications of this soil behaviour for plant growth and soil erosion are not difficult to imagine, and such land can become completely unproductive.

Humification

Dead plant and animal tissue is broken down by larger organisms called 'macro-organisms' (such as earthworms and termites) and also by 'micro-organisms' (such as bacteria and fungi). The end product of this complex series of breakdown processes is 'humus', another important soil colloid. Humus plays a vital role in storing ions and water and in improving the physical characteristics of soil for plant growth. Bearing a negative electrical charge, like many clays, humus attracts cations (Fig. 1.6(a)). During the humification process vital plant nutrients are released. Since humification is particularly active at the soil surface (Fig. 1.5), these released nutrients can move down with infiltrating water into the soil profile and root system.

The speed of humification and form of its products depend on the nature of the decaying organic material, the moisture content, the temperature and the populations of macro- and micro-organisms. The techniques of composting are designed to accelerate humification. The woody tissue of trees and shrubs is much more resistant to humification than is soft tissue such as tree leaves and grass. Higher temperatures, as in the tropics or in summer in temperate regions, accelerate humification if water is available.

Soil-profile development as a physicochemical system

Figures 1.4 and 1.5 show developed soil profiles, this development involving the movement or translocation in all directions of chemicals released by humification, weathering and other physicochemical processes. It is the spatial gradation in the intensity of the various soil-forming processes, combined with the translocation or transport of products, which leads to the different layers or 'horizons' typically found in soil profiles (Figs. 1.3–1.5).

The storage and exchange of plant-nutrient cations by the colloidal clay–humus complex is a dynamic process. A plant-nutrient cation such as calcium (Ca^{2+}) can be displaced by other cations such as the hydrogen ion (H^+). Hydrogen ions can be supplied in abundance during the decomposition of organic matter, acids being released during this process. With such an abundant supply, these hydrogen ions can come to dominate the negatively charged clay–humus complex. This gives an acid reaction to the soil. This characteristic is measured in terms of the concentration of hydrogen ions, expressed in logarithmic form, and so written pH, where

$$pH = \log_{10}(1/[H^+]) \tag{1.1}$$

with $[H^+]$ the hydrogen-ion concentration. The reciprocal form adopted in the expression for pH in Eq. (1.1) is in order to give the term pH a positive value. A solution is defined as being acid, neutral, or alkaline, depending on the relative abundance of hydrogen ions (H^+) and hydroxyl ions (OH^-). In water, some of the molecules split up or 'dissociate' into the two oppositely charged ions H^+ and OH^-. When the the numbers of H^+ and OH^- ions are equal, the pH is found to be 7, and the solution is defined as neutral in its reaction. If pH < 7 in soil solutions, it is acid in reaction, with H^+ ions dominating OH^- ions. However, if pH > 7, the soil solution is alkaline, and OH^- ions dominate.

The displacement of plant-nutrient cations such as calcium (Ca^{2+}) and ammonium (NH_4^+) by H^+ ions in acid soils means that these nutrients are more readily moved down through the soil profile by infiltrating water. This process is called 'leaching'. Significant leaching of plant nutrients to depths below the root zone can lead to nutritional deficiency in vegetation.

If soils become sufficiently acidic (say pH < 5), then even the normally much less soluble or mobile ions, such as those involving aluminium, iron and manganese, can become sufficiently soluble to move through the soil profile. In sufficient concentrations the ions of these metals are toxic to many plants. Thus the growth of vegetation can be much affected by the pH of the soil.

These processes of ion exchange and subsequent translocation through the soil profile add to the formation of distinguishable soil horizons (Figs. 1.4 and 1.5). Such downwardly directed 'translocational processes' are obviously more active in humid rather than arid environments. The loss by leaching of the common negatively charged nitrate ion (NO_3^-) is particularly significant, since nitrate is the major source of the essential plant nutrient nitrogen, and nitrate can also accumulate in groundwater to an extent that makes it unfit for human consumption.

By contrast, in arid environments, soil water that is not taken up by vegetation commonly moves towards the site of evaporation at or close to the soil surface. This upward movement can lead to the surface accumulation of the most mobile ions, which are sodium and calcium. Hence areas of naturally occurring soil salinity occur in arid environments, even though aridity retards the rate of rock weathering, which is one source of saline salts. The processes whereby areas of land can fall prey to the slow degradation caused by salinity are described in Chapter 12.

Aggregation processes

Yet another soil-forming process involves the aggregation or bringing together of soil constituents. In this process clay and humus colloids in particular stick together, and also bind the less-weathered soil components and rock fragments into recognisable units called 'aggregates'. Humus and other organic residues play a particularly important role in gluing soil components together into structural units, although oxides and hydroxides can also form the major basis of aggregate formation in some soils.

Soil material within soil aggregates is somewhat protected from translocation processes. It is the pore space between aggregates that provides the main channels for the ready movement of water, dissolved chemicals, or 'solutes', and also colloidal material. Larger pores can also allow the transmission of larger-sized solids.

In the undisturbed state, voids in the soil 'fabric' commonly take on a readily recognisable pattern, giving rise to a 'pedal structure'. The voids between soil peds can be somewhat extensive, and, if such voids are present, they play a substantial role in water and solute transport through the soil profile.

Erosional and surface transport processes

As previously noted, Figs. 1.3–1.5 could give the erroneous impression that soil formation always takes place as a building process vertically above the underlying weathering rock which provides the initial source material to commence soil formation. Whilst this can be so, it is by no means always the case.

Lateral movement under gravity is common and can be particularly rapid and dramatic in steep and mountainous topography. In cold regions or at higher altitudes, glacial action may be the dominant cause of lateral movement of sediment, bulldozing off prior sediment, resulting in deposits that are then worked on by erosional and transport processes involving water and wind. For example, repeated glaciations during the ice ages produced vast arrays of lithospheric material in North America and Eurasia, where soil-formation processes are still in relatively early stages of development. The lateral airborne spread of ash from explosive volcanic eruptions is another special example of lateral spreading of material on which soil-forming processes can take place. The Hawaiian and Philippine islands, and the region surrounding the volcanic eruption at Mt. St. Helens in the USA provide notable examples of soils forming on such disturbed material.

Sediment can be transported over large distances by water. Streams and rivers are one efficient means of long-distance transport of sediment, such 'alluvial processes' resulting in extensive deposition of transported sediment, for example when river banks are overtopped in flooding. Aeolian (or wind-driven) processes can transport fine sediment, such as clays, over continental and intercontinental distances. The very deep deposits of wind-transported 'loess' in China and elsewhere illustrate the effectiveness of this mechanism. When the fine fractions of soil are winnowed over geological time scales by wind erosion, what is ultimately left behind is deserts of coarse sand, examples being the sand deserts of Africa and Australia. The fine component removed can be discovered as a significant soil component in surrounding downwind regions of its deposition, though the soils may be quite different from loess deposits.

The nature of the transported sediment is a major clue to understanding soil formation where such deposits are a significant contributor to the soil material. Where the rate of sediment-transport processes is greatly accelerated by human action, land degradation can result, as is mentioned in the following subsection. Water erosion processes are considered in Chapter 8, and, when these occur at relatively high rates, they can result in land degradation and impairment of stream-water quality as discussed in Chapter 9.

Environmental concerns involving soils

Environmental concern has arisen in relation to the world's soils because of recognition that degradation of the quality of our land resource is commonly occurring. In areal terms, much, though not all, such degradation is the result, usually unintended, of agricultural and pastoral land use, amplified by pressure from growing population demands for food, fibre, fuel, housing and other needs. The biophysical sciences help in understanding the processes involved in land degradation, and the social sciences in elucidating the broader context and social reasons why human action can so exaggerate natural processes as to affect soil quality deleteriously. Whilst this understanding is of potential assistance, translation of that potential into improved or more appropriate land-management practices by land users involves a complex web that includes society as a whole, or at least governments, and aspects of national and international trade.

Erosion of the landscape is a natural and necessary geophysical process (Fig. 1.2), but some practices lead to such an acceleration of these natural erosional processes that continued productive use of the land is threatened. In more humid environments degradation of soil by profile truncation can be accelerated by water-driven erosion and cultivation. In more arid regions both wind- and water-driven erosion processes can play such a role.

These degradation processes can be further accelerated by a breakdown in the stability of soil structure or aggregation, which can accompany continuous mechanical cultivation. This form of soil degradation is a decline in the ability of soil to cohere together in aggregates or crumbs. This decline in the stability of aggregates is commonly associated with, or due to, a corresponding decline in the level of the soil's organic matter, in particular its humus content. This rundown in humus level can be due to restricted return of organic material following harvesting operations, which is a possible outcome in large-scale mechanised farming practices. It is the larger pores between stable soil aggregates which allow water to enter or infiltrate quickly into the soil surface. Hence any soil-management practice that leads to a reduction in the ability of the soil to form and sustain a suitable level of aggregation leads to a decrease in the rate of infiltration of water, thus increasing the likelihood of water running off or over the soil surface.

In some soils a rather compact 'surface seal' can form, greatly restricting the possible rate of entry of water into the soil surface. The tendency to form a seal can be either a natural characteristic of the soil, or the end result of degrading soil-management practices. Irrespective of whether entry of water into the soil profile is restricted by surface sealing or by a reduction in the size and frequency of water-stable pore spaces between aggregates, less water is stored in the profile for subsequent growth of vegetation, a factor of considerable importance in sub-humid climates. In turn this may lead to less return of organic matter to the soil surface, setting up a spiral of land degradation that is evident in some regions of the world.

In those parts of the world where the flatness of the landscape limits the drainage of water away to the sea, the concentration of naturally occurring salts can rise to levels that are toxic to most vegetation, or at least severely restrict plant growth. Whilst this situation can develop naturally in certain locations, extensive land clearing and irrigation have made salt accumulation at the soil surface, or 'salinity', a growing and intensifying threat in susceptible regions.

Whilst better quantification of the varied consequences of soil degradation is sorely needed, there is no doubt that the consequences of soil degradation are much more widespread and diverse than the direct effects on the sustainability and productivity of land-based activities such as agriculture, animal production and forestry. Soil degradation is also involved in the widespread decline in the quality of surface waters as well as the contamination of groundwater (Lal *et al.*, 1998). Understanding and managing such adverse effects involve considerations at watershed scales. Where appropriate, this book seeks to contribute to this needed expansion in scale of consideration of soil and water resource and environmental issues.

World-wide concern with the effects of the increase in concentrations of greenhouse gases in the earth's atmosphere has drawn attention to the role of soil with its biotic and vegetation components as a major sink of carbon dioxide (CO_2), the major greenhouse gas. The extent and types of land and vegetation activities and the management practices employed in them can determine whether such activities produce a net source or sink of carbon.

Increasing the organic content of soil in well-managed land can improve soil quality by improving nutrient cycling and stimulating biological activity in soil. It is fortunate not only that all such soil-improving and -conserving activities improve soil quality, but also that the consequent sequestration of carbon from the atmosphere can be of significant importance in moderating the currently continuing increase in the earth's atmospheric carbon dioxide concentration (IPPC, 2000).

Land degradation brought about by any cause provides a threat to the ability of terrestrial ecosystems to sustain their biological productivity. The basis for understanding some of these environmental problems of soil and water management will be given later in this book.

1.5 Energy- and water-exchange systems at the earth/atmosphere interface

The flows of energy towards and away from the earth's surface are described as exchanges of energy at this interface. This energy exchange is a basic determinant of natural ecosystems and is intimately linked to the distribution of salt or fresh water (the 'hydrosphere'). Life on earth depends on the basic form of energy which is radiation emitted from the sun in our solar system by virtue of its extremely high temperature, approximately 6000 K (where absolute temperature measured in K or degrees Kelvin is equal to temperature in °C + 273 very closely). Radiation emitted by any substance or body because of its temperature is called 'thermal radiation'. Thermal radiation, so called because of its heating effect on absorption, is just a part of the much more extensive 'electromagnetic spectrum'. Electromagnetic radiation can be characterised either by its frequency or by the length of its repetition pattern, known as its wavelength. X-rays are electromagnetic radiation with a higher frequency (or shorter wavelength) than that of thermal radiation. In contrast, the electromagnetic waves employed in radio and television communication are of lower frequency (or longer wavelength). Light visible to human eyes forms a part of the thermal radiation spectrum. However, electromagnetic radiation with wavelengths both shorter and longer than the visible waveband also has a thermal effect.

Thermal radiation increases very rapidly with the temperature of the emitter of that radiation. Thus, even though our sun is a long way away from the earth in terms of human experience, its radiation received on earth is still the major energy source, with heat flow from within the earth's hot mantle being trivial in comparison. All electromagnetic radiation travels at the same enormous speed of light. A physics text such as Giancoli (1991) describes the basic physical characteristics of such radiation.

The hotter the body, the shorter the wavelength of its thermal radiation, which is distributed over a range of wavelengths, peaking at a wavelength characteristic of the emitting body's temperature. Because of the sun's high temperature, thermal radiation from the sun is called 'short-wave' radiation, and its peak power is in the range of wavelengths the human eye can see. Figure 1.7 seeks to illustrate this feature.

Under full power the heating element of an electric stove, or the metal filament of a bar radiator, can be seen glowing a red colour. When the power source is switched off, the heated element cools, and the colour fades from the red end of the visible spectrum. Even when the element has cooled sufficiently for emitted radiation to be no longer visible, thermal radiation can still be felt on one's hand in proximity to the element. Such radiation is called 'infrared radiation' and is of wavelength longer than the visible spectrum.

It is not just bodies we would regard as hot that emit thermal radiation. Indeed, all components of the earth around us, whether solids, liquids, or gases, emit thermal radiation, referred to as 'long-wave' radiation

Figure 1.7 Illustrating the sun as the source of short-wave thermal radiation in our planetary system because of its high temperature. The peak wavelength for solar radiation is about 0.5 μm. The long-wave thermal radiation from the earth peaks at close to 10 μm, some twenty times longer in wavelength than that from the sun.

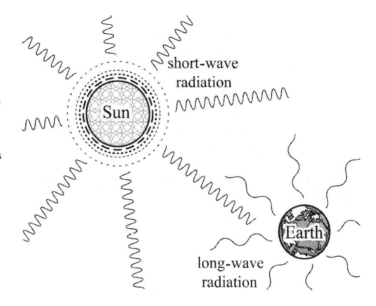

since its wavelength is considerably longer than than of visible radiation (Fig. 1.7).

The energy budget at the earth's surface by day

A fraction of the solar short-wave radiation is either scattered or absorbed as it passes through the earth's atmosphere and some is reflected back to outer space, in a form described as 'diffuse' rather than like reflection from a mirror. Figure 1.8 illustrates the various fates of this extraterrestrial solar radiation, resulting finally in what is called the 'global radiation' input to the earth's surface.

A small, but obviously vital, fraction of the global radiation input is used in photosynthesis (Fig. 1.8). If water is available at the earth's surface, or if there is vegetation that can draw on water throughout its root profile, then the major fraction of the radiation input is used to provide the energy needed to evaporate water. This energy is called the 'latent heat of vaporisation' of water – *latent* because the heat involved is used, not to heat the water, but to break apart the bonds holding liquid water molecules together, allowing water to escape as a vapour or gas.

Evaporation can take place freely from open water surfaces, such as oceans and lakes, and from moist terrestrial surfaces. As such surfaces dry out, the site of evaporation can retreat, perhaps into the soil surface, or into the inside surfaces of plant leaves. Evaporation of water that has been transported up through vegetation is called 'transpiration'.

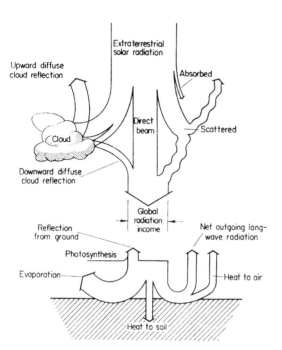

Figure 1.8 Components of the daytime heat-energy exchange at the earth's surface. (From Rose (1979).)

This distinction from other evaporative fluxes can be helpful since, unless the surfaces of vegetation are wet, vegetation has direct control over most of its water loss to the atmosphere. This control of the transpiration flux is achieved by partial or complete closure of the myriad number of small ports in leaves through which the evaporative flux passes, the 'stomata' being the structure of these small ports, which are visible under a microscope. Controlled reduction of the stomatal opening allows vegetation to increase the resistance to vapour flow from the evaporation source within the leaf leading to its passage out through the stomatal pore to the atmosphere. When the amount of water within the vegetation is restricted, closure of the stomata allows the vegetation to avoid dehydration, which might otherwise be sufficiently severe for growth to be retarded, or even lead to death of the plant or tree.

Whilst water loss to the atmosphere from the land surface can be measured reasonably accurately, in practice it is commonly difficult to partition the water loss accurately between transpiration and other sources of evaporation. For this reason the total flux is commonly called 'evapotranspiration'. The term 'evaporation' or 'evaporation flux' can be used in slightly different ways in different literature sources. In some literature, the term evaporation is used to describe the flux of water vapour up into the air resulting from a change in phase from liquid to gas, regardless of the site at which that phase change occurred. This site may be at a water surface or moist soil surface, or within a leaf, as in transpiration. In other literature, the term evaporation can be applied more restrictively, implying that the phase change involved occurred at a free water surface.

How we see the land surface by day is through solar radiation reflected from land surfaces into our eyes. This reflected fraction is higher from white sand, for example, than from a dark-coloured soil, or from vegetation. In addition to this radiation reflected upwards from the earth's surface and radiation used to evaporate water, there are other radiant fluxes, and we will now discuss these other fates of the global radiation income shown pictorially in Fig. 1.8.

- Some solar radiation is absorbed by the earth's land or water surfaces. If that surface is soil, then heat can flow by 'thermal conduction' from the heated soil surface down to lower layers of the soil. In solids, transport of heat energy by thermal conduction can be imagined as the energetic molecular vibration associated with high temperature being passed on to adjacent cooler and less-energetically vibrating molecules. Since this process takes place at the molecular level, it can be relatively slow to penetrate, but the energy involved by day in heating soil can still be substantial, especially if the soil is bare (Fig. 1.8).
- Some of the radiation absorbed by the earth's surface can also be used to heat the air above it. Heat transfer to the air also commences by thermal conduction at the earth/atmosphere interface. However, eddies caused by wind blowing over the land or water surface then largely take over from thermal conduction as the dominant mechanism of movement of heat energy up through the air. Even in the absence of wind, since hotter air is less dense than cooler air, cooler air can flow downwards and displace hotter air at the earth's surface, and the resultant somewhat-organised circulation transports heat up into the air, the process being referred to as heat transfer by 'convection'.
- Finally, radiation is emitted upwards by the earth's surface in the long-wavelength waveband typical of its range of terrestrial temperatures (around 300 K). However, certain gases in the atmosphere such as water vapour and carbon dioxide, though small in volume percentages, significantly absorb this outgoing long-wave radiation, becoming warmed thereby. These gases also radiate at the long wavelengths appropriate to their temperature, and the downward component of this radiation reaching the earth's surface is nearly all absorbed by it. As the result of these outgoing and incoming long-wave radiation fluxes to the earth's surface there is a net outgoing radiation flux, which is commonly positive by day because the earth's surface is usually warmer than the effective radiative temperature of the atmosphere above it.

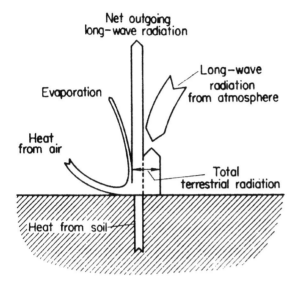

Figure 1.9 Components
of the *night-time* heat
exchange at the earth's
surface. (After Rose
(1978).)

Night-time energy exchanges at the earth's surface

Figure 1.9 shows the heat exchange components at night time. At night
there is no short-wave radiation of energetic significance, despite the
visually significant solar radiation reflected from the moon (or moon-
light). Thus, during the night, long-wave radiation plays a dominant role
in the energy exchange (Fig. 1.9).

Whilst Fig. 1.8 shows only the net result of outgoing over incoming long-wave-radiation streams by day, Fig. 1.9 indicates that, as is the
case by day as well as by night, this net outgoing long-wave radiation
results from the difference between the outgoing emitted terrestrial radiation and the downwardly directed or incoming long-wave radiation from
the atmosphere. At night this net long-wave flux can also be outgoing,
especially so if the sky is cloud-free, a major reason why nights can be
cold under these conditions. However, especially with heavy low cloud
at night, it is possible for this net long-wave-radiation exchange to be
incoming to the earth.

At night transpiration is greatly restricted, since without sunlight
most plants close their stomata through which water vapour escapes. As
the surface cools at night, heat will be directed upwards from depth in
the soil towards the cooler surface (Fig. 1.9). Also, if air near the ground
is warmer than the earth's surface, then heat will be transported from the
warmer air to that surface as shown in Fig. 1.9. Of course, this heat flux
can also be in the reverse direction, from surface to air.

Quantitative distinction between short- and long-wave thermal radiation

The general nature of short- and long-wave radiant heat exchanges has been described in connection with Figs. 1.7–1.9. In this subsection, the distinction between thermal radiation of solar and terrestrial origin will be explicitly defined. More exact radiation terminology will also be introduced.

The flux density of any emitted radiation is called the 'radiant emittance', whereas the flux density received on a surface is called the 'irradiance'. Both are measured in watts per square metre ($\mathrm{W\,m^{-2}}$). (Such units are discussed in Section 1.6).

The concept of wavelength of any wave motion, including thermal radiation, is defined in Fig. 3.14 on page 113 and denoted by λ. The wavelength of thermal radiation is commonly expressed in micrometres, μm (or $\mathrm{m} \times 10^{-6}$). The thermal radiation emitted by all material occurs over a range of wavelengths referred to as a 'waveband'. The wavelength distribution of emittance depends strongly on the temperature of the emitting body, as we know from kitchen experience. We cannot see thermal radiation from an electric-stove element when it is at room temperature because of its long wavelength; but, as the element heats up, a dull red colour becomes visible, rising further to a bright cherry red as its temperature increases. The white-hot filament of a light bulb operates at such a high temperature that it emits radiation over a wide spectrum of short wavelengths.

A body that absorbs radiation of all wavelengths falling upon it is called a 'black body', a term following from our visual experience. (Another descriptor used is a 'full emitter', since a full emitter is also a full absorber of radiation.) The sun emits as a black body in this special sense of the term. Quantum theory in physics was initially developed to provide a theoretical interpretation of the measured distribution of emittance from a black body as a function of the wavelength of that emittance. The radiant emittance per unit wavelength is called the 'spectral radiant emittance', Φ_λ. The unit of spectral radiant emittance is $\mathrm{W\,m^{-2}\,\mu m^{-1}}$. 'Planck's law' of radiation emission mathematically describes this energy distribution as a function of emitted radiation wavelength, λ (μm) and emitting body temperature T (K):

$$\Phi_\lambda = c_1 \Big/ \left\{ \lambda^5 \left[\exp\left(\tfrac{c_2}{\lambda T} \right) - 1 \right] \right\} \tag{1.2}$$

where constant $c_1 = 3.7 \times 10^8$ W $\mu\mathrm{m}^4$ m^{-2} and $c_2 = 1.44 \times 10^4$ μm K. The term 'exp' in Eq. (1.2) refers to the exponential function, which is accessible on a calculator or computer.

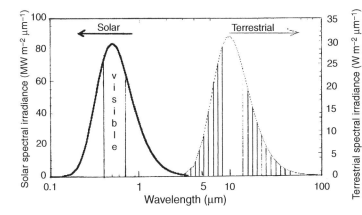

Figure 1.10 A wavelength comparison of the spectral radiant emittance of the sun at about 5800 K (use the left-hand ordinate scale) and that from a typical terrestrial surface temperature of 300 K (or 27 °C) (use the right-hand scale).

Example 1.1

Using Eq. (1.2), calculate the spectral radiant emittance as a function of wavelength for a black body with a temperature of 5800 K (approximating the sun's surface temperature, though its interior temperatures are much higher). Compare the results of this calculation with that for thermal radiation from a body with a surface temperature of 300 K (which is in the range of the earth's surface temperatures).

Solution

The results of calculations of Φ_λ for various values of λ, using Eq. (1.2), and the two very different values of T are given in Fig. 1.10. Note that, because of its enormous magnitude, the spectral radiant emittance of the sun is expressed in millions of watts or megawatts, the unit being $MW\,m^{-2}\,\mu m^{-1}$. The comparatively very much lower spectral radiant emittance at terrestrial temperatures is expressed in $W\,m^{-2}\,\mu m^{-1}$. The wavelength scale is logarithmic in nature.

Figure 1.10 shows that there is very little overlap between the short-wave and long-wave radiation from the solar and terrestrial sources. This separation at a wavelength of 3–4 μm is very useful since there are typical differences between the ways that radiation from these two sources interacts with the climate system.

Figure 1.10 also shows that the maximum in spectral radiant emittance from the sun is for a wavelength of about 0.5 μm. This peak in emittance is within the waveband visible to the human eye, which is from about 0.4 to 0.7 μm, corresponding to from yellow to red in colour, and is delineated in Fig. 1.10. The corresponding maximum for terrestrial radiation is about 10 μm. All long-wave radiation is invisible, as is also the case for the ultraviolet and infrared wavebands below and beyond the visible part of the solar spectrum.

The line-filled segments of the earth's emittance distribution in Fig. 1.10 are a simplified representation of where there are extensive bands of absorption in the atmosphere of the earth's long-wave radiation. Water vapour and carbon dioxide are two important atmospheric sources of this radiation absorption, though other trace gases also absorb (and emit) long-wave radiation. Thus most of the earth's radiation is lost to space only in the less-absorptive waveband or 'window' from about 8 to 13 μm (Fig. 1.10). Even this window is much less transparent if heavy cloud is present. It is re-radiation back to earth from such long-wave-absorbing trace gases in the atmosphere that is the source of the normal warming or 'greenhouse' effect.

Though they are more transparent to short-wave than to long-wave thermal radiation, atmospheric gases, particularly water vapour and ozone (O_3), absorb some solar radiation. The global radiation income at the earth's surface, shown in Fig. 1.8, is only some 50% on average of the radiation incident on the outside of the earth's atmosphere. This is due to scattering, absorption and reflection by water droplets in cloud masses; selective absorption and scattering by atmospheric gases; and scattering and diffuse reflection by small particles such as dust and smoke.

The flux density of solar radiation on a horizontal surface is readily measured by a 'solarimeter', an instrument that is responsive to this flux, and whose surface is covered by a glass dome to protect it from dust and the variable effects of wind. However, the burn record obtained by focusing solar radiation through a glass sphere onto a recording paper strip (as in the simpler Campbell–Stokes recorder) can also be empirically related to the daily global radiation income.

Not all short-wave radiation from the sun is effective in photosynthesis. The photosynthetically active radiation is believed to be largely that which is visible to human eyes, with wavelengths between 0.4 and 0.7 μm. This waveband straddles the median wavelength of solar radiation (0.51 μm).

Ozone plays the vital role of absorbing much of the solar radiation in the short 'ultraviolet' wavelengths that can be lethal to biological organisms. Hence the serious concern over the depletion in concentration, or so-called 'hole', in the ozone layer. The protective ozone layer resides mostly in the stratosphere, some 15–30 km above the earth. Despite international agreement by most countries to cease production of chlorine-containing gases responsible for destruction of ozone, it is expected that recovery of effective ozone protection will take at least several decades, possibly longer.

The enhanced greenhouse effect

Discusion will focus on carbon dioxide, even though this is by no means the only atmospheric gas contributing to the greenhouse effect. (Houghton (1997) provides a much fuller discussion.)

Whilst the greenhouse effect has been effective and somewhat benign over the time scale of recent development of life on earth, evidence of past variations in the atmospheric carbon dioxide concentration and in the earth's temperature brings with it a serious warning to current human generations. But first, the good news.

Without long-wave-radiation-absorbing gases in the atmosphere the earth's temperature would be something like that of the moon, that is some 33 °C colder on average than that which we currently experience on earth. The analogy between this warming and that which occurs in a glass greenhouse is far from complete. Nevertheless, there is value in the analogy since the sun's radiation can readily penetrate the glass of a glasshouse, but the long-wave radiation is largely trapped by it. So we should be thankful for the 'greenhouse effect' provided by the limited transparency of the atmosphere to long-wave radiation.

Evidence for global temperature variation over geological time has shown that low concentrations of carbon dioxide (CO_2) correlate with low temperatures in the last ice age. Furthermore, support has been obtained for the hypothesis that there has been a general correlation between CO_2 concentration and the earth's temperature over recent geological time, higher temperatures correlating with higher CO_2 concentrations. Also measurement has demonstrated that there has been an accelerating increase in atmospheric CO_2 concentration over the last 200 years, associated particularly with the burning of fossil fuels, though burning of timber on cleared land has also been a contributor. Burning fossil fuels such as coal, oil and gas releases CO_2 taken from the earth's atmosphere long ago when the atmospheric CO_2 concentration was very much higher than it is at present. Continued burning of fossil fuels, even at current rates, is predicted to lead to substantially higher CO_2 concentrations. Human activity is also manufacturing or releasing other gases that behave somewhat similarly to CO_2 in terms of absorption of long-wave radiation when they enter the atmosphere.

This evidence has quickened interest in much earlier investigations of the greenhouse effect, which go back to 1896 when Arrhenius, a Swedish chemist, developed theoretical considerations suggesting that an increase in CO_2 concentration in the atmosphere would warm the earth. This prediction was supported by the scientist and mathematician Joseph Fourier, who recognised that, without atmospheric absorption of radiation, the earth's surface would be some 30 °C cooler than it is. Fourier recognised that long-wave radiation (which he termed 'non-luminous' radiation) was much more heavily absorbed in the earth's atmosphere than was light (or short-wave radiation generally). Thus increasing the concentration of gases in the atmosphere that absorb long-wave radiation would be expected to increase atmospheric temperature and reduce the net outgoing long-wave radiation from the earth's surface (Fig. 1.8).

Fourier's explanations of the expected consequences of the presence of radiation-absorbing gases in the atmosphere for global warming still provide the basis for concerns that the increasing atmospheric concentrations of CO_2 and certain other gases may be sufficient to cause climatic or related changes that could be disruptive, possibly necessitating rapid human and ecological adaptation. Research stimulated by such concerns has shown that, though the processes involved are more numerous than Fourier could have imagined, the issues are of great importance as well as of considerable complexity. Texts on this topic, such as Houghton (1997), should be consulted in order to follow physical aspects of this rapidly developing area of research.

Governments, industry and society as a whole are also studying the possibly widespread nature of the response needed if predictions, which currently are largely based on global climatic modelling, prove to be realistic. Improvements in such modelling are leading to better agreement between predictions and observed global increases in temperature, and there is now an almost universal consensus that global warming due to human activities is an issue the world must face up to. A variety of evidence apparently in line with global warming is emerging. This evidence ranges from direct global temperature measurement, through general retreat of glaciers, to polar and ecological observations.

Modelling indicates that, even to limit the increase of CO_2 concentration to two or three times its current level, a reduction in CO_2 emission to some one third of its present rate would be required. Thus, in order to modify the predicted climate change significantly, it seems that quite major changes in energy generation and use are needed. Much more efficient use of energy, in particular by energy-intensive nations, and a shift from widespread dependence on fossil fuels to their replacement by renewable energy forms are commonly suggested as necessary steps to be undertaken as quickly as is feasible. Because of our current high dependence on fossil fuels, such change is a major challenge, to which the response has hardly begun.

However, attention must also be given to the issue of non-atmospheric sinks and storage of carbon, as well as emission of carbon into the atmosphere. On land, the important effects in this complex question of carbon balance at planetary scale of the extent and types of land use, and of improved land management, were alluded to earlier.

Improvements in agricultural and forestry operations can at the same time improve soil quality and increase the capacity of the land to take up, or sequester, carbon dioxide from the earth's atmosphere (IPCC, 2000).

The global climatic models which allow these predictions to be made seek to represent the wide range of physical processes that take place in the atmosphere. These processes are themselves tightly coupled to what happens on the earth's land and ocean surfaces through the exchange of energy, water and carbon dioxide. In dealing with environmental matters we also need to understand the movement over and into land surfaces and water bodies of plant nutrients, pesticides and other chemical contaminants. In order to measure all such quantities, we need to be familiar with the units in which they are measured. Also, especially in dealing with environmental matters, it is quite common that useful data are available, but in a variety of units. For example, industry is a common source of information for use in environmental analysis, but industry uses a great variety of units to express quantities of interest to it.

A major source of error in quantitative analysis commonly arises in converting the magnitudes of quantities measured or expressed in one system of units into another needed for calculation. Thus it is very helpful, before going further into quantitative expression and calculation, to be confident in such conversion of units. That is the purpose of the following final section of this chapter, Section 1.6.

If you are familiar and confident with the conversion of quantities from one system of units to another, a skill you can check with the exercises at the end of the chapter, then the material in Section 1.6 can simply be used for later reference if required. However, Section 1.6 also introduces some basic physical quantities and their definitions, so it is suggested that you have a look at it to check that you are comfortable with the concepts introduced.

1.6 How things are measured – units and conversions

An introduction to units

All environmental issues involve a question of magnitude. Suppose, for example, that you wish to obtain information on the levels of a soil or water contaminant thought to be a problem in the area where you live.

To obtain data that will stand up to criticism, measurements with some kind of instrument that indicates the concentration of the contaminant in question will be required. As is typical in environmental issues, once measurements have been obtained, their significance has to be assessed. Judgement is required as to whether the contaminant level is considered too high for human safety, and such judgement involves many types of science and a range of areas of knowledge, including human perceptions, attitudes and values. The development of actions to help manage or mitigate the problem can involve yet further types of knowledge and understanding.

But despite all that – and 'all that' is very important – the proper recognition and definition of potential or actual problems arising from soil or water contaminants still depends on measurement.

Also, when we seek to develop models that we believe may describe or help us to understand some process of environmental interest, the application and testing of such models depends on measurement.

Measuring any quantity, such as the height of a person, involves adoption of some standard unit for length or distance; then the process of measurement involves comparison of the height of the person with that standard to give, say, 5 foot 7 inches or 1.7 m or 170 cm. So, clearly there is more than one standard of measurement in use – as is true of all physical measurement!

Even though there is now general agreement in scientific communication on using a particular metric system of measurement, data are not always available, and measurements not always made, in that agreed system. This agreed system is the Système Internationale (or International System), abbreviated to SI. This system is based on the *metre* (for length), *k*ilogram (for mass) and *second* (for time), and so was once referred to as the MKS system of units. Another metric system sometimes used is based on the *c*entimetre (a hundredth of a metre), the *gram* (a thousandth of a kilogram), and the *second*, so being referred to as the cgs system of units.

An important group of quantities is needed in order to describe how things move. The most familiar is 'average speed', defined as the ratio of the distance travelled to the time which has elapsed during that motion. However, the speed of flight of a bird, for example, is continuously changing. This change during its flight can be captured by a sports camera with rapid exposures automatically triggered, say, each hundredth of a second, yielding a record of the distance moved during each period. In principle we can make the elapsed time and so the distance travelled as small as we like, and there will still be a value for the speed given by the ratio of these two small quantities. The limiting value of this speed as the time interval becomes smaller and smaller is called the 'instantaneous speed'.

As well as the ability of the bird to change its instantaneous speed continuously, equally important if it is to catch an insect on the wing or negotiate its way through a thicket, is its ability to change its direction of flight. Both these factors of speed and direction are taken into account in the term 'average velocity', defined as the ratio of a displacement Δx *in a specified direction* to the time period, Δt, taken to complete that displacement. Thus, for practical purposes, we can say that velocity is speed in a specified direction. Again, if we imagine the elapsed time and displacement to become indefinitely small, the ratio $\Delta x / \Delta t$ is called the 'instantaneous velocity', v. The instantaneous velocity of a bird can be changed either by a change in instantaneous speed or by a change in the direction of motion.

Many will remember the childhood experience of the periodic motion provided by a swing. During the swinging motion the speed varies periodically from a maximum at the bottom of the swing to that position at maximum displacement where the direction of motion changes, and you come to rest for an instant. The direction of motion also continually alters during the swing, so the instantaneous velocity varies for both these reasons.

The rate of change of velocity is called 'acceleration'. Thus, if there is a change Δv in velocity v during a time period Δt, then the 'average acceleration' over this time period, a, is defined as

$$a = \Delta v / \Delta t \qquad (1.3)$$

If v is measured in metres per second (m/s or m s^{-1}), then a would be expressed in metres per second per second, written as m/s^2 or m s^{-2}.

We arrived at the concept of instantaneous velocity as the limit of average velocity as both the displacement Δx and the time interval Δt became indefinitely small. An instantaneous acceleration, a, can be similarly thought of as the limiting value of the average acceleration, a, as the time interval Δt (and so Δv also) becomes smaller and smaller.

Those who have enjoyed the experience of swinging on a swing may remember the varying pressure on your bottom during the arc of motion. Also, at least to get the motion started, it was helpful if someone gave you a push or a pull from behind. A push or a pull describes our intuitive idea of what, in physical science, is called a 'force'. A force may but need not, cause something to move. For example, sitting at rest on a swing, the downward force due to the weight of your body is resisted by an equal upward force from the swing seat, which is supported in turn by the force in the chains or ropes on which the swing seat hangs. Hence, at rest, there is no net force.

Isaac Newton (1642–1727) expanded on this concept of force in his 'first law of motion' by stating that a 'body continues in its state of rest, or in uniform motion in a straight line [i.e. uniform velocity], unless it is compelled to change that state by a net force acting on it'.

In this space age, we are familiar with the fact that a person in orbit around the earth appears to be weightless, and it must be quite frustrating when your can of beans for lunch floats away out of reach. What has changed in space is not the 'quantity of matter' in the can of beans, but the net gravitational attractive force we call its weight. The intuitive concept of 'quantity of matter' can be made more precise using Newton's first law by introducing the concepts of 'mass' and 'inertia'. The mass of a body can then be conceived in dynamic terms as related to the tendency of a body to continue with a constant velocity, which is called its inertia. This inertial concept of mass relates to everyday experiences such as the difficulty in suddenly reversing our direction of motion.

Newton's celebrated 'second law of motion', which is depended upon in planning the launch of satellites, by an agile monkey jumping from tree to tree, or in understanding the erosion of river banks, brings together the concepts of force, mass and acceleration. This law is usually described in terms of what happens to an 'object'; however, an object can be anything, even a volume of water with only an imaginary surface enclosing it. Newton's second law states that 'The acceleration of an object is directly proportional to the net force acting on it and is inversely proportional to its mass. The direction of the acceleration is in the direction of the applied net force.'

Thus, denoting net force by F and mass by m,

$$a \propto F/m$$

With a suitable choice of units this proportionality can be changed into the equality

$$F = ma \tag{1.4}$$

Using Eq. (1.4), the unit of force in the SI system of units is defined as the force required to accelerate a mass of one kilogram (abbreviated kg) by one metre per second per second (written m s^{-2}); and this unit of force is called a newton (written N). As in the case of force, most SI units are named after past historically famous physicists, and a list of some of the important units used is given in Table 1.1. The units given in Table 1.1 will be introduced in this and later chapters.

Vector and scalar quantities

Force, velocity and acceleration are examples of a large class of physical quantities for which we require information on their direction as well as magnitude in order for them to be adequately defined. Members of this class of quantities are called 'vectors' or vector quantities. A moment's thought will show that temperature and humidity are not vector quantities, because, after being told their magnitude, it does not make sense

Table 1.1 *A list of fundamental and some dynamic units*

	Unit	Symbol	Equivalent
Fundamental units			
Length	metre	m	
Mass	kilogram	kg	
Time	second	s	
Dynamic units			
Force	newton	N	
Pressure, stress	pascal	Pa	$N\,m^{-2}$
Work, energy	joule	J	$N\,m$
Power	watt	W	$J\,s^{-1}$

to ask 'and in what direction?'. These directionless quantities are called 'scalars'.

Since velocity is defined as speed in a particular direction, velocity is a vector, whereas speed is a scalar quantity. This distinction is important in understanding Newton's laws of motion. Though Newton's laws will be considered more fully in Section 2.2, Newton's first law states that any body will continue either at rest or in motion with uniform speed and direction (i.e. constant velocity) unless it is acted on by a net force.

The term 'net force' in this first law of Newton assumes that, if a number of forces are acting on a body, these can be added up in some way. Force is a vector, and the addition of vectors is a generalisation of arithmetic addition. The generalisation is required because vector quantities have direction as well as magnitude. The addition of the vectors **A** and **N** acting on a body is carried out by completing the parallelogram as shown in the figure below, the sum **A** + **N** being given by the diagonal.

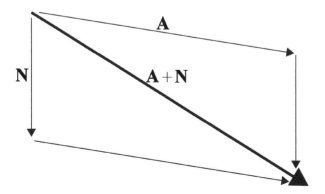

Displacement from one location to another is also a vector. Displacement from E to F in the next figure, and then from F to G, has the net effect of moving from E to G, which is the sum of the two component vectors. (Note that there is an implicit parallelogram EFGH.)

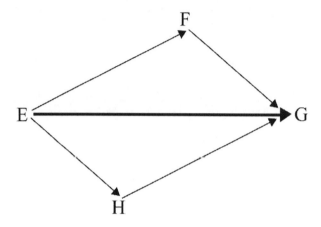

Weight and gravitational units

A particularly important value of acceleration is that of a body falling freely under gravity, denoted by g. (Strictly g applies to a body falling in a vacuum or at low speeds at which air-resistance forces are negligible.) The gravitational field exerted by planet earth provides a particular value for g at any given altitude, being about $9.8 \, \text{m s}^{-2}$ at modest altitudes. The gravitational force acting on a mass is called its weight.

The value of F in Eq. (1.4) when $a = g$ is called the weight of the body, W. For a body of mass m the weight is obtained by putting $a = g$ in Eq. (1.4) to give

$$W = mg \tag{1.5}$$

In contrast to the standard SI system of units, 'gravitational' systems of units are defined in terms of unit force instead of unit mass. In the 'metric gravitational' system the kilogram force or weight of a kilogram mass is chosen as a fundamental unit. This weight is written kg wt., and it is defined as the force required to support a unit kilogram mass. On putting $m = 1$ kg in Eq. (1.5), it follows that 1 kg wt. $= g$ newtons $= 9.8$ N on the earth's surface.

A system of units commonly referred to as the 'British gravitational' or 'English' system is widely used in engineering, especially in the USA.

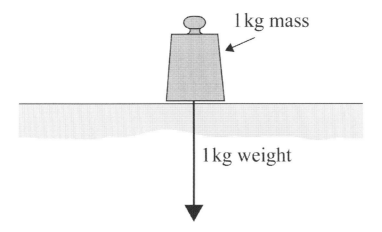

The unit of length in this system of units is the foot (ft), and that of time is the second (s); the other basic unit is the unit of force, taken as the pound weight (lb wt., or just pound, p). (Since the word 'pound' is sometimes used for force and sometimes for mass, you have to be clear of the context in order to avoid ambiguity!)

In gravitational systems of units the unit of mass follows from what mass experiences under unit acceleration and under unit force; hence the unit of mass is g times the absolute unit of mass, and is called the 'slug' and 'metric slug' in the British and metric gravitational systems, respectively. Any engineering-mechanics text will clarify gravitational units further, as will any engineering handbook (which will also provide a source of information that is very useful in physical environmental investigations). Dingman (1984) gives a very useful review of units and conversions, including gravitational units.

We have seen how more complex physical quantities, such as force, can be defined in terms of more fundamental quantities such as mass, length and time. Letting M, L and T stand for the fundamental 'dimensions' of mass, length and time, respectively, it follows from Eq. (1.4) that force is defined in terms of MLT^{-2}. All terms in a correct physical equation obviously must have the same 'dimensions'. Some quantities, such as the number π, are dimensionless or non-dimensional.

Conversion of units of measurement of any quantity

Because data come in various units, it is important to be able to convert quantities correctly from one system of units to another. Errors commonly occur in making such conversions, so some examples in which

Table 1.2 *Some common conversions from non-SI to SI units*

Unit	Abbreviation	Conversion factor
Gram	g	$\times 10^{-3} = \text{kg}$
Pound mass	lb^a	$\times\, 0.454 = \text{kg}$
Foot	ft	$\times\, 0.3048 = \text{m}$
Inch	in	$\times\, 0.0254 = \text{m}$
Kilometre	km	$\times\, 10^3 = \text{m}$
Centimetre	cm	$\times\, 10^{-2} = \text{m}$
Millimetre	mm	$\times\, 10^{-3} = \text{m}$
Acre	ac	$\times\, 4047 = \text{m}^2$
Hectare	ha	$\times\, 10^4 = \text{m}^2$
Litre	l	$\times\, 10^{-3} = \text{m}^3$
Gallon (US)	gal (US)	$\times\, 0.003\,785 - \text{m}^3$
Hour	h	$\times\, 3600 = \text{s}$
Minute	min	$\times\, 60 = \text{s}$
Pound force	lb wt. or lb^a	$\times\, 4.448 = \text{N}$
Pound force per square inch	p.s.i	$\times\, 6894 = \text{N m}^{-2b}$ (or pascal, Pa)
kg wt. per square centimetre	kg wt. cm^{-2}	$\times\, 9.8 \times 10^4 = \text{N m}^{-2b}$ (or pascal, Pa)

^{a} Note possible confusion.

^{b} $1\ \text{N m}^{-2}$ is called a pascal (Pa).

we make use of the brief list of conversions in Table 1.2 follow. (A further list is given in the appendix in Table A.1).

Example 1.2

A small lake or pond is being examined because of its eutrophication problem, being filled with green algae. A survey gave its area as 2.7 acres (ac) and average depth about 5.5 ft. The investigating scientist needs to use this information to calculate the volume of water in the pond in cubic metres (m^3). Can you help by doing this calculation?

Solution
From Table 1.2,

$$2.7\ \text{ac} = 2.7\ \text{ac} \times 4047\frac{\text{m}^2}{\text{ac}} = 10\,927\ \text{m}^2$$

and the depth is

$$5.5\ \text{ft} = 5.5\ \text{ft} \times 0.3048\frac{\text{m}}{\text{ft}} = 1.676\ \text{m}$$

Thus the pond volume is $18\,318\ \text{m}^3$.

Question. Suppose that there is an uncertainty of ± 0.4 ft in the estimated pond depth in the earlier example, how many significant figures would be warranted in the calculated pond volume?

$$\% \text{ error in pond depth } = \frac{0.4}{5.5} \times 100 = 7.3\%$$

and 7.3% of $18\,318$ m^3 $= 1332$ m^3.

So rounding off the reported volume to $18\,000$ m^3 would provide a better indication of likely accuracy in the figure. The number 18 to the left of the zeros is called the 'significant figures'.

It is commonly regarded as unprofessional to quote more figures than those which are significant as judged from uncertainty or error. Rather, the number of significant figures should indicate the accuracy of the result, though this advice is not always followed in publications.

We can generalise the way in which units were converted in the above pond-volume example as follows. Suppose that a quantity is expressed as Q_a in the 'A' system of units, and the conversion factor (CF) gives the ratio of 'B' units to 'A' units. This CF is to be interpreted as the number of 'B' units in one 'A' unit, the number of m^2 in an acre, or the number of metres in a foot in Example 1.2. Then the same quantity in 'B' units, Q_b, is given by

$$Q_b \text{ (in B units)} = Q_a \text{ (in A units)} \times CF \text{ (B units/A units)} \qquad (1.6)$$

Considering Table 1.2 in relation to Eq. (1.6), suppose that 'B' refers to SI units, and so the conversion factors in the table are for the ratio (SI unit/non-SI unit).

Note that, in Example 1.2, it is as though you can algebraically cancel units out. Error is less likely if you write out the conversion of units in this way. Make sure that you have the conversion factor in the required ratio – not its inverse!

Reports indicate that errors in conversion of units may have contributed to failure in a NASA Mars lander!

Example 1.3

Problem setting
After secondary treatment, sewage sludge can be discharged to evaporation bays for drying by evaporation to take place. How large should these evaporation bays be to allow evaporative drying with a known rate of input volume of wet treated sewage?

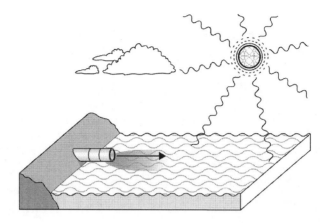

Data

The daily output of a small treated sewage plant is 750 000 US gallons per day, where 1 US gallon = 231 in³.

At the site the average daily rate of evaporation from an open water surface for the design period is equivalent to a 3.5-mm depth of ponded water per day.

Design problem

Assume that evaporation from the sewage sludge takes place at the same rate as from water. What area of evaporation bays is needed in order to maintain an evaporation rate equal to the rate of input of treated sludge? This area is needed in hectares, where 1 hectare (ha) = 10^4 m².

Solution

Suppose that we don't have information such as Table 1.2 to hand and need to use the conversions provided. Then the input of 750 000 US gal d^{-1} is equivalent to

$$750\,000\,\frac{\text{gal}}{\text{d}} \times 231\,\frac{\text{in}^3}{\text{gal}} \times (0.0254)^3\,\frac{\text{m}^3}{\text{in}^3} = 2.84 \times 10^3\,\frac{\text{m}^2}{\text{d}}$$

(Note that we have used a generalisation of Eq. (1.6).)

Let A be the required area of evaporation pond (m²). Then the loss of water by evaporation is $A \times 3.5 \times 10^{-3}$ m³ d⁻¹, which must equal the input, 2.84×10^3 m³ d⁻¹. Thus

$$A = 8.11 \times 10^5\,\text{m}^2$$
$$= 81.1\,\text{ha}$$

(*Note.* Exercise 1.7 carries this example a little further.)

Example 1.4

Problem setting

The amount of soil eroded from a cultivated field plot can be measured by collecting the soil lost from the plot in a low-slope collecting channel at the downslope end of the plot as shown in the figure that follows. Because water moves slowly in this low-slope channel, the larger soil components drop out or are deposited in it. The smaller or finer fraction of soil flows from the collecting channel with the runoff water. A sub-sample of this runoff water is collected to measure the concentration of sediment in it (i.e. the kg sediment per m^3 of runoff). The total amount of soil lost from the plot is the sum of that deposited in the collecting channel and that suspended in the runoff water.

Calculate the mass of soil eroded in kg given the following data, and then calculate the soil loss per unit area of the plot in tonnes ha^{-1}, where 1 tonne (a metric ton) $= 1000$ kg.

Data

The area of the cultivated plot is 0.5 acre. The length of the collecting trough is 20 ft, and its width is 1 ft. The average depth of eroded soil collected in the trough (measured after it had dried) was 3 inches. The soil density (mass/volume) was found to be 1.1 g cm^{-3}. The concentration of suspended sediment sub-sampled from the runoff water was 10 g per litre ($10 \, \mathrm{g} \, \mathrm{l}^{-1}$), the total volume of runoff being 3000 litres.

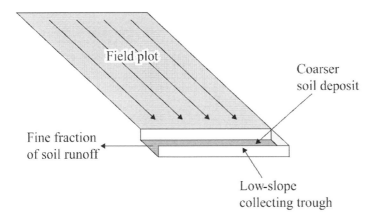

Solution

Volume of soil in collection trough

$$= 20 \, \text{ft} \times 1 \, \text{ft} \times (3/12) \, \text{ft}$$

which, from Table 1.2,

$$= 5\,\text{ft}^3 \times (0.3048)^3\,\frac{\text{m}^2}{\text{ft}^3}$$

$$= 0.142\,\text{m}^3$$

Mass of soil in collection trough

$$= \text{volume} \times \text{density}$$

$$= 0.142\,\text{m}^3 \times 1.1\,\text{g} \times 10^{-3}\,\frac{\text{kg}}{\text{g}} \times \frac{1}{\text{cm}^3} \times \frac{\text{cm}^3}{(10^{-2})^3\,\text{m}^3}$$

$$= 156\,\text{kg}$$

Mass of suspended soil

$$= \text{sediment concentration} \times \text{total volume of runoff}$$

$$= 10\,\text{g} \times 10^{-3}\,\frac{\text{kg}}{\text{g}} \times \frac{3000\,\text{l} \times 10^{-3}\,\text{m}^3\,\text{l}^{-1}}{11 \times 10^{-3}\,\text{m}^3\,\text{l}^{-1}}$$

$$= 30\,\text{kg}$$

(Note that $1\,\text{g}\,\text{l}^{-1} = 1\,\text{kg}\,\text{m}^{-3}$.)
So total soil loss

$$= 30 + 156 = 186\,\text{kg}$$

Thus soil loss per unit area of plot

$$= \frac{186\,\text{kg}}{0.5\,\text{acre} \times 4047\,\text{m}^2/\text{acre}}$$

$$= 0.0918\,\text{kg}\,\text{m}^{-2}$$

$$= 0.0918\,\text{kg} \times 10^{-3}\,\frac{\text{tonnes}}{\text{kg}} \times \frac{1}{1\,\text{m}^2 \times 10^{-4}\,\text{ha/m}^{-2}}$$

$$= 0.918\,\text{tonnes}\,\text{ha}^{-1}$$

Data limitations and error

There are limitations in all data. Apart from outright mistakes, errors can arise from many sources, including faulty or inadequately calibrated instrumentation, or from inadequacies in sampling the environmental variable or pollutant of interest. It is always worth checking the repeatability or stability of measurement systems as well as their accuracy, which is judged by calibration against a standard instrument. However, in many environmental studies resource limitations commonly restrict the number of sampling sites or the suitability of their location, and this is a major source of data limitation. Such limitations may lead to erroneous conclusions, or at least seriously restrict the validity of conclusions that can be drawn from the data collected. For example, in investigating city air pollution, where and how frequently should measurements be taken, at how many sites and at what height? Provided that some data are collected, there are methodologies that can help in answering such questions, but limitations on resources available to meet the requirements of a scientifically adequate study may provide the chief source of uncertainty.

In seeking to deal with many environmental problems it is not uncommon for there to be few or no data of the type really required. If so, the best way forwards may be to seek physically based relations to surrogate or alternative variables on which some data are available. For example, there may be no data on solar radiation received on the earth's surface at some location of interest; however, there could be data recorded nearby on the hours of sunshine received, and there is an empirical relation between these two variables that was originally found by Ångström.

Whatever the cause of error or uncertainty in a measured or estimated variable, some estimate of the magnitude of the error can be made. With multiple measurement of the same quantity, statistically based measures of such uncertainty can be made. However, if such data are not available, careful consideration usually allows us to make a useful possible error estimate, perhaps of the likely maximum error δA in some measured or estimated quantity of magnitude A. Very often the quantity of interest is not just A, but involves the addition, subtraction, multiplication, or division of A with another quantity or with other quantities. There are simple rules for calculating the error in such compound calculations that depend on the particular arithmetical operation used. These simple rules are as follows.

(1) In addition or subtraction of quantities the errors are added to obtain the error in the resulting quantity, i.e.

$$(A \pm \delta A) \pm (B \pm \delta B) = A \pm B \pm (\delta A + \delta B)$$

(2) When multiplying or dividing quantities, the relative or percentage error in the compound quantity is the sum of the relative or percentage errors in the component quantities. The reason for this is illustrated for multiplication (and you may wish to justify that the same rule holds for division):

$$(A \pm \delta A) \times (B \pm \delta B) = AB \pm (B\,\delta A + A\,\delta B)$$

Thus the relative error

$$= \frac{B\,\delta A + A\,\delta B}{AB}$$

$$= \frac{\delta A}{A} + \frac{\delta B}{B}$$

(3) When a quantity is raised to a power p, the relative error in A raised to the power p is p times the relative error in A. (This follows from the above multiplication rule for errors.)

Following the symbols are some exercises or questions you can try for yourselves to ensure mastery of the material in this chapter and expand your experience. Answers are on page 421.

Main symbols for Chapter 1

a Acceleration (Eq. (1.3))
F Force (Eq. (1.4))
m Mass
pH   A measure of hydrogen-ion concentration (Eq. (1.1))
t Time
T Absolute temperature (in K = °C + 273)
W Weight (Eq. (1.5))

λ Wavelength of thermal radiation
Φ_λ Spectral radiant emittance (Eq. (1.2))

Exercises

On units and conversions

1.1 A small watershed or catchment basin has an area of $0.5\,km^2$. Convert this area to m^2, ha and ft^2.

1.2 A particular irrigation farmer normally receives a yearly total of 100 ac ft of water, where 1 ac ft is the volume of water of an area of 1 acre ponded to a depth of 1 ft. If this water were supplied by a uniform continuous flow of water over the whole year, what would be the volumetric flow rate expressed in units of $ft^3\,min^{-1}$, $m^3\,s^{-1}$ (a cubic metre per second sometimes being called a 'cumec') and litres h^{-1}?

1.3 Rain fell at an average rate of $10\,mm\,h^{-1}$ for 4 h, and 20% of rainfall received on an area of 20 ha ran off the area. What is the total volume of runoff from this area during the 4-h period expressed in m^3 and ft^3?

1.4 Pressure is force per unit area. A hoofed animal can exert a pressure on the soil of 15 kg wt. cm^{-2}. Convert this pressure into units of $N\,m^{-2}$, lb wt. ft^{-2} and pound weight per square inch (written as p.s.i, a commonly used unit in some areas of engineering).

1.5 Estimate the pressure (in any appropriate unit) which you exert by standing with both feet on the soil.

1.6 The period, P, of the swing of a pendulum of length L is given by the equation

$$P = 2\pi(L/g)^{0.5}$$

where g is the acceleration due to gravity. Show that the terms on either side of the equation have the same dimensions.

1.7 Reread Example 1.3 in the text.

At one stage of the works development the area of the evaporation bay receiving this flow of sewage was only 10 ha. If the rate of sewage input was as given in the example, what volume rate of sewage, in $m^3 \, d^{-1}$, must have been lost from the area to an adjacent stream or to ground water beneath the soil?

Assuming that all this excess sewage is lost to the nearby stream, which has a mean annual stream flow of 48×10^6 US gallons per day, what is the percentage of excess effluent to the mean flow of the stream?

If the seasonal minimum flow of the stream is only 0.3×10^6 US gallons per day, what relation does the excess sewage flow bear to this minimum stream flow? Comment on possible implications of this calculation.

1.8 The air-quality standard of the US Environmental Protection Authority sets 90 micrograms per m^3 as the maximum annual mean concentration of total suspended particulates. (A microgram is 10^{-6} g.) Convert this particulate concentration into $kg \, km^{-3}$.

1.9 Because survey equipment was not available at the time, a rectangular field was measured by a person stepping, and found to be 500 yards long and 220 yards wide (a yard being 3 ft). The person who made the measurement is able to step a distance within 2% of the correct distance. Would the area of this field be estimated with an accuracy to the nearest 0.01, 0.1, 0.5, or 1.0 acre?

1.10 In Example 1.2 of the earlier text, the uncertainty in the depth of the pond was estimated to be 0.4 ft. If the possible error in measuring the pond surface area was 0.2 acre, what is the percentage possible error in the calculated volume of the pond? Also, what is this possible error expressed in m^3?

On thermal radiation

1.11 List three differences between thermal radiation from the sun and thermal radiation from the earth or its atmosphere.

1.12 The total radiant emittance of any body, Φ, is given by the Stefan–Boltzmann law:

$$\Phi = \varepsilon \sigma T^4 \ (W \, m^{-2})$$

where ε is the emissivity (or emission coefficient) of the body, σ is the Stefan–Boltzmann constant,

$$\sigma = 5.67 \times 10^{-8} \, W \, m^{-2} \, K^{-4}$$

and T is the absolute temperature of the body in degrees Kelvin (K), where $K = °C + 273$.

Using this Stefan–Boltzmann equation for radiant emittance, calculate the ratio of the radiant emittance of the sun to that of the earth, assuming that the effective temperature, T, of the sun is 6000 K and that that of the earth is 290 K. Also assume that the emissivities of the sun and the earth are the same.

1.13 (a) It can be shown to follow from Planck's law of radiation given by Eq. (1.2) that, at least to a good approximation, the maximum in emitted radiant energy occurs at a wavelength, λ_m, given by

$$\lambda_m T = 2897$$

when λ is in microns (μm) and T is in kelvins (K). This relationship is known as Wien's law. For solar energy, $\lambda_m = 0.5 \ \mu$m.

What is the equivalent emissivity temperature of the sun calculated using Wien's law?

(b) Assuming that the earth's average global temperature is 15 °C, what is the wavelength of its maximum radiation, λ_m?

References and bibliography

Australian Academy of Science (1994). *Environmental Science*. Canberra: Australian Academy of Science.

Brady, N. C., and Weil, R. R. (1999). *The Nature and Properties of Soils*, 12th edn. Englewood Cliffs, New Jersey: Prentice-Hall International, Inc.

Christopherson, R. W. (1997). *Geosystems: An Introduction to Physical Geography*, 3rd edn. New Jersey: Prentice-Hall International, Inc.

Dingman, S. L. (1984). *Fluvial Hydrology*. New York: W. H. Freeman and Co.

Giancoli, D. C. (1991). *Physics: Principles with Applications*, 3rd edn. Englewood Cliffs, New Jersey: Prentice-Hall International, Inc.

Houghton, J. (1997). *Global Warming: The Complete Briefing*, 2nd edn. Cambridge: Cambridge University Press.

IPCC (International Panel on Climate Change) (2000). *Land Use, Land Use Change, and Forestry*. New York: Cambridge University Press.

Lal, R., Blum, W. H., Valentin, C., and Stewart, B. A. (eds.) (1998). *Methods for Assessment of Soil Degradation*. Boca Raton, Florida: CRC Press.

McTainsh, G. H., and Boughton, W. C. (1993). *Land Degradation Processes in Australia*. Melbourne: Longman Cheshire.

Murphy, B., and Nance, D. (1999). *Earth Science Today*. Pacific Grove, California: Brooks/Cole Publishing Company.

Paton, T. R. (1978). *The Formation of Soil Material*. London: George Allen & Unwin.

Robinson, P. J. and Henderson-Sellers, A. (1999). *Contemporary Climatology*, 2nd edn. Harlow: Pearson Education Limited.

Rose, C. W. (1979). *Agricultural Physics*. Oxford: Pergamon Press.

Strahler, A. N., and Strahler, A. H. (1973). *Environmental Geoscience: Interactions between Natural Systems and Man*. Santa Barbara, California: Hamilton Publishing Company.

2
Soil and soil strength

2.1 Introduction

Significant environmental problems involving soils can occur when the soil material moves. An earthen river bank may be weakened by heavy rain and slide into the river, where it is broken up and transported down the flooding river, the added sediment impairing water quality and finally reducing the storage capacity of any downstream dam, where most sediment deposits. More dramatically, many of the inhabitants of a village on a steep mountain slope could be buried in an avalanche of soil and rock (see chapter frontispiece). World-wide news indicates that such disasters occur frequently, and loss of life can be considerable. Whilst landslides occur naturally on sloping or mountainous land, human action, such as clearing vegetation, can also be a trigger, possibly due to alteration in the hydrology of the site. Cutting roads into hillsides has also led to landslides with serious consequences. The force causing the mass movement of soil is the weight of the soil itself, this being due to gravity. Whether the river bank collapses, or the avalanche develops, depends on whether the strength of the soil is sufficiently great to sustain the forces acting on it.

The increase in rate of soil erosion often associated with mechanised soil tillage provides a threat to the long-term sustainability of agriculture in many parts of the world. Whilst an objective of tillage is to provide a suitable seed bed and soil medium for the rapid root development of subsequently growing crops, this objective is achieved by reducing the strength of soil, making it more vulnerable to erosion processes, as discussed in Chapter 8. Mechanised cultivation methods have sometimes been so aggressive as to weaken the soil more than is necessary to enhance plant growth. Hence there is now growing adoption of practices that minimize the amount and severity of tillage, one objective of this change being to leave the soil with adequate strength to resist erosion, either by water or by wind.

Thus the strength of soil is a determining factor in whether or not it moves in surface erosion or mass-movement processes. The basic physical concepts which determine whether or not material moves are described in terms of forces acting on a 'body', conceived of as some rigid lump of matter. We will briefly introduce these quite basic ideas of how it is that material can move using this simple physical abstraction of a 'rigid body'. This abstraction is not irrelevant to soil, even though soil is more a continuum of bodies; for example, the sand in a sand hill can be considered to consist of a large collection of individual sand grains. Even when an entire slope of sand moves downhill in a landslide, the individual sand grains can still be adequately described as small rigid bodies rolling and sliding as they interact with each other in the overall mass movement of the bulk of the soil.

Thus, although at the small scale of the sand grain it has edges and there will be voids or air spaces between the grains (perhaps partly water-filled), at a larger scale we can consider the sand hill to be a continuum with bulk characteristics such as strength and density (mass per unit volume) that may vary through space and with time.

However, before considering the mechanical behaviour (or 'mechanics') of such a continuum, we shall extend the brief introduction to the mechanics of a point mass or rigid body given in Section 1.6. These concepts were first codified by Newton (1642–1727).

2.2 Introductory mechanics

The area of physics known as 'statics' deals with the analysis of forces in situations of static equilibrium, where no body under consideration is moving relative to the earth. For such bodies to be at equilibrium, there must be no net or resultant force acting on them. However, if there is no net force acting on a body, it doesn't have to be at rest; it could be moving with a constant velocity (i.e. with constant speed in a constant direction). This is explained in Newton's first 'law' of motion that can be written as

> Every body continues in its state of rest, or in uniform motion in a straight line [i.e. uniform velocity], unless it is compelled to change that state by a net force acting on it.

Admittedly, 'uniform motion in a straight line' is not something commonly observed because of the ubiquitous presence of frictional forces that tend to retard motion. The motion of a billiard ball on its table might be the best approximation to this type of motion we have seen. (Strictly this first law makes assumptions about the spatial reference plane which we will ignore, but see Giancoli (1991), for example.)

Newton's first law also assumes that we know what is meant by a force, and his second law can be interpreted as meeting this need. However, his second law involves another concept, that of momentum, which is defined (for a body) as the product of its mass (m) and velocity (v). Momentum is thus a vector quantity (see Section 1.6), and this is made explicit in Newton's second law:

> The rate of change of momentum of a body is proportional to the net force acting on it, and takes place in the direction of the net force applied.

For a body of constant mass, the rate of change of momentum is equal to the product of mass and rate of change of velocity, or the product of mass and acceleration, a. Hence the net force, F, can be written as

$$F \propto ma \tag{2.1}$$

The SI unit of force is chosen so that the proportionality in Eq. (2.1) becomes an equality. Thus the unit of force, one newton (N), is defined as the force which accelerates a mass of one kilogram (kg) by one metre per second per second, written as either $1\,\text{m/s}^2$ or $1\,\text{m s}^{-2}$. Using such units, Newton's second law can be written

$$F = ma \tag{2.2}$$

Example 2.1

A rock of mass 500 kg perched on an incline of 30 degrees becomes dislodged by an earth tremor. Calculate the initial acceleration of the rock assuming that there is no initial resistance to its movement.

Solution

The force causing the rock's acceleration is the component of the rock's weight which acts in the down-plane direction, given by $mg \sin \theta$ (as shown in the figure below).

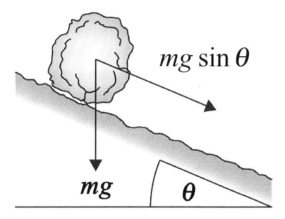

From Eq. (2.2),

$$
\begin{aligned}
a &= F/m \\
&= mg \sin \theta / m \\
&= g \sin \theta \\
&= 9.8 \sin 30^\circ \\
&= 4.9 \, \text{m s}^{-2}
\end{aligned}
$$

Note that, since the mass m cancels out, this acceleration is the same for all masses under the stated assumption.

Newton's third law can be stated as follows:

Whenever one body exerts a force on a second body, the second exerts a force equal in magnitude but opposite in direction on the first.

Thus, in walking, for example, the force our feet exert on the ground is equal and opposite to the force the ground exerts on our feet. If so, why is the acceleration of the earth due to our force on it so completely negligible?

Figure 2.1 The overhanging
section of river bank exerts a
stress on the surface ABCD.
The 'riparian zone' is the strip
of land adjacent to the river
bank.

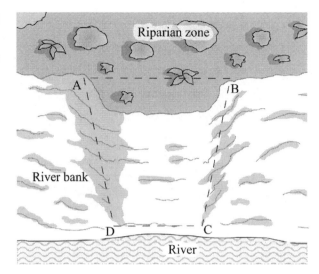

Figure 2.1 The overhanging section of river bank exerts a stress on the surface ABCD. The 'riparian zone' is the strip of land adjacent to the river bank.

In this chapter we consider the effect of forces on solids, focusing particularly on soil because of its importance as an environmental component of our world. The weight of soil itself is a force, caused by the attraction of the earth to everything in its proximity, including our moon. Since the gravitational attractive force acts on every part of a body equally, this kind of force is called a 'body force'. This type of force can be distinguished from a 'surface force', such as the frictional force that exists between any body and another over which it is sliding.

The acceleration of a body falling under gravity, or the acceleration due to gravity, g, is in practice reduced by forces that resist any motion on earth; however, the gravitational force acting on a body of mass m is still given by replacing the acceleration a in Eq. (2.2) by the acceleration g that would occur in the absence of any resistive force. Thus the weight of a body, mg, is a force not to be confused with its mass, m. The clear distinction between mass and weight was one of Newton's great achievements.

The weight of soil itself is a force. Consider the weight of soil in the overhanging section of river bank shown in Fig. 2.1, where the flowing river has undercut the bank and some has already fallen into the river. The weight of the overhanging bank exerts a stress (a surface force per unit area) on the vertical surface shown as ABCD in Fig. 2.1.

If the 'strength' of the soil is not adequate to overcome the 'stress' (or stresses) produced by the overhanging bank then failure will occur, and this section of the bank too will slip into the river. This type of movement of soil is referred to as 'mass movement' since the soil moves in bulk. Other examples of mass movement are landslides, rock falls and soil failure in trench works or other excavations. Mass movement of soil is usually contrasted with the transport of soil in fluids as in soil erosion by wind or water. The distinction between these two forms of sediment transport is not always absolutely clear-cut, however. For example, when a saturated region of soil starts to slide downslope it would be regarded as mass movement; however, if this sliding soil enters a river, there is a transition whereby the sediment is not moving under gravity, but is being carried by flowing water.

In following chapters we will consider the role of forces in the behaviour of fluids and the interacting forces between fluids and solids, for example between water and soil. Another type of interaction between water and soil is the decrease in the strength of soil as water infiltrates into it, so that soil can be damaged or grossly deformed by walking on it, or by cultivation when it is too wet and weak. In contrast, dry beach sand is weak and cohesionless, but, if some (not too much) water is added, the sand can be moulded and will become more cohesive and stronger than when dry. We will seek to understand this apparent paradox later in this chapter.

2.3 Physical characteristics of soil

An overview of the processes involved in the development of soil from the weathering of rock material was given in Sections 1.3 and 1.4. Weathering involves more than simply the disintegration of rock material into inorganic fractions. Rather, it is also a transformation into new and varied types of minerals.

The development of plant and animal life on earth provided the source of another vitally important component of soils, the organic fraction, this being typically less than some 5% of the total soil mass. Because most organic matter accumulates on the earth's surface, it is usually the upper layers of the soil profile (the topsoil) where organic components are most abundant.

The breakdown of accumulated plant and animal matter is carried out by a vast range of soil organisms, somewhat arbitrarily divided into larger macro-organisms and smaller (usually microscopic) soil micro-organisms. Organic matter is the main energy source for the extensive array of soil organisms, and the processing of organic material by these organisms releases vital plant nutrients such as nitrogen and phosphorus.

The end result of the breakdown and decomposition of organic material is a more resistant product, described in Section 1.4 as humus. This colloidal material, usually dark or black in colour, can store even more water and nutrient ions per unit mass than can its inorganic colloidal counterpart, the clay minerals. Humus also has most important desirable effects on physical characteristics of soil, making it more friable in nature, facilitating the penetration of plant roots. Thus the positive contribution made by humus to the soil's capacity for plant and tree production is far greater than might be expected from its small percentage of soil composition.

Continuous cultivation with little or no return of organic matter to the soil inevitably leads to a decline through time in this important humus fraction. The consequences of this decline are reflected in a corresponding drop in the natural fertility or availability of nutrients of the soil, but the physical characteristics of soil also suffer. The deleterious effects on physical characteristics of soil include the following:

- a reduction in capacity to form coherent complex soil units, called soil 'aggregates' or crumbs;
- a decrease in the rate at which water can infiltrate into the soil; and
- a decrease in ability of the soil to store water in the profile in an amount and form available for subsequent use by vegetation.

Thus, though this book focuses particularly on the physical characteristics of soil and their consequences for environmental management, the biological and chemical characteristics of soil are intimately involved in affecting these physical characteristics, as well as being of enormous environmental significance themselves. One example of the great importance of the soil's chemistry is that soils with significant amounts of sodium associated with them can be very susceptible to erosion by water, making them quite unsuitable for plant growth.

It is desirable to be able to describe the variety of soil characteristics in a repeatable and quantitative way. In addition to colour, a commonly employed and useful way to describe soil involves firstly separating the soil into its individual prime particles using standardized techniques to disrupt the bonds holding soil together in aggregates. The proportions by weight in each of various 'fractions' or size ranges are then determined, larger fractions being determined by standard sieves. Separation and measurement of the finer soil fractions makes use of the strong effect of particle size on the velocity with which a particle settles in water – the 'settling velocity'. Fine sediment falling through water very quickly achieves a 'terminal' (or maximum) velocity.

The arbitrary but standard size limits and names for the corresponding soil fractions adopted by the International Soil Science Society are given in Table 2.1. Information on the proportions of soil minerals in each of these size fractions is called a 'mechanical analysis'. There is general international agreement on the upper size limit for the clay fraction ($2\ \mu m$) and the lower size limit for gravel (2 mm). However, in the

Table 2.1 *Fractions in mechanical analysis used by the International Society of Soil Science*

Name of fraction size limits expressed as particle diameters	
Gravel	Above 2 mm
Coarse sand	2.0–0.2 mm
Fine sand	0.2–0.02 mm
Silt	0.02–0.002 mm
Clay	<0.002 mm (2 μm)

USA, the upper size limit for silt is taken as 0.05 mm, and the sand fraction is more finely subdivided than in Table 2.1 (Marshall, Holmes and Rose, 1996).

Material forming the clay fraction, defined as less than 2 μm in size, has undergone intensive chemical weathering, and much of the chemical and colloidal behaviour of soil is associated with this finest soil fraction. Presentation of the relative proportions of material in the various fractions defined in Table 2.1 is one way used to provide a descriptive name of the soil. Thus a sandy clay soil could describe a soil rather low in silt compared with a silty clay. The term 'loam' is often used in describing a soil not dominated by the strongly cohesive behaviour of high-clay soils, or by the non-cohesive characteristics of sand. Further refined, mechanical analysis is the basis of what is called 'soil texture'. This characteristic of soil is distinct from, but somewhat related to, the 'consistency' of soil discussed later in Section 2.7.

The properties of soil are strongly influenced by the characteristics of the rock which, on weathering, provided the parent material on which the soil subsequently formed. This influence is usually quite evident in the soil texture (the proportions of sand, silt and clay in the soil – Table 2.1). For example, those rocks such as granites and sandstones which are rich in quartz weather to form sandy soils. Weathering of rocks poor in quartz (such as basalt) often leads to the formation of clay soils. The originating rock material also has vital implications for the chemical and plant-nutritional characteristics of soil.

Such broad relationships between original rock and subsequent soil can be evident even when soils are not developed from rock *in situ*, but rather on alluvium (sediment deposited from flowing water), or on colluvium (weathered sediment transported downslope mainly by gravity). However, even such broad relationships between originating rock and subsequent soil can be significantly modified by other factors such as climate, topography, soil organisms and the duration of soil formation.

Figure 2.2 Soil aggregates or crumbs, with water within and between adjacent aggregates. (After Rose (1979).)

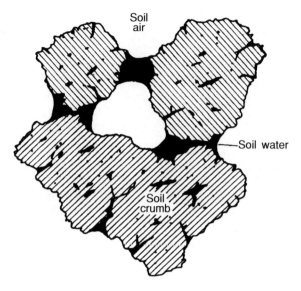

Soil air

Soil water

Soil crumb

The mineral and organic soil components form a matrix enclosing irregularly shaped pores or voids. This 'pore space' is commonly filled partly with air and water vapour and partly with water in the liquid phase, as illustrated in Fig. 2.2.

If in play you have ever been buried in sand at the beach, you will know that you are subject to a pressure in supporting the sand above you. Pressure is defined as the force divided by the area on which the force is acting, and Fig. 2.3 shows how, at any depth z beneath the soil surface, the soil has to support the weight of soil above it. The resulting pressure is called an 'overburden pressure' (p), or 'normal stress', σ, written $\sigma(z)$ since its magnitude depends on z. Figure 2.3 shows that the normal stress is given by the weight (W) of the overburden above that depth divided by the area (A) of the overburden. Just like pressure, stress is a force per unit area, but stress and pressure are formally the same only if the force is at right angles (or normal) to the surface (as in normal stress).

Thus, referring to Fig. 2.3,

$$\sigma = \text{weight of overburden in prism}/A$$

Now, from Eq. (1.2),

$$W = \text{mass} \times g$$

and

$$\text{mass} = \gamma \times \text{volume of prism}$$
$$= \gamma A z$$

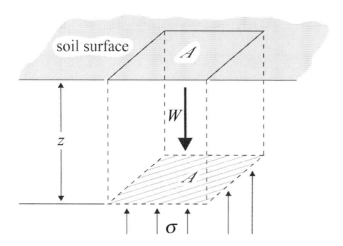

Figure 2.3 The overburden pressure or normal stress σ at depth z in the soil profile.

where γ is the average density of soil in the prism shown in Fig. 2.3. Thus, finally,

$$\sigma(z) = \gamma gz \qquad (2.3)$$

This overburden pressure is a major contributor to soil strength. In excavating a hole in soil to insert a fence post, for example, an increase in soil strength with depth into the soil is experienced. One reason for this obvious increase in soil strength with depth is the increase in overburden pressure described in Eq. (2.3).

Example 2.2

Assuming that the average density of wet soil is $2000 \, \text{kg m}^{-3}$, what would the normal stress due to the weight of overburden at a depth of 3 m be? Express this stress in the SI unit (N m^{-2} or Pa) and an alternative gravitational unit used in the engineering literature, kg wt. cm^{-2}.

Solution
From Eq. (2.3),

$$\sigma(z = 3) = 2000 \times 9.8 \times 3 = 5.88 \times 10^4 \, \text{N m}^{-2} \text{ or Pa}$$

$$= 5.88 \times 10^4 \frac{\text{N}}{\text{m}^2} \times 10^{-4} \frac{\text{m}^2}{\text{cm}^2} \times \frac{1}{9.8} \frac{\text{kg wt.}}{\text{N}}$$

$$= 0.60 \, \text{kg wt. cm}^{-2}$$

Overburden pressure tends to compress or consolidate soil, squeezing the solid components closely together and thus reducing the volume occupied by the pore space between the solids which can contain air or

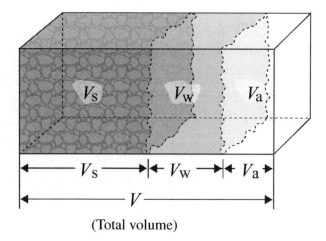

Figure 2.4 Illustrating the various volume fractions of soil.

(Total volume)

water. It is for this reason that the 'bulk density' of soil, ρ_b, typically increases with depth, where ρ_b is the (oven-dry) mass of soil solids (m_s) per unit volume of space (V) from which the soil mass was obtained. Thus

$$\rho_b = m_s/V \tag{2.4}$$

As shown in Fig. 2.4, any volume (V) of soil is made up of the volume V_s of the solids, volume V_w of water and volume V_a of air and water vapour so that

$$V = V_s + V_w + V_a \tag{2.5}$$

The ratio of pore volume ($V - V_s$) to V is called the 'porosity', ε. Thus

$$\varepsilon = (V - V_s)/V \tag{2.6}$$

An alternative measure of pore space is the 'voids ratio' given by $(V - V_s)/V_s$.

The 'water content' of soil plays an important role in determining soil strength. Water content can be expressed on a mass basis, w, where

$$w = m_w/m_s \tag{2.7}$$

and m_w and m_s are the masses of soil water and dry solids in the volume (V). However, there are advantages in some contexts of expressing water content on a volumetric basis, where the volumetric water content, θ, is given by

$$\theta = V_w/V \tag{2.8}$$

It follows from Eqs. (2.4), (2.7) and (2.8) that

$$\theta = w\rho_b/\rho \qquad (2.9)$$

where ρ is the density of water, its mass per unit volume (m_w/V).

The gravimetric water content w (Eq. (2.7)) is simply determined by weighing an amount of soil whose moisture content is to be determined, oven drying at 105 degrees Celsius (105 °C) (adopted as a standard temperature) and reweighing to give the mass of oven-dry soil (m_s). The mass of water, m_w, comes from the drop in mass on drying.

The bulk density of soil *in situ* can be determined in various ways, but one of the simplest is to excavate a volume of soil and determine its oven-dry mass m_s. The volume, V, from which the mass comes (Eq. (2.4)) is then determined by filling the excavation with dry sand of known bulk density. The volume V is calculated from the mass of sand required to fill the excavation.

With ρ_b and w determined as above, the volumetric water content, θ, can be calculated using Eq. (2.9). Whilst this methodology for determining θ may be adequate near the soil surface, there are other ways of measuring the water content of soil that are more convenient and suitable in some contexts. The principles of some of these other methods using equipment such as the neutron moisture meter and time-domain reflectometry are described in Marshall, Holmes and Rose (1996).

Example 2.3

In determining the bulk density of surface soil, 2.5 kg of moist soil was excavated with a spade. On oven-drying to 105 °C this soil lost 0.5 kg in mass. The excavation was filled level to the top of the soil with 3.0 kg of dry sand of standard bulk density 1600 kg m^{-3}. Calculate the bulk density of the soil.

Solution
The dry mass of soil is $m_s = 2.0$ kg, and

volume of sand fill = mass/density = $3.0/1600 = 1.88 \times 10^{-3}$ m^3

Thus

$$\rho_b = m_s/V$$
$$= 2.0/1.88 \times 10^{-3}$$
$$= 1070 \text{ kg m}^{-3}$$

Though the nature of the soil itself and its water content can contribute to soil strength, in the next subsection we shall examine how overburden pressure experienced by soil commonly plays an even larger role in determining its strength. The consolidation, or decrease in porosity, produced by overburden pressure in clay soils can remain unrelieved over geological time scales, even if natural erosion processes over that time scale have removed great depths of soil so that the current surface soil experiences negligible overburden pressure. Such clay soils are said to be in an 'overconsolidated' condition, a condition characterised by high strength and bulk density and low porosity.

Overconsolidated clay contrasts with 'normally consolidated' clay, which has never been subjected to a greater overburden pressure than that which it currently experiences. Disturbance, such as cultivation, largely removes the effect of overconsolidation. The bulk densities of soils low in clay minerals are rather independent of the pressure applied to them, so overconsolidation does not occur.

Example 2.4

What is the relation between the bulk density of a particular soil (ρ_b) and the volumetric water content θ if the pore space is always completely filled with water?

Solution
If there is no air in the pore space, $V_a = 0$, so

$$V = V_s + V_w \quad \text{(Eq. (2.5))}$$

Thus

$$\theta = V_w/V = 1 - V_s/V$$

and, if the density of soil solids is ρ_s,

$$\rho_b = m_s/V = \rho_s V_s/V$$
$$= \rho_s - \rho_s \theta$$

This relationship is shown in Fig. 2.5.

The slope of the dashed line is $-\rho_s$, and the intercept, when $\theta = 0$, is ρ_s (see the equation above). Water content corresponding to A on the solid line in Fig. 2.5 yields the maximum bulk density, and compaction at this water content is used in road construction, for which a high bulk density of soil is desired. Air in the pore space is essential for the growth of most plants (paddy rice being an exception, for example). Without continuity in air space within soil the carbon dioxide evolved by plant roots and other soil-living organisms could reach levels harmful to plants. Figure 2.5 also shows the relationship between ρ_b and θ if the air volume fraction V_a/V is 0.1 or 10%.

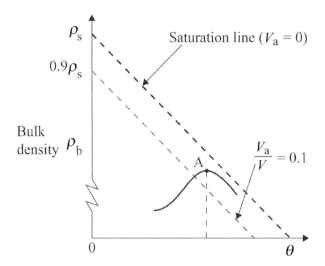

Figure 2.5 The relation between bulk density and θ if the soil pore space is filled with water (dashed line). The solid line shows the form of the relationship between ρ_b and θ obtained if a soil at different water contents is subjected to a constant amount of compaction.

'Sediment' is an even more general term than soil, at least in terms of the size of component units involved. The term sediment is used to describe any material on earth that moves, or can move, due to environmental forces, and so may range from great rocks that can topple or be moved in major floods down to the finest microscopic soil component.

This brief introduction to soil and physical characteristics of soil is expanded in texts such as Marshall, Holmes and Rose (1996).

2.4 Strength and behaviour of sediments

If dry sand is poured onto a surface, it tends to form an approximately conical pile as illustrated in Fig. 2.6. Each particular dry non-cohesive granular material, such as sand or gravel, forms a characteristic angle of repose when poured onto a horizontal surface. A surface at that angle is evidently stable, higher slopes being unstable, material being redistributed by sliding and rolling downhill until a surface with the stable angle of repose is achieved.

Consider the forces acting on an individual sand grain resting on an inclined plane surface (Fig. 2.7). The vertically downward acting weight of the sand grain, W, can be resolved into two components, one at right angles to the plane, namely $W\cos\alpha$, and the other parallel to the plane, $W\sin\alpha$. As discussed in Section 1.2, W can be regarded as the sum of these two components, so that these two components can be regarded as replacing W.

The component force $W\cos\alpha$ and the normal reaction R to it are action and reaction forces referred to in Newton's third law, and are

Figure 2.6 Showing the angle of repose, ϕ, formed by dry sand poured onto a horizontal surface.

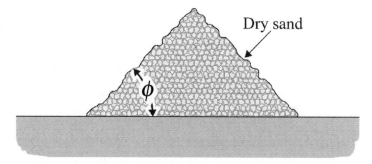

Dry sand

Figure 2.7 Forces acting on a sand grain on an inclined surface.

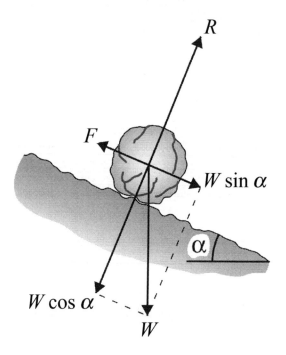

R

F

$W \sin \alpha$

α

$W \cos \alpha$

W

therefore equal in magnitude. If the sand grain doesn't slide down the plane in response to component force $W \sin \alpha$, it is because there is an equal and oppositely directed force up the plane, F, this being a frictional force between the sand grain and the plane surface. No motion of a body implies that there is no net force acting on it, and this is satisfied if there is no net force either parallel or perpendicular to the plane (Fig. 2.7).

However, as the angle α is increased, the sand grain will commence to slide at some critical angle, ϕ, which depends on the nature of the materials involved. Movement of the sand grain also indicates that there is an upper or critical limit to the value of F, denoted F_c. It is found

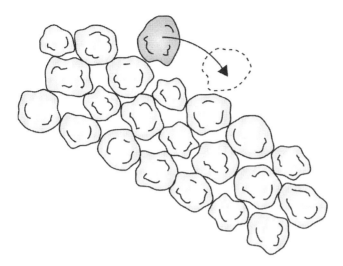

Figure 2.8 A sand grain rolling down an inclined bed of other sand grains.

experimentally that, for any particular pair of materials, the ratio F_c/R is approximately constant, and this ratio is called the static coefficient of friction, μ. Thus

$$\mu = F_c/R \tag{2.10}$$

(Once the sand grain is sliding, the frictional interaction force F_c will decrease a little, but this is of less concern to us here.)

If a sand grain is sliding, not over a planar surface, but over the rough surface provided by many other sand grains, then a component of F_c could be the force required to lift the sand grain over its neighbour (Fig. 2.8). If so, the source of F_c is not entirely frictional in nature. Indeed, the source of frictional forces is complex, especially in natural materials such as soil, sand and rock.

Frictional forces quite generally act to oppose those forces tending to produce movement, such as occurs in the slippage of one layer of material over another. As illustrated in Fig. 2.7 and Eq. (2.10), a force normal to the plane of slip is involved in frictional forces, and this force can be due to the weight of the material itself.

From Fig. 2.7, assuming that the angle α is at its maximum or critical value, ϕ (the angle of repose), then

$$\mu = F_c/R = W \sin\phi / W \cos\phi$$
$$= \tan\phi$$

or

$$F_c = R \tan\phi = \mu R \tag{2.11}$$

Children standing near the edge of a landslide in East Timor, 1999. Reproduced by permission of Lisette Wilson and Kevin Austin.

Experimentally it is found that $\phi = 27$–$40°$ for sands, corresponding to $\mu = 0.5$–0.8. For coarser non-cohesive material such as gravel (commonly regarded as material of size >2 mm), ϕ is greater and can be of order 40–$50°$ ($\mu = 0.8$–1.2). Values of ϕ for soils can vary significantly depending on composition, water content and prior stress history.

So far we have been considering isolated grains of material. However, in mass movement such as landslides and avalanches, vast numbers of grains are involved. For granular material, such as snow, sand, sandy soils, or rock debris, landslides can be crudely represented as an upper mobile layer of material sliding over immobile base material, with sliding taking place at a 'plane of failure' (Fig. 2.9).

At the plane of failure we can imagine one sheet of grains sliding over another stationary sheet of material, the angle of the failure plane being ϕ (Fig. 2.10).

For any number of grains in contact with others, n, we can write, using average values for the quantities involved;

$$nF_{\mathrm{c}} = nR\tan\phi \qquad (2.12)$$

where ϕ is called in this context the angle of internal shearing resistance, or simply the friction angle, somewhat akin to the critical angle of friction described earlier. Cancelling out n from both sides of Eq. (2.12) shows that Eq. (2.11) can still be expected to hold, even with an assembly of particles, and, indeed, it has been used for at least 200 years.

This frictional characteristic of sediment is one important source of its strength, another source being its common ability to stick together or 'cohere'. This cohesive source of strength is noticeable in soils of

Plane of failure

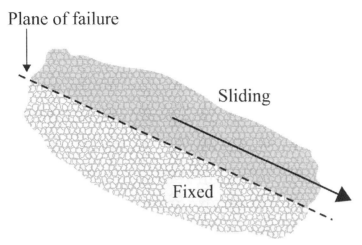

Sliding

Fixed

Figure 2.9 Illustrating a landslide.

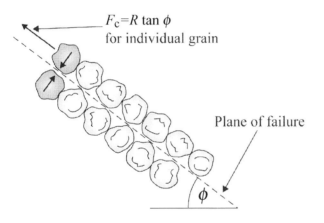

$F_c = R \tan \phi$
for individual grain

Plane of failure

ϕ

high clay content and, in extreme form, in rock. Before we consider the strength of soil and sediments in general any further, we shall discuss the ideas of stress, strength and strain.

2.5 Stress, strength and strain

Whatever the magnitude or kind of force, its effect depends on the area over which the force acts. The ratio of force to area is called a 'stress'. There are various kinds of stress. Perhaps the simplest kind of stress to think about is 'tensile stress' as in a crane cable or a rope supporting a rock climber. The tensile stress is the force in the cable or rope divided by its cross-sectional area. As the tensile stress in the cable is increased,

Figure 2.11 Three types of stress and strain: (a) tensile, (b) compressive and (c) shear.

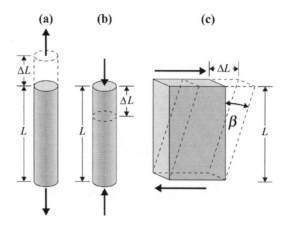

its length increases slightly, and the increase in length divided by the original unstressed length is called the 'tensile strain'.

The weight of a person standing on soil divided by the area of their shoes is the magnitude of the compressive stress exerted on the soil. All stresses can be measured in the SI unit of newtons per square metre or $N\,m^{-2}$, which is called one pascal (Pa). Stresses in soil are commonly expressed in kPa, and stresses in rocks in MPa (a million Pa).

Both tensile and compressive stresses in solids create a change in dimension, the resulting strain being defined by

$$strain = \Delta L / L$$

where ΔL is the change in length of the original length, L (Figs. 2.11(a) and (b)). For some solids, such as metals, there is is a range of stress over which the ratio $E = $ stress/strain is constant, and E is called 'Young's modulus', a characteristic of the material.

In shear stress (Fig. 2.11(c)), the forces are parallel to a surface, rather than perpendicular to it as in tension and compression. For shear stress the strain is $\Delta L / L = \tan \beta$, where β is the 'shear angle'. It is shear stress that is most important for the failure of particulate materials such as sediment.

The maximum resistance that a solid can offer to an applied stress is called the 'strength' of the material. A stress beyond the strength of the material commonly produces a very large increase in strain or complete rupture, described as 'failure'. Such failure in a part of the earth's surface is referred to as 'rapid mass movement', landslides and rock failures being examples. A less dramatic example of inadequate

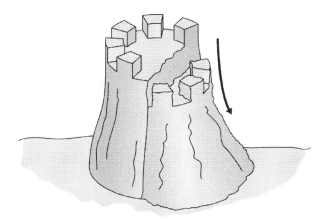

Figure 2.12 Failure indicated by sliding in response to shear stresses in a wet 'sand castle'.

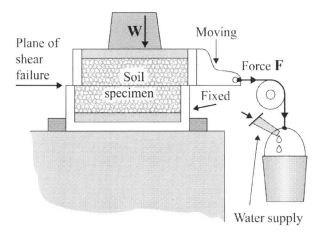

Figure 2.13 The principle of the shear box used to measure the shear strength of soil. The plane on which shear failure will ultimately take place is indicated.

shear strength is the failure of a section of a wet 'sand castle' as shown in Fig. 2.12.

However, even without failure or rapid mass movement occurring, stresses in the earth's surface can result in rather continuous or ongoing strain at such a slow rate that it may be almost imperceptible. The movement of glacier ice and 'soil creep' are examples of such slow mass movement.

A common method of measuring the shear strength of soil in the laboratory is to use the 'shear box' illustrated in Fig. 2.13. Soil is placed in the shear box and subjected to a 'normal stress', which in Fig. 2.13 is due to the weight W. The normal stress in Fig. 2.13 is the stress normal (or at right angles) to the plane of failure. The shearing force F (Fig. 2.13) is then gradually increased (for example by adding water

Figure 2.14 (a) The principle of the shear-vane equipment used to measure the shear strength of soil in the field. (b) Showing the cylindrical surface of soil sheared by the shear vane.

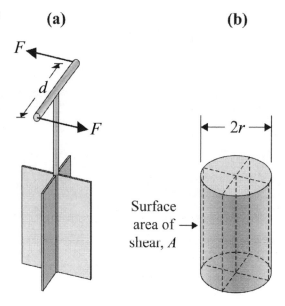

(a)   **(b)**

F

d

F

Surface area of → shear, A

2r

to the receptacle as in Fig. 2.13) until failure occurs. If the cross-sectional area of soil in the box is A, then the shear stress τ is given by

$$\tau = F/A \qquad (2.13)$$

and, at failure, when τ is at its maximum value, this stress is called the 'shear strength' of the soil, s. Thus,

shear strength, s = maximum value of τ

= (maximum value of F)/A

= shear stress at failure

One way of measuring shear strength in the field is to use a 'shear vane' as shown in Fig. 2.14. This equipment consists of two thin crossed blades, which are pressed into the soil. Then two equal and opposite forces, F, are applied to the handle of the equipment, these forces being separated by a distance d. The product Fd is called a 'torque', and, as illustrated in Fig. 2.14(a), this torque tends to shear a cylindrical surface of area A (neglecting the shear on the bottom of the cylinder) and radius r. Denote the shear stress on this cylindrical soil surface by τ (N m^{-2} or Pa). Then the force applied tangentially around the entire cylindrical surface is τA (N), and the moment of this force (i.e. the product of the force and the radius at which it operates) is τAr. This moment must equal

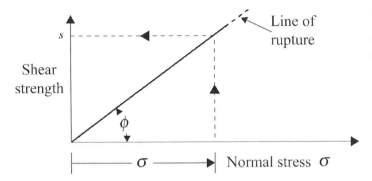

Figure 2.15 The dependence of the shear strength of non-cohesive sediment on the normal stress.

the applied torque, Fd, so

$$\tau\, Ar = Fd$$

At failure, $\tau = s$, the shear strength of the soil, so

$$s = F_c d/(Ar) \qquad\qquad (2.14)$$

where F_c is the maximum magnitude of F which occurs at failure.

2.6 Soil strength

For an assembly of soil particles, it follows from Eq. (2.12), if there are n particles in an area of surface A, that

$$n F_c / A = (n R / A)\, \tan \phi$$

or

$$s = \sigma \tan \phi \qquad\qquad (2.15)$$

where σ is the normal stress due to overburden pressure as defined in Eq. (2.3). Equation (2.15) defines the source of soil strength which is entirely due to friction and the existence of an overburden pressure σ. This simple equation is found to describe quite well the strength of non-cohesive sediment such as sand and gravel, and is illustrated in Fig. 2.15.

Right at the soil surface, where $\sigma = 0$, a sandy soil surface, for example, is indicated as having no strength, and, if the sand is dry, this agrees with experience. An implication of this is the well-known tendency of desert sands to shift under the shear stress exerted on the sand by wind. (Moist sand does have some strength, for reasons discussed in Section 2.7.) If you have excavated a hole by hand in beach sand, you

Figure 2.16 The increase in
shear strength with depth for
a non-cohesive soil.

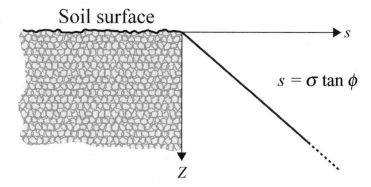

will be well aware of the increase in strength of the sand with depth,
which is illustrated in Fig. 2.16.

Combining Eqs. (2.3) and (2.15) gives

$$s = \gamma g z \ \tan \phi \qquad (2.16)$$

this increase in shear strength with depth being shown in Fig. 2.16.

Cohesive soils

Soils with a significant fraction of finer materials, clay and silt
(Section 2.2), tend to stick together, thus having some strength even
in the absence of any overburden pressure. Lithified materials such as
rocks and cemented material can have very great strength. The 'cohesive strength', C, is the shear strength of sediment in the absence of
overburden pressure, thus providing a source of strength that is lacking in non-cohesive sediments. Following a French physicist, Coulomb
(1736–1806), the more general expression for the strength of cohesive
sediments, called 'Coulomb's law', is

$$s = C + \sigma \ \tan \phi \qquad (2.17)$$

Coulomb's failure law is illustrated in Fig. 2.17.

Both the cohesive strength C and the friction angle ϕ can vary substantially with the water content of the soil, θ. If soils have any clay or
silt component then C and ϕ decrease with increasing water content, as
illustrated by the slippery and weak nature of a wet clay soil surface.

Example 2.5

An undisturbed sample of soil was cut into two pieces, each piece being
placed in a shear box (see Fig. 2.12). The soil in one shear box was

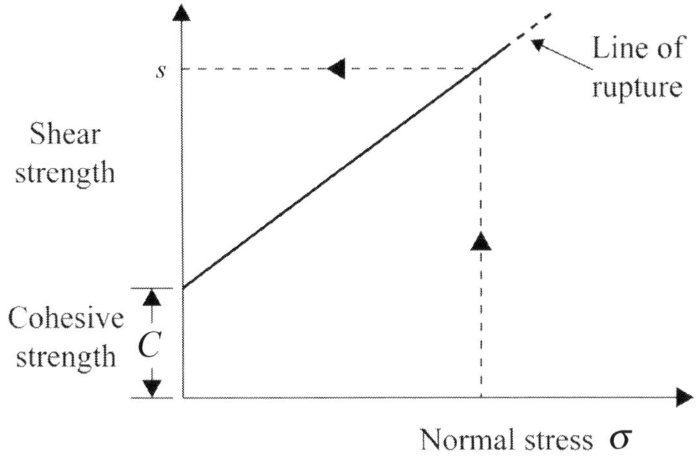

Figure 2.17 Illustrating Coulomb's failure law.

loaded with a mass of 10 kg, and the force required to shear the sample was 100 N. The soil in the second shear box was loaded with a mass of 30 kg, and a force of 200 N was then needed to cause failure.

If the plan area of soil in the box is $0.2 \, \text{m}^2$, calculate the cohesive strength and the angle of friction for this soil.

Solution

$$s = C + \sigma \tan \phi \quad (\text{Eq. (2.17)})$$

For the first box,

$$\sigma_1 = 10 \times 9.8/0.2 = 490 \, \text{N m}^{-2}$$
$$s_1 = 100/0.2 = 500 \, \text{N m}^{-2}$$

For the second box,

$$\sigma_2 = 1470 \, \text{N m}^{-2}$$
$$s_2 = 200/0.2 = 1000 \, \text{N m}^{-2}$$

Substituting these values into Eq. (2.17) and solving the two resulting equations simultaneously gives $C = 259 \, \text{N m}^{-2}$ and $\tan \phi = 0.510$, so $\phi = 27°$.

Example 2.6

(a) A trench with vertical walls is excavated in soil whose physical characteristics are assumed uniform to the depth, H, of the trench (see the figure). The cross-hatched section of the bank of the trench can slip into the trench if shear failure occurs on the potential failure plane, which can be shown to be at an angle of $(45 + \phi/2)$ degrees to the horizontal.

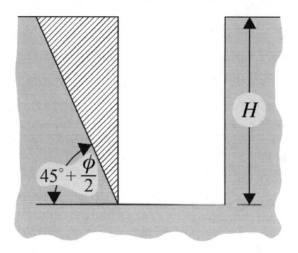

Firstly show that the weight, W, of the cross-hatched section per metre of trench length is given by

$$W = (\gamma/2)H^2 g \tan(45 - \phi/2) \quad (\mathrm{N\,m^{-1}}) \tag{i}$$

where γ $(\mathrm{kg\,m^{-3}})$ is is the wet density of the soil, and g $(\mathrm{m\,s^{-2}})$ is the acceleration due to gravity.

Solution
Consider a 1-m length of the cross-hatched section of the trench wall. The weight of this section is

$$W = \mathrm{mass} \times g$$
$$= \gamma \times \mathrm{volume} \times g$$
$$= \gamma g \times \tfrac{1}{2} \times (\mathrm{base} \times \mathrm{altitude}) \times 1 \, (\mathrm{m})$$

or

$$W = \gamma g \times \tfrac{1}{2}H^2 \tan(45 - \phi/2) \quad (\mathrm{N\,m^{-1}})$$

(b) If $\gamma = 1600 \, \mathrm{kg\,m^{-3}}$, $H = 5 \, \mathrm{m}$, $g = 9.8 \, \mathrm{m\,s^{-2}}$ and $\phi = 38°$, what is the magnitude of W?

Solution
By substitution of these values into (i), $W = 95.6 \times 10^4 \, \mathrm{N\,m^{-1}}$.

(c) Draw a diagram showing the forces acting on a 1-m length of the cross-hatched section of the trench wall shown in the sketch in part (a). Next to it draw a similar diagram, but showing forces resolved perpendicular to and parallel with the plane where shear failure could occur.

Solution

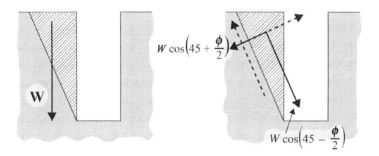

In the figure above, solid lines represent the force due to weight W of unit length of the crosshatched section of the trench wall (or its components) and dashed lines represent opposing forces (oppositely directed and equal in magnitude if in equilibrium).

(d) Derive an analytical expression for the average shear stress, τ, acting on the potential plane of failure.

Solution

$$\text{shear stress } \tau = \frac{\text{force/m acting parallel to the plane of failure}}{\text{area of 1-m length of failure plane}}$$

This area of the failure plane is $[H/\cos(45 - \phi/2)] \times 1$ (m). The right-hand figure in part (c) shows that the force parallel to the failure plane is $W\cos(45 - \phi/2)$. Thus

$$\tau = W \cos^2(45 - \phi/2)/H \qquad \text{(ii)}$$

(e) If the average strength of the soil in the trench wall is 100 kPa (or 100 000 Pa), would shear failure of the trench bank be expected if the values of the parameters are as in part (b)?

Solution
Substituting $W = 95.6 \times 10^4 \, \text{N} \, \text{m}^{-1}$ and other values from (b) into equation (ii) gives $\tau = 155 \, \text{kPa}$. Since this shear stress exceeds the average shear strength, failure of the wall of the excavated trench would be expected. (Thus the wall of the trench would need to be supported to prevent failure.)

Figure 2.18 (a) The rise of water inside a small-diameter capillary tube. (b) Water held between two sand grains by capillary forces.

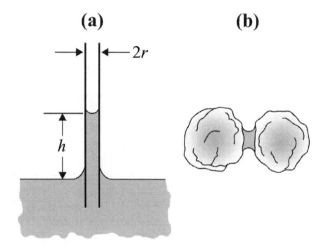

Figure 2.18 (a) The rise of water inside a small-diameter capillary tube. (b) Water held between two sand grains by capillary forces.

2.7 Effects of water content on soil strength

As soil absorbs water its strength is generally reduced because of the decreases in cohesive strength C and friction angle ϕ. Yet, as those who have made 'sand castles' on the beach have learnt, the cohesionless weak behaviour of dry sand can be strengthened enough to form stable shapes by the addition of some water – provided that the amount of water added is not enough to saturate the sand. How does water give dry sand some cohesive strength?

The answer to this question is related to the commonly observed phenomenon of 'capillarity' – the fact that water is seen to rise up inside a small-diameter glass 'capillary' tube as shown in Fig. 2.18(a).

Since the pressure on the horizontal water surface outside the capillary tube is atmospheric, it is clear that the pressure under the curved surface of water in the capillary tube of Fig. 2.18(a) must be less than atmospheric. Water rises in the capillary tube because the strong attractive (or 'adhesive') forces between the molecules of water and glass are stronger than the attraction of water molecules for each other.

When a drop of liquid is placed in contact with a solid surface there are two possible types of behaviour.

- The liquid drop may remain as a compact drop that does not expand over the solid surface, mercury on a glass surface being an example (Fig. 2.19(a)).
- Alternatively the liquid may spread out on the surface, at least to some extent, or even spread over the surface to 'wet' it. Water on glass can be an example of this type of wetting behaviour, depending on the cleanliness of the glass (Fig. 2.19(b)).

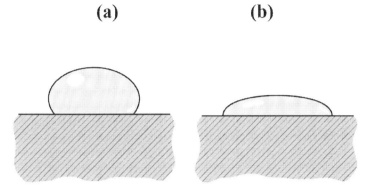

(a)   (b)

Figure 2.19 The two possible types of interaction of a liquid and a solid surface.

If the liquid remains as a drop, the liquid and solid surfaces meet at a fairly definite 'contact angle', but this angle is effectively zero if the liquid wets the solid, as is assumed in Fig. 2.18(a). If so, it is as though a force, T, acts on each unit length of the water/glass interface in an upward direction, given an upward force on the water in the capillary of $2\pi r T$, where r is the radius of the capillary tube (Fig. 2.18(a)). This must support the weight of the water column inside the capillary tube, which is the product

$$(\text{water density } \rho) \times (\text{volume of water}) \times g$$

or $\rho \pi r^2 h g$, where h is the capillary elevation (Fig. 2.18(a)). At equilibrium these two oppositely directed forces must be equal, from which it follows that the pressure less than atmospheric, or 'suction', τ_s, at A in Fig. 2.18(a) is given by

$$\tau_s = \rho g h = 2T/r \qquad (2.18)$$

Notice that the smaller the value of r the larger the pressure deficit or suction.

Assuming that the water between the two sand grains in Fig. 2.18(b) wets the grains, then there will also be a pressure deficit or suction, τ_s, in this geometrically complex form of capillary water. This suction will obviously tend to attract the two sand grains to each other, acting in some respects like a weak glue, so adding some cohesive strength to the otherwise non-cohesive sand grains. On extending this concept to a three-dimensional array of sand grains, it is possible to understand why adding some water to sand (though not enough to saturate the pore volume) gives it some cohesive strength, so that the sand behaves in a qualitatively similar manner to a cohesive soil.

Suppose that a volume of sand is saturated and excess water is then allowed to drain. The water can then be at a suction (or pressure less than

Figure 2.20 Effects of
suction and degree of
saturation on the effective
stress of a beach sand drying
from saturation. The
experimental data points for
effective stress agree closely
with the suction until the
fractional saturation falls to
about 0.6. (From Marshall,
Holmes and Rose (1996).)

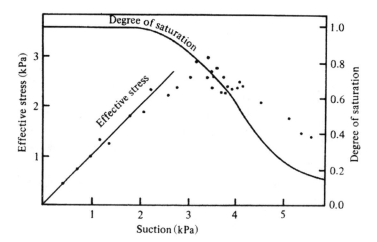

Figure 2.20 Effects of suction and degree of saturation on the effective stress of a beach sand drying from saturation. The experimental data points for effective stress agree closely with the suction until the fractional saturation falls to about 0.6. (From Marshall, Holmes and Rose (1996).)

atmospheric) τ_s, with the pore space remaining essentially saturated. Then the 'effective stress', σ_e, at any depth in the sand is greater than the normal stress, σ, by the amount of that suction, so that

$$\sigma_e = \sigma + \tau_s \tag{2.19}$$

Coulomb's failure law then requires modification from its form in Eq. (2.17) to become

$$s = C_e + \sigma_e \tan\phi_e \tag{2.20}$$

where the suffix e is added to C, σ and ϕ to indicate effective values whose magnitudes depend on the suction. From Eqs. (2.19) and (2.20) it is clear that the suction τ_s increases the strength of the sand, in line with common experience.

If more water is allowed to drain from the pore space between the sand grains, so that the sand becomes 'unsaturated', then water appears as rings such as shown in Fig. 2.18(b), which cover only a fraction of the area of any plane within the sand. Hence the effectiveness of suction in increasing effective stress is reduced by a factor related to the fractional degree of saturation, χ, which for sand is virtually the same as the volumetric water content, θ. Then Eq. (2.19) is modified to

$$\sigma_e = \sigma + \chi\tau_s \tag{2.21}$$

where $0 \le \chi \le 1$, where $\chi = 1$ at saturation, and $\chi = 0$ for dry sand.

Figure 2.20 illustrates the dependence of σ_e on suction τ_s for a range of values of χ, or degrees of saturation. This figure shows that, until χ falls below about 0.6, the increase in effective stress is closely equal to the suction, as expected from Eq. (2.21) if $\chi = 1$. However, as χ

decreases the effectiveness of the increasing suction on increasing the effective stress declines, as will the soil strength.

Consider a soil profile that is saturated with water. At depth z beneath the soil surface the total normal stress is due to the weight of soil and water above that depth (Fig. 2.16). At depth in saturated soil the pressure of water in the soil pores is positive (or greater than atmospheric). When this pore water pressure is positive we will represent it by the symbol u, to distinguish it from a negative pore water pressure, or suction, denoted by τ_s. Thus the total overburden load expressed as the normal stress, σ, will be borne partly by effective inter-aggregate stresses in the soil, represented by σ_e, but partly by the positive pore water pressure, u. Thus

$$\sigma = \sigma_e + u \qquad (2.22)$$

where σ_e is the effective stress. It is only this effective inter-granular stress which provides soil with strength due to friction, and, from Eq. (2.22),

$$\sigma_e = \sigma - u \qquad (2.23)$$

From Eqs. (2.23) and (2.20) it follows that a positive pore water pressure will reduce the shear strength of soil. Pore water pressure can also affect the cohesive strength and friction angle, as indicated in the modified version of Coulomb's failure law given in Eq. (2.20). Thus Eq. (2.20) applies irrespective of whether the pore water pressure is positive or negative.

A positive pore water pressure can arise due to the accumulation of infiltrated rainfall or snow melt, especially when drainage is impeded by a soil layer less permeable than surface layers. Such an increase in pore water is the most common cause of landslides because of the consequent reduction in soil strength (Eqs. (2.23) and (2.20)).

Example 2.7

Suppose that the angle ϕ of internal friction of a clay soil is related to the volumetric water content θ by the equation

$$\phi = 27 - 45\theta \quad \text{(degrees)}$$

Also, the cohesive strength C of this soil is given by

$$C = 4000 - 8000\theta \quad (\text{N m}^{-2})$$

and the wet density γ is given by

$$\gamma = 1200 + 1000\theta \quad (\text{kg m}^{-3})$$

Calculate the shear strength, s, of the soil at a depth 2 m beneath the surface of this soil, (a) when $\theta = 0.1$ and (b) when $\theta = 0.4$.

Solution

At depth $z = 2$ m, from Eq. (2.3),

for $\theta = 0.1$

$\sigma = (1200 + 100) \times 9.8 \times 2$
$\quad = 2.55 \times 10^4 \, \mathrm{N\,m^{-2}}$
$\phi = 22.5°$
$C = 3200 \, \mathrm{N\,m^{-2}}$
$s = 3200 + 2.55 \times 10^4 \tan 22.5$
$\quad = 13.8 \times 10^3 \, \mathrm{N\,m^{-2}}$
$\quad = 13.8 \, \mathrm{kPa}$

and for $\theta = 0.4$

$\sigma = (1200 + 400) \times 9.8 \times 2$
$\quad = 3.14 \times 10^4 \, \mathrm{N\,m^{-2}}$
$\phi = 9°$
$C = 800 \, \mathrm{N\,m^{-2}}$
$s = 800 + 3.14 \times 10^4 \tan 9$
$\quad = 0.801 \times 10^3 \, \mathrm{N\,m^{-2}}$
$\quad = 0.801 \, \mathrm{kPa}$

2.8 The consistency of soil

The term 'consistency', when used in this context, describes the state of soil, meaning whether it behaves more like a solid, a plastic material, or a liquid. The type of behaviour of soil can be similar to any of these three possible states, depending on its water content.

The application of a stress to a solid causes a change in dimension as illustrated in Fig. 2.11. For metal solids there is an upper limit to stress, called the 'elastic limit', below which the strain is proportional to the stress, and the strain is reversible, the original length being regained on removal of the stress. If stress is increased beyond the elastic limit, the relationship between stress and strain is typically non-linear, and, after some degree of irreversible strain, failure will occur.

The types of relationship between stress and strain for two types of soil are shown in Fig. 2.21.

As the water content of soil increases substantially, its behaviour changes gradually from that of a solid to that of a liquid, for which the concept of strength is meaningless, the relevant property for a liquid being 'viscosity', which is discussed in Chapter 3.

As water is added to cohesive soils (i.e. soils with a significant clay content), the consistency changes from that of a solid, through a plastic stage, to a slurry with similarities in behaviour to a liquid. The water contents at which there is an arbitrary but experimentally defined transition between these consistency states are called the 'Atterburg limits'. The lower of these two limits is called the 'plastic limit' defining the lower limit to the plastic state. This limit is the gravitational water content at which soil rolled out into a thin thread between the palms of one's hands starts to crumble. The higher of the two limits, which defines the upper

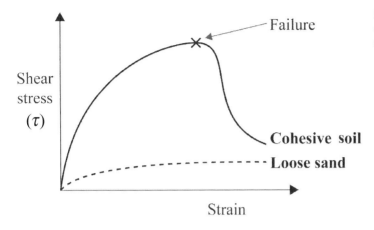

Figure 2.21 Stress–strain relationships for a cohesive soil and loose sand.

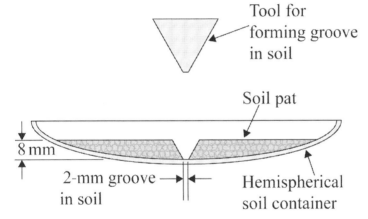

Figure 2.22 The equipment used to determine the liquid limit of a cohesive soil (shown in cross-section).

limit to the range of water content within which the soil exhibits plastic behaviour, is called the 'liquid limit'. To determine this limit, a pat of soil is prepared in an open spherical metal cup as shown in Fig. 2.22. Using a special metal tool, a groove of standard shape and size shown in Fig. 2.22 is produced in the wet soil. The cup containing the prepared soil is allowed to fall through a standard height in a standard mechanical device, and the number of blows required for the two separated parts of the soil to move together and just touch is recorded.

There are general correlations between the texture of soil as indicated by its mechanical analysis (Section 2.2) and the magnitude of the Atterburg consistency limits. An example of this correlation is given for two soils in Table 2.2.

The strength of an unconfined cylindrical or prismatic sample of cohesive soil under compression is another measure of a soil's

Table 2.2 *Illustrating the general correlation between mechanical analysis and Atterburg consistency limits for soils*

Texture	Clay percentage	Consistency limits (gravimetric water content, %)	
		Plastic limit	Liquid limit
Sandy loam	12	16	21
Clay	51	36	83

Figure 2.23 Different forms of failure of a cylindrical sample of a cohesive soil in an unconfined compression test.

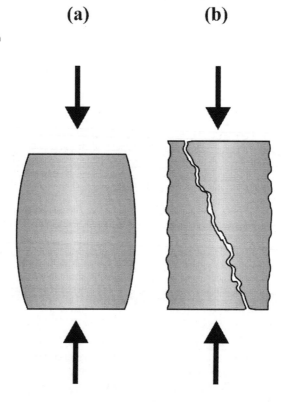

(a)   **(b)**

consistency (Fig. 2.23). In the field any volume of soil is supported or laterally confined by adjacent soil. However, a cylindrical or prismatic sample of cohesive soil can be extracted and tested in compression without such lateral support. The stress at failure is called the 'unconfined compressive strength' of the soil. As shown in Figs. 2.23(a) and (b), failure of such an unconfined soil sample can occur in different forms.

Figure 2.23(a) illustrates failure of soil of more plastic or 'soft' consistency, whereas the form of failure in Fig. 2.23(b) indicates a more brittle or stiff consistency. Of course, the same soil sample could exhibit either type of consistency at different water contents.

A clay soil with an unconfined compressive strength of up to approximately 5×10^4 N m^{-2} (or 50 kPa) can be said to have a 'soft' consistency, whereas a soil with strength several times greater than this may be called 'stiff'.

2.9 Some environmental implications of the mechanical characteristics of soil

The mechanical analysis of soil described in Section 2.3 is one important factor affecting the soil's ability to form structural aggregates, and the stability of these aggregates in turn has a large influence on the ability of soil to absorb rainfall of any significant rate. The form of structural organisation of soil in the soil profile also plays an important role in modifying the rate of movement of water through the soil profile and the accompanying translocation of chemicals necessary for the growth of vegetation (Fig. 1.5).

Table 2.2 illustrated the strong interactions between clay content (a dominant component in mechanical analysis) and the consistency of soil. These characteristics play an important role in determining the response to, or consequence of, soil disturbance by cultivation in agriculture, for example. All these factors also affect the strength of soils, which is also strongly modified by soil water content. The strength of soil can be sufficiently great that plant roots cannot penetrate the soil matrix. Also, for germinating seeds to emerge through the soil surface, the strength of the soil layer above the seed must not exceed certain limits.

The whole mechanical behaviour of the earth's mantle, most obvious in terms of earth-surface processes and landscape formation, is largely mediated by the mechanical characteristics of soil, driven by climatic characteristics, and modified by biological and land-use factors. Earth-surface processes also involve sediment transported by the two fluids water and wind. The rapid soil movement in landslides and the less rapid movements in soil heaving or soil creep all have consequences for human safety and the integrity of the foundations of homes, buildings, roads and other built structures. Failures of earth in road and trench construction and in mining operations have all taken a substantial toll on human life.

Klute (1986) and McKenzie, Coughlan and Cresswell (2002) describe a much wider range of methods of soil analysis and physical measurement than is covered in this chapter. The latter reference pays particular attention to how such measurement can assist in evaluating the suitability of land for possible alternative uses. Using soil analysis to warn against unsuitable use of land can help avoid land degradation.

Main symbols for Chapter 2

a Acceleration

A Area

E Young's modulus

C Cohesive strength (Eq. (2.17))

F Force (Eq. (2.2))

F_e Critical value of frictional force

g Acceleration due to gravity

L Length

m Mass

n Number of grains in contact with others

R Reaction force

s Shear strength

u Pore water pressure

V Volume (with component volumes V_s, V_w and V_a – Eq. (2.5))

w soil water content on a mass basis (Eq. (2.7))

W Weight

z Depth

γ Average wet soil density

ε Porosity (Eq. (2.6))

θ Angle, or volumetric water content (Eq. (2.8))

μ Coefficient of friction (Eq. (2.10))

ρ_b Bulk density of soil

$\sigma(z)$ Normal stress at depth z

σ_e Effective stress (Eq. (2.3))

τ Shear stress (Eq. (2.13))

τ_s Soil water suction (Eq. (2.18))

ϕ Critical angle

Exercises

2.1 (a) Draw three diagrams illustrating the three types of stress that can occur in solids (singly or in combination). These three types of stress are

- tensile stress,
- compressive stress and
- shear stress.

(b) In describing the relationship between force and the resulting deformation (i.e. between stress and strain), what is meant by the terms 'elastic region' and 'plastic region'?

2.2 Discuss the concepts of pressure, normal stress, shear stress and shear strength. What are the physical dimensions of each of these terms?

2.3 On hillslopes it is possible for the stress arising from the downslope component of the weight of overburden to exceed the strength of the soil at some depth, in which case failure occurs, resulting in a layer of the slope sliding downhill. This may occur, for example, when soil strength is reduced by increasing water content after heavy rain or melting of snow, especially when the pore water pressure increases.

(a) The following figure shows a segment of hillside of slope angle α, slope length y, vertical depth z and weight W, being of unit width normal to the figure.

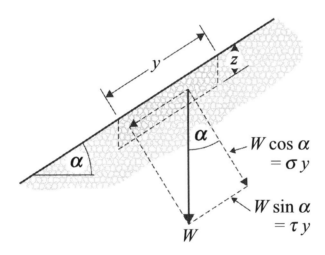

Assuming that the soil profile is homogeneous and of wet density γ, show that the weight of the hillside segment, W, is given by

$$W = \gamma gyz \sin(90 - \alpha)$$
$$= \gamma gyz \cos \alpha \qquad \text{(i)}$$

(b) Using the expression in (i) for W, show that the normal stress, σ, on the potential plane of failure at depth z is given by

$$\sigma = \gamma gz \cos^2 \alpha$$

and that the shear stress on the same plane, τ, is given by

$$\tau = \gamma g z \cos \alpha \sin \alpha$$

(c) Hence, using Coulomb's law (Eq. (2.17)), show that failure of the slope at depth z will occur (i.e. when $\tau = s$) if

$$\tan \alpha = C/(\gamma g z \cos^2 \alpha) + \tan \phi \qquad (ii)$$

(Equation (ii) expresses the limiting condition for a landslide to occur.)

(d) Calculate α, the angle of slope at which a landslide could occur, if $C = 7\,\text{kPa}$, $\gamma = 1500\,\text{kg m}^{-3}$, $z = 1.5\,\text{m}$ and $\phi = 12°$.
(Note. Solving equation (ii) for α is an example of a very common class of equations to which an analytical solution either is not obtainable or is very complex. The solution of equation (ii) can readily be obtained by trial and error, by plotting both the left-hand side and the right-hand side of equation (ii) against the angle α, or by using a numerical solving routine such as 'solver' in EXCEL™ software. Note that two solutions for α can be obtained, the lower value, $34°$, being the physically relevant solution.)

2.4 The following data have been published for a shear-box test measuring the strength of a remoulded sandy clay soil sample. During these tests sufficient time was allowed for full drainage of water to occur. Like much data from soil-mechanics sources, the system of units employed is gravitational and based on the unit of mass of the pound (lb), where $1\,\text{lb} = 0.454\,\text{kg}$.

Normal load (lb wt.)	20	40	60	80	90
Shearing resistance (lb wt.)	44	51	60	67	70

The area of the shear box used in the tests was $5.59\,\text{in}^2$ (or square inches) where $1\,\text{in}$ (or inch) $= 2.54\,\text{cm} = 0.0254\,\text{m}$. The shear strength is the shear resistance divided by the area of the shear box. The normal stress (σ) is the normal load divided by the area of the shear box.

Plot the two sets of data given in the table against each other, using their given units of lb wt. Hand fit a straight line to the data.

(i) From this figure calculate the cohesive strength of the soil in units of lb wt./in^2.
(ii) Convert this cohesive strength in lb wt/in^2 to the SI unit of N m^{-2} or Pa, and express it in kPa (10^3 Pa).
(iii) Obtain the angle of shearing resistance, ϕ, from the diagram you have plotted.

(iv) Indicate whether or not you believe that the angle ϕ obtained from the plotted diagram would be the same as that which you would have obtained had you first converted the data in lb wt. into newtons (N). Explain the reason for your belief.

2.5 An undisturbed sample of soil removed from the field was divided and placed in two direct-shear boxes (such as are shown in Fig. 2.13). The soil in one shear box was loaded with a mass of 10 kg, and the force needed to shear the sample was 100 N. The soil in the second shear box was loaded with a mass of 30 kg, and the force at shearing was 200 N.

The plan area of soil in the shear box is $0.2\,\mathrm{m}^2$. Calculate the cohesive strength and angle of friction assuming the two sub-samples to be identical.

References and bibliography

Barnes, G. E. (1995). *Soil Mechanics: Principles and Practice*. Basingstoke: Macmillan.

Giancoli, D. C. (1991). *Physics: Principles with Applications*, 3rd edn. Englewood Cliffs, New Jersey: Prentice-Hall International, Inc.

Hillel, D. (1982). *Introduction to Soil Physics*. New York: Academic Press.

Klute, A. (ed.) (1986). *Methods of Soil Analysis*, Part 1, 2nd edn. Madison, Wisconsin: American Society of Agronomy.

Marshal, T. J., Holmes, J. W., and Rose, C. W. (1996). *Soil Physics*, 3rd edn. Cambridge: Cambridge University Press.

McKenzie, N. J., Coughlan, K. J., and Cresswell, H. P. (2002). *Soil Physical Measurement and Interpretation for Land Evaluation*. Collingwood, Victoria: CSIRO Publishing.

Rose, C. W. (1979). *Agricultural Physics*. Oxford: Pergamon Press.

Whitlow, R. (1995). *Soil Mechanics*, 3rd edn. New York: Wiley.

3
The behaviour of liquids

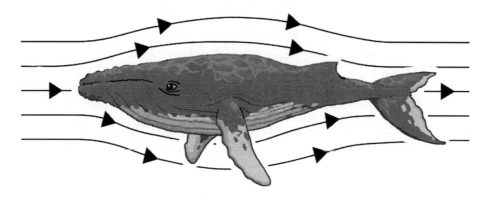

3.1 Introduction

Liquids (such as water) and gases (like air) share the common character-istic of being able to flow quite readily, but, unlike a solid object, liquids and gases do not have a shape of their own. Because of their ability to flow, the liquid and gaseous phases of matter are collectively referred to as 'fluids'.

The bulk or macroscopic properties of all matter, which includes solids, liquids and gases, are interpreted in terms of the strength of the forces between the atoms or molecules involved. In solids the forces of attraction between the atomic building blocks are sufficiently strong that they are almost fixed in position with respect to each other, though they vibrate about their mean positions. The energy of this vibration increases with temperature. At the melting point of ice ($0\,^\circ$C), addition of heat energy can overcome the strength of the binding forces between the water molecules in the solid ice phase, freeing the molecules of water to roll around each other in the liquid phase. Application of an even greater amount of energy removes water molecules far apart from each other into the gas phase – the process of 'evaporation'. Water molecules in the gas phase are referred to as water vapour. Gas-phase molecules move extremely rapidly at ordinary temperatures, and collisions between molecules are so violent that repulsive forces are invoked rather than the attractive forces typical of the liquid and gas phases. This is because attractive forces between atoms and molecules can turn to repulsion at close range.

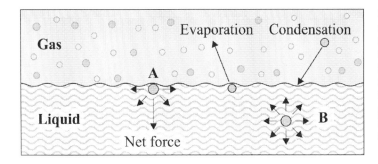

Figure 3.1 Illustrating the net inward attraction on a molecule at A located near the liquid surface, which is the source of surface tension. Evaporation of liquid molecules into the gas phase and their return in condensation is also illustrated.

Liquid surfaces have energy associated with them due to the net inward attraction of other liquid molecules for any molecule near the surface (see A, Fig. 3.1). This energy is expressed in the phenomenon of surface tension considered in Section 2.7 as a factor capable of adding some strength to non-cohesive sediments such as sand. The liquid surface behaves as if it is in tension – hence the name surface tension. It is a surface phenomenon, the molecule at B in Fig. 3.1 away from the surface not experiencing any persistent net force.

The ability of small insects such as pond skaters (a type of water beetle) to walk on the surface of water is due to the effect of surface tension. Depression of the water surface by the leg of a beetle increases the surface area and so the surface energy, resulting in a force opposing that exerted on the water surface by the beetle's leg. The role of surface tension in capillarity was considered in Section 2.7.

3.2 The environmental significance of liquids

Many environmental issues of concern to society crucially involve the behaviour and characteristics of liquids and the alteration of these characteristics by human behaviour. Surface water in rivers and dams is a major source of drinking water, so the amount and quality of such water is important to human health. In some countries toxic algal blooms have occurred in streams whose flows have been reduced. Low flows in some of Australia's inland rivers have resulted in increases in salinity due to seepage into the river of saline groundwater, which reduces stream turbidity, so improving the penetration of sunlight conducive to algal growth, given adequate nutrients.

However, too high a turbidity in streams, due to excessive soil erosion in river catchments, or to river dredging, can also impair water quality. As will be discussed in later chapters, the flow of water over unprotected land surfaces is a major cause of soil erosion.

Liquids of all kinds can move through the pore space of soil. Whilst this is necessary in order to recharge water stored underground (an important source of pumped water), it also means that potential pollutants from petroleum spills, industrial chemical leakage, or pesticides may also find their way into underground aquifers. Naturally occurring saline salts are the most extensive 'pollutant' of inland soil and water in some countries, and the movements of these salts are related directly to movement of water through the soil.

3.3 Fluid pressure and buoyancy

The first characteristic considered, density, has the same meaning and definition for liquids and gases (i.e. fluids) as it does for solids. The density of a fluid, ρ, is the mass of fluid in any particular volume, divided by that volume. The SI unit of density is kg m^{-3}.

If we dive beneath the sea surface, or go high into the air, our ears experience the resulting change in fluid pressure. The pressure, p, at any point in a fluid is defined as the ratio of the force F which acts at right angles (or perpendicular) to an area A, so that

$$p = F/A \tag{3.1}$$

Since the pressure must be the same whatever the orientation of the plane area A, the fluid pressure at any given point is the same in any direction. This implies that, for a fluid at rest and in contact with a solid surface, the pressure exerted by the fluid must be perpendicular to that surface, as shown in the accompanying figure, since, if this were not so, there would be a component force parallel to the solid surface, which would result in fluid motion.

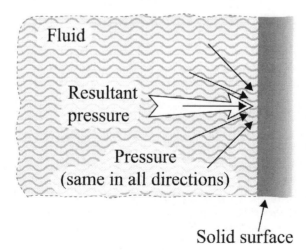

At the earth's surface the weight of the gaseous atmosphere above is experienced as a ubiquitous atmospheric pressure. Most instruments used to measure pressure, such as a gauge used to measure pressure in a vehicle or bicycle tyre, measure only the pressure in excess of the atmospheric pressure. The atmospheric pressure must be added to this excess or 'gauge pressure' in order to give the absolute magnitude of the pressure.

It follows from Eq. (3.1) that the unit for measuring fluid pressure is $N m^{-2}$, and this SI unit is called a pascal, Pa (see Table 1.1).

Example 3.1

If atmospheric pressure is 101.3 kPa, what is the absolute pressure at a depth 50 m beneath the surface of a freshwater lake, with water density $1000 \, kg \, m^{-3}$?

Solution
The weight of lake water produces a force on the area A shown in the figure given by

$$F = \rho \times \text{volume} \times g$$
$$= \rho A h g$$

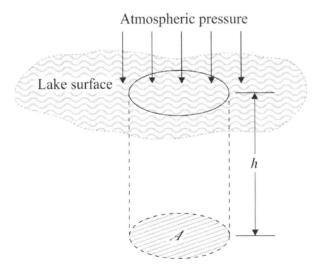

Atmospheric pressure

Lake surface

h

A

Thus the pressure is

$$p = F/A = \rho g h \qquad (3.2)$$

Figure 3.2 (a) Showing the forces acting on a volume of fluid at rest and completely enclosed by an imaginary surface. W_f is the weight of the fluid volume and B the buoyancy force. (b) The forces acting on a solid object of exactly the same shape and size replacing the volume of water defined in Fig. 3.2(a). W_s is the weight of the solid.

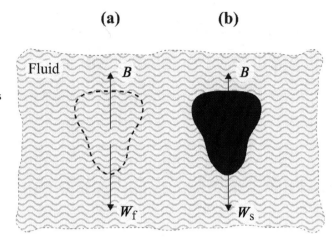

In this case

$$p = 1000 \times 9.8 \times 50$$
$$= 490 \, \text{kPa}$$

Thus the absolute pressure is $490 + 101.3 = 591.3$ kPa.

In the above example it is assumed that the water density did not increase with depth or pressure, i.e. that water is incompressible. For all but great depths in the ocean this assumption is very closely satisfied. Note that Eq. (3.2) indicates that the pressure at any given depth beneath a liquid surface will be the same.

Consider the arbitrary volume of water shown in Fig. 3.2(a) to be enclosed by an imaginary surface. Since the fluid is at rest, the downward-acting weight of this volume of water must be exactly counterbalanced by an equal upward force. This upward force, equal but opposed to the weight of fluid in the volume, is called the 'buoyancy force', B.

Now imagine this volume of water to be excluded by a solid object of exactly the same size and shape as the volume enclosed in Fig. 3.2(a). Since the solid object (shown in Fig. 3.2(b)) excludes an exactly equal volume of water, it follows that the buoyancy force on the immersed object must be exactly the same as acted on the displaced volume of water. This insight is expressed in '*Archimedes' principle*', namely that

the upward buoyancy force acting on any object immersed in a fluid is equal to the weight of the fluid displaced by that object.

If, in Fig. 3.2(b), $W_s > B$ then the solid object will sink; if $W_s < B$ then the solid object will float in the fluid. These alternatives are governed by whether the density of the solid is greater than or less than that of the fluid.

Example 3.2

A rectangular barge floats in water to a depth h. Show that Archimedes' principle applies also to a floating object.

Solution

The upward buoyancy force acting on the barge

$$= \rho ghbl$$
$$= \rho (hbl)g$$

where hbl is the volume of the displaced fluid. So

$$\text{buoyancy force} = (\text{mass of displaced fluid})g$$
$$= \text{weight of fluid displaced}$$

Thus Archimedes' principle applies also to objects that float.

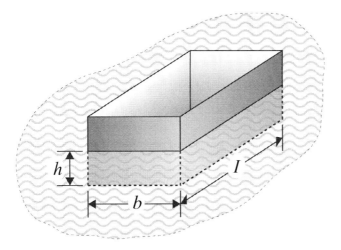

Example 3.3

The ratio of the density of any material to that of pure water at a temperature of $4\,^{\circ}\mathrm{C}$ is called its 'specific gravity'. The specific gravity of ice (Sg_i) is 0.917, and that of sea water (Sg_s) is 1.025. Hence show that the fraction of an iceberg which is above the surface of the sea is 0.105.

Solution

Let V_a be the volume of iceberg above the sea, and V_b be the volume below the sea (see the figure). Then

$$\text{density of iceberg} = \text{Sg}_i \times (\text{water density})$$
$$= \text{Sg}_i \times \rho$$
$$\therefore \text{mass of iceberg} = \text{Sg}_i \times \rho(V_a + V_b)$$

From Archimedes' principle,

$$\text{weight of iceberg} = \text{weight of seawater displaced}$$

or

$$\text{Sg}_i \times \rho(V_a + V_b)g = \text{Sg}_s \rho V_b g$$

Thus

$$V_b/(V_a + V_b) = \text{Sg}_i/\text{Sg}_s$$

Thus the required ratio is

$$V_a/(V_a + V_b) = 0.105$$

Hence some 90% of the volume of a floating iceberg is beneath the sea surface.

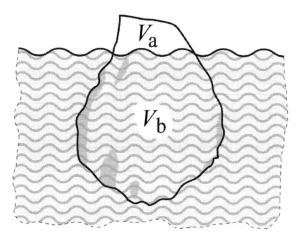

3.4 Liquids in motion

So far we have been considering liquids that are at rest, a static condition. Many issues of environmental concern arise because liquids can move, and the study of this topic is called 'hydrodynamics'. We all have some experience of fluids in motion – the effects of wind and flowing water being part of our experience. Despite the impact of the enormous

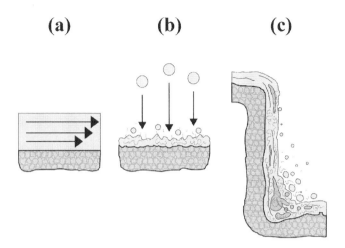

(a)   **(b)**   **(c)**

Figure 3.3 (a) Uniform laminar flow of shallow water over a gently sloping plane bed. (b) Rapid turbulent shallow flow with raindrop impact. (c) Water plunging down a waterfall into a pool results in strong turbulent mixing.

difference in density between water ($1000 \, \text{kg m}^{-3}$) and air ($\approx 1.2 \, \text{kg m}^{-3}$ at sea level and $20\,^{\circ}\text{C}$), there are some similarities in behaviour between the flows of water and air. The flow can be gentle and orderly at sufficiently low speeds, a gently flowing stream being an example. Alternatively, flow can be erratic and disorderly at higher speeds, a turbulent mountain stream or strong gusty wind being examples. A difference between water and air is that water can present a surface of its own, whereas the atmosphere just gradually decreases in density with height above the earth's surface.

Thus flowing liquids and gases share common characteristics of occurring either in orderly layers, called 'laminar flow', or in erratic flows or eddies characteristic of 'turbulent flow'. Whilst there is some gradual transition between these two types of flow, which of the two types of flow dominates has very important environmental implications, for example in the degree of mixing of a pollutant and hence the concentrations experienced.

Figure 3.3 illustrates these two different types of flow regime. If a volume of fluid is accelerating, then, from Newton's second law (Eq. (2.2)), there must be a net force acting on it. What are the kinds of forces that act on a fluid? In considering solids in Chapter 2 we distinguished between forces due to weight and those due to mechanical pressure or stress. (Furthermore, we distinguished between normal stresses and shear stresses.) In addition to these types of force associated with solids, for fluids in motion a third type of force, which arises from that property of fluids called 'viscosity', can be important. The forces that arise from viscosity are called 'viscous forces'. In terms of everyday experience, we are aware that more effort is required to stir a jar of honey than to stir a jar of water. Since the viscosity of a fluid is a measure of the resistance it

Figure 3.4 Illustrating the principle of operation of a rotating viscometer. The measurement liquid fills the space between the inner cylinder, which is rotated at a constant speed, and a fixed outer cylindrical shell.

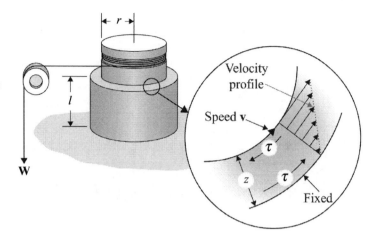

offers to a change of shape being brought about by shear stresses, we would say that honey is more viscous than water.

We shall discover that fluid viscosity is one of the factors involved in determining whether the flow of a fluid is laminar or turbulent. However, before we go into that, how can the viscosity of a fluid be measured more quantitatively than by stirring with a spoon?

Viscosity is measured with equipment called a 'viscometer'. The principle of operation of one form of viscometer is illustrated in Fig. 3.4.

The weight W applies a 'torque' given by the product Wr to the inner cylinder, and this accelerates to reach a constant speed of rotation. At this dynamic equilibrium the applied torque results in an equal and opposite torque due to the shear stress τ arising from the viscosity of the measurement fluid between the two cylinders (Fig. 3.4):

$$\text{resisting torque} = \text{force} \times \text{radius}$$
$$= \text{stress} \times \text{area} \times \text{radius}$$
$$= \tau \times (2\pi rl) \times r$$

Equating the applied and resisting torques,

$$\tau = Wr/(2\pi r^2 l) = W/(2\pi rl)$$

The rate of shearing of the liquid is given by the increase in fluid velocity, v, divided by the distance over which this increase in velocity is gained, z (Fig. 3.4), and so is given by the velocity gradient v/z. In laminar flow this increase in velocity is linear.

Provided that the flow of liquid in the viscometer is laminar, it is experimentally found that there is support for a hypothesis originally proposed by Newton, namely that

$$\tau \propto \text{rate of shearing}$$

so

$$\tau \propto \frac{v}{z} \quad \text{or} \quad \frac{\Delta v}{\Delta z} \quad \text{in general}$$

where Δv is a small increase in v and Δz is the corresponding increase in z. This proportionality can be written as an equality by the introduction of a constant, η, known as the 'coefficient of dynamic viscosity'. Thus

$$\tau = \eta \frac{\Delta v}{\Delta z} \tag{3.3}$$

from which equation the fluid property η can be calculated.

In liquids, the adhesive forces between the liquid molecules and the cylinder walls ensure that the velocity of the liquid in contact with the wall is equal to the velocity of the wall (i.e. the fluid velocity at either cylindrical wall is the same as that of the wall in Fig. 3.4). The velocity gradient shown in the inset indicates how the moving cylinder pulls the fluid with it, and each layer pulls the layer further away from it due to cohesive forces between the molecules.

Since, from Eq. (3.3), $\eta = \tau/(\Delta v/\Delta z)$, dimensional analysis shows that the SI unit of η is Pa s (or pascal seconds). At $20\,^{\circ}\text{C}$, the dynamic viscosity of water is 10^{-3} Pa s, decreasing as temperature increases. It follows from Eq. (3.3) that the velocity gradient $\Delta v/\Delta z$ must be considerable before the viscous shear stress, τ, becomes appreciable. As fluids flow over a fixed surface the velocity gradient decreases with distance from the surface, as shown in Fig. 3.5.

For flow over a submerged object the value of δ can be quite small. For example $\delta \approx 5$ mm after water flowing at 0.1 m s^{-1} has moved 1 m over a relatively smooth solid surface. Beyond a distance δ from a solid surface the velocity gradient is essentially zero, and thus viscosity has no effect (Eq. (3.3)). The distance δ over which viscosity has an effect in laminar flow has been called the 'boundary layer'. The role of the boundary layer is of fundamental importance to understanding all interactions involving fluids moving relative to a solid surface.

Whilst boundary layers can be thin for flow over discrete objects, for the flow of water in a river, or the atmospheric flow over the earth's surface, boundary layers can be very much thicker. The spatial development of a boundary layer as turbulent water flows over a submerged solid surface, such as a flat-bottomed boat in a turbulent stream, is illustrated in Fig. 3.6.

Figure 3.5 (a) Showing how fluid velocity decreases at distances closer to a solid surface than δ. (b) The change with distance z in the value of the velocity gradient $\Delta v/\Delta z$, given by the reciprocal of the slope of the velocity profile in (a).

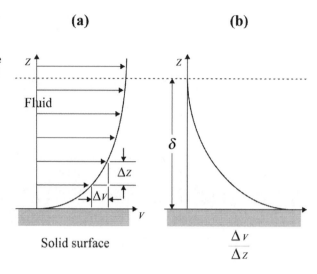

Figure 3.6 The growth in thickness of an initially laminar boundary layer in a turbulence-free stream, which undergoes transition to a turbulent condition. (The vertical scale is greatly magnified compared with the horizontal.)

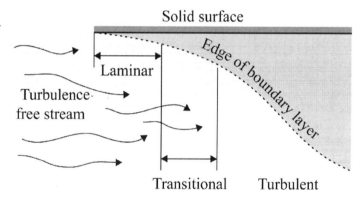

As illustrated in Fig. 3.6, even in a turbulence free stream, the boundary layer is initially laminar in nature, but transition to a turbulent boundary layer will take place if the scale of the solid surface is sufficient.

Turbulent flow is characterised by three-dimensional eddies in which pockets of fluid move in irregular curved motions. These eddies have a rotational characteristic, which can be roughly conceived of as erratic whirls of fluid of a wide range of sizes or frequencies. The looping motion of smoke emissions from a chimney stack and the irregular whirls of motion in a rapidly moving mountain stream are indicators at their particular scales of turbulent eddies. The initial energy for turbulent motion often seems to come from relatively large eddies generated by chaotic shearing and flow separation from solid surfaces. This initial energy is then dissipated by cascading down in scale through a spectrum of eddy sizes or frequencies, ultimately being dissipated into thermal motion of molecules or heat. However, the scale of turbulence is characteristically much greater than the molecular scale, and the ability to generate shear stress due to turbulence is commensurably much greater than the viscous shear stresses of laminar flow described by Eq. (3.3).

Equation (3.3) for viscous shear stresses is commonly written in equivalent form as

$$\tau = \frac{\eta}{\rho} \frac{\Delta(\rho v)}{\Delta z}$$

or

$$\tau = v \frac{\Delta(\rho v)}{\Delta z} \tag{3.4}$$

where $v = \eta/\rho$ is the 'kinematic viscosity', a property of the fluid that is important in laminar flow.

In turbulent flow, a velocity gradient results in shear stresses that are much greater than the viscous shear stresses described by Eq. (3.4). Shear stresses in turbulent flow are referred to as 'eddy shear stresses', and these can be described by an equation of form similar to Eq. (3.4), namely

$$\tau = K \frac{\Delta(\rho v)}{\Delta z} \tag{3.5}$$

where K is commonly called a 'kinematic eddy viscosity' because its role is similar to that of v for laminar flow (Eq. (3.4)). K is given different names in different areas of science or engineering, and is sometimes called an 'eddy diffusivity' or 'turbulent transport coefficient'.

In principle, molecular-scale processes can still take place, even in the presence of turbulence, so that the term K in Eq. (3.5) is strictly $K + v$. However, K is commonly so much greater than v that, in practice, v is ignored and Eq. (3.5) employed. The reason why $K \gg v$ is because of the vastly greater mixing in turbulent flow compared with that which takes place (only at a molecular scale) in laminar flow.

An engineer, Osborne Reynolds (1842–1912), asked the following interesting and important question: 'What are the factors that determine whether any particular fluid flow will be laminar or turbulent?' The answer to this question can be expressed in terms of a non-dimensional number, known as the 'Reynolds number' (Re), which is the ratio of two relevant types of force involved in fluid motion, namely inertial and viscous forces. (For the flow of liquids in which motion of the liquid surface is important, as in rivers and ocean waves, another non-dimensional number, called the Froude number, is also important.)

Consider approximately spherical grains of sand of a range of diameters, d, settling in water (Fig. 3.7). The situation is exactly the same irrespective of whether the liquid is stationary and the sand grain moving or vice versa.

Figure 3.7 Spherical sand grains of two different sizes settling in water. In (a) all flow is laminar. In (b) the sand grain is larger, and settles with a higher velocity, v. Separation of the boundary layer leads to turbulent flow behind the grain.

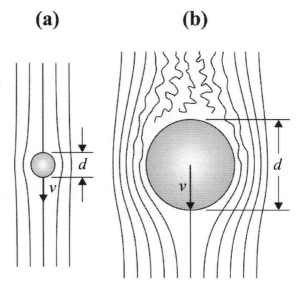

In either of Figs. 3.7(a) and 3.7(b) the flow of fluid around the spherical grain is at least initially spherical in form. The magnitude of any inertial force is the product of mass and acceleration (Eq. (2.2)). The acceleration experienced due to moving in a circle of radius r with uniform speed v is shown later in Example 3.4 to be v^2/r. Thus the inertial force, F_I, acting on a unit volume of liquid (of density ρ) around a spherical object of diameter d (Fig. 3.7) is given by

$$F_I \propto \rho d^3 \times v^2/r$$
$$\propto \rho d^2 \times v^2$$

The viscous force, F_v, acting on the same volume of liquid, using Eq. (3.3), is such that

$$F_v \propto \eta \frac{\Delta v}{\Delta z} \times \text{area}$$
$$\propto \eta \frac{v}{d} d^2$$

or

$$F_v \propto \eta v d$$

Thus the ratio of these two types of forces is

$$\frac{F_I}{F_v} = \frac{\rho d^2 v^2}{\eta v d} = \frac{\rho v d}{\eta} = \frac{v d}{\nu} \tag{3.6}$$

The non-dimensional force ratio in Eq. (3.6) is what is defined as the Reynolds number, Re, so

$$Re = \frac{\rho v d}{\eta} = \frac{v d}{\nu} \tag{3.7}$$

In this derivation, d is interpreted as the size scale of the object immersed in the liquid (Fig. 3.7). However, the Reynolds number is similarly defined in different situations in which the scale factor d might be the diameter of a pipe through which liquid is flowing at mean velocity v, or it might be the distance flow has moved from the leading edge over the plate shown in Fig. 3.6.

The magnitude of the Reynolds number can vary over a very wide range. For example, $Re \approx 10^{-3}$ for a small silt particle settling in water, whereas $Re \approx 10^7$ for a feeding shark or dolphin.

The great value of Re lies not so much in its exact value, but in the fact that its magnitude gives a good indication of whether flow will be laminar, turbulent, or in some transitional type of situation in which flow has some of the characteristics of both these types of flow. Especially in the transitional region, whether flow is dominantly laminar or turbulent can depend on some rather specific detail of the type of surface or object over which the fluid is flowing. Despite uncertainties of that kind, flow is generally laminar when $Re < 2000$ and turbulent if $Re > 10^4 - 10^5$.

Example 3.4

Show that, for uniform circular motion with speed v in a circle of radius r, the acceleration is v^2/r, directed towards the centre of the motion.

Solution

Though the speed is constant, figure (a) shows velocities as v_1 at point A and v_2 at point B. Figure (b) shows the change in velocity Δv in movement from A to B. It may be shown from geometry that the isosceles triangles in figures (a) and (b) with equal angles shown as θ are geometrically similar.

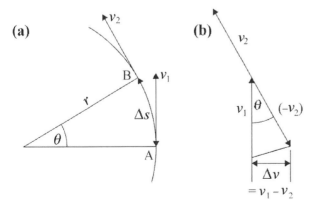

From the geometrical similarity,

$$\Delta v/v = \Delta s/r$$

Thus the acceleration is

$$a = \mathrm{limit}\left(\frac{\Delta v}{\Delta t}\right) = \mathrm{limit}\left(\frac{v\,\Delta s}{r\,\Delta t}\right)$$

$$= v^2/r$$

since the limit of $\Delta s/\Delta t$ is v.

Example 3.5

To investigate the flow of water round a dolphin, a scale-model dolphin is constructed (a fifth full size) and lowered into a model-testing tank through which water is circulated. For water flow round the model dolphin to have similar characteristics to that around the real dolphin moving at any speed relative to the water, what should the ratio of the speed of water in the model-testing tank (where the model is stationary) to that of the dolphin moving in still water be?

Solution

Note that only the relative speed of dolphin and water is relevant, not which is moving. Similarity of flow in both cases requires the Reynolds number to be the same. Considering Eq. (3.7), the kinematic viscosity v of water in the testing tank and the sea will be similar (though v varies a little with temperature and sea-salt concentration). Neglecting such minor variation in v, from Eq. (3.7) the requirement of flow similarity, on which the validity of the model testing rests, is therefore that

$$vd = v_m d_m$$

where the subscript m refers to the model dolphin, and the real dolphin is implied by vd, where v is velocity and d refers to any significant scale dimension of the dolphin, such as its length or maximum diameter.

The geometrical scale $1 : 5$ implies that $d_m/d = 1/5$. It follows that $v/v_m = 1/5$ or $v_m/v = 5$, implying that the relative velocity between the model dolphin and water must be five times higher than for a corresponding speed of a real dolphin in the sea. Provided that this is observed, measurements made on the model, such as the drag force experienced, can be a useful guide to what occurs in reality.

Note. This example illustrates the principles involved in using scale models in experimentation.

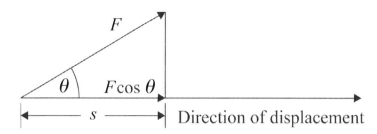

Figure 3.8 The work done by force F is given by the product of displacement s and the force component ($F\cos\theta$) in the direction of the displacement s.

3.5 Energy and fluid flow

The words 'work' and 'energy' are used in everyday speech. However, in the physical sciences these words are used in a quite restricted and specific sense, although this sense is still somewhat related to our every-day experience of exertion. You may have tried and failed to pull up a particularly resistant weed from the garden: but, despite all your effort, no work may have been done using the physical definition of 'work', which is as follows.

Work is done when a force moves in the direction in which it is acting, and the magnitude of the work, W, is the product of the force, F, and the distance moved, s, so that

$$W = Fs \tag{3.8}$$

If the displacement, s, is in the opposite direction to the force, F, the work will be done *on* the force, not *by* the force, and the work done is regarded as negative (i.e. $F \times (-s) = -Fs$).

The direction of movement caused by a force doesn't always coincide with the direction of the force, in which case the work done is given by the product of the displacement and the component of the force in the direction of displacement. Thus (see Fig. 3.8)

$$W = F\cos\theta \times s \tag{3.9}$$

Although F and s are vector quantities, work is given only by the magnitude of their product (Eqs. (3.8) or (3.9)), so W is a scalar quantity. From these equations and Table 1.1 the unit of work is N m (i.e. newton metre). The unit N m is called a 'joule' (after James Joule, 1818–1889, who clarified the relationship between work and heat).

The outcome of work done on any volume of liquid depends on the nature of the force, F. If the force is due to the weight of the volume of liquid, work done in lifting the volume is stored, since the work done can be recovered when the volume of liquid returns to its original position. The work done in such circumstances is said to be stored as 'potential energy' (PE).

Figure 3.9 During the motion in which water in a fountain is squirted up into the air its velocity gradually declines.

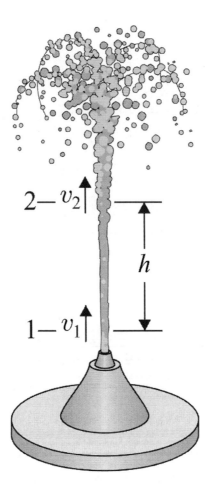

However, if the force involved is frictional in nature, the work done is ultimately lost by conversion into heat, just as the work done in accelerating a motor vehicle can be dissipated into heat by application of the brakes. As discussed in Section 3.4, frictional forces in liquids are associated with viscosity in laminar flow and with turbulent eddies in turbulent flow.

Thus work done in fluids can either end up as an increase in energy, or be dissipated by frictional forces. In practice frictional forces of some kind are always present, so the energy gained is always less than the work done. However, there are systems in which the dissipation of energy by frictional forces is sufficiently small that its neglect still allows a good approximation to result. The jet of water in the fountain shown in Fig. 3.9 is such a system. As the water in this jet rises into the air, work is done against the downward-acting weight on each volume of liquid. Thus the potential energy of liquid in the jet increases as it moves from level 1 to level 2 in Fig. 3.9.

The potential energy is easily calculated since the force doing the work needed to lift each volume of water is the force opposing the downward-acting weight of the liquid. The weight W of mass m is mg, and, since the mass of unit volume of liquid is its density, ρ, the weight of unit volume of liquid is ρg. From Eq. (3.7) the work done in lifting unit mass of water from one level to another through a height h (Fig. 3.9) is

$$W = \rho gh \quad \text{(per unit volume)}$$

or

$$W = mgh \quad \text{(for mass m)} \qquad (3.10)$$
$$= gh \quad \text{(per unit mass)}$$

This work W is equal to the gain in potential energy, ΔPE, by water rising through a height h. Thus

$$\Delta PE = \rho gh \quad \text{(per unit volume)} \qquad (3.11)$$

Neglecting any frictional forces experienced by the jet, this gain in PE is at the expense of a decrease in motion energy (KE), the velocity v_2 at level 2 (Fig. 3.9) being less than the velocity v_1 at level 1. How can the energy due to motion be expressed quantitatively?

In seeking to answer this question we make use of an important physical concept known as the 'principle of conservation of energy'. This principle states that, whenever energy is converted from one form into another form, the energy lost from one form appears in exactly equivalent amounts in the transformed form (or forms). (When, as in nuclear reactions, mass is converted into energy, this principle has to be extended to include mass as well as energy.)

If we neglect frictional losses in the water jet of Fig. 3.9, then the increase in potential energy per unit volume of water rising from level 1 to level 2, given by Eq. (3.11), must come from some exactly equivalent loss of some other form of energy. This source of energy is due to the velocity of water in the jet, and is called 'kinetic energy' (KE). The velocity v_2 at height 2 in Fig. 3.9 will be less than v_1 at height 1, and the velocity eventually falls to zero at the top of the jet. Thus the loss in KE due to the decrease in velocity of water moving from level 1 to level 2 must equal the gain in PE, written ΔPE, given by Eq. (3.11).

Let us now derive an expression for the loss in KE, and so obtain a definition of kinetic energy in terms of velocity.

The term g in $\Delta PE = \rho gh$ is the acceleration due to gravity, a uniform acceleration which doesn't alter significantly over modest fall heights. Thus this acceleration can be written

$$g = (v_1 - v_2)/t$$

where t is the time taken for a particular volume of water to rise from level 1 to level 2. This calculation gives the magnitude of the acceleration g as a positive quantity. However, acceleration is a vector quantity, and acts downwards. Thus, if we take the (upward) direction of motion of the water to be the positive direction, the acceleration of the water, a, will be given by $a = -g$, indicating its deceleration. Thus, since

$$a = (v_2 - v_1)/t = -g$$

we have

$$v_2 = v_1 + at$$

Since the acceleration (or deceleration) is constant, velocity decreases linearly with time. So the average velocity, h/t, is also given by $(v_2 + v_1)/2$. Thus $h = (v_2 + v_1)t/2$. Now replacing t by its expression available from the earlier definition of a, it follows that

$$h = (v_2 + v_1)(v_2 - v_1)/(2a)$$
$$= \left(v_2^2 - v_1^2\right)/(2a)$$

from which

$$a(= -g) = \left(v_2^2 - v_1^2\right)/(2h)$$

Consequently,

$$\Delta PE = \rho gh \quad \text{(from Eq. (3.11))}$$
$$= \text{decrease in KE}$$
$$= -\rho ah$$
$$= -\rho h(v_2^2 - v_1^2)/(2h)$$

or the decrease in KE $= \frac{1}{2}\rho v_1^2 - \frac{1}{2}\rho v_2^2$.

Hence the KE of a unit volume of fluid of density ρ moving with a velocity v is given by

$$KE = \tfrac{1}{2}\rho v^2 \quad \text{(per unit volume)} \tag{3.12}$$

or

$$KE = \tfrac{1}{2}mv^2 \quad \text{(for mass } m\text{)}$$

Returning to Fig. 3.9, there is an exchange of kinds of energy as the water jet rises from level 1 to level 2. The work done *by* the force due to the weight of water acting downwards when the movement is upwards is *negative*, so the equation

work done by the weight force = gain in KE

may be expressed, using Eqs. (3.11) and (3.12), as

$$-\rho g h = \frac{\rho}{2}\left(v_2^2 - v_1^2\right)$$

So

$$\frac{\rho}{2}v_1^2 = \frac{\rho}{2}v_2^2 + \rho g h$$

or

$$(KE + PE)_{level\ 1} = (KE + PE)_{level\ 2} \qquad (3.13)$$

Equation (3.13) is a particular example of the energy-conservation principle.

It should be noticed that the term 'conservation' in the energy-conservation law is used to mean that the total energy remains absolutely constant in any process. This meaning is quite different from that when a policy of energy conservation is recommended. In this context to 'conserve' energy really means to minimise fuel use by using energy efficiently and wisely.

Equation (3.13) is also a special case of a more general equation of energy conservation in fluid flow known as 'Bernoulli's equation', after Daniel Bernoulli (1700–1782). This equation recognises that pressure in a fluid can change, and that there is a change in energy associated with any change in pressure. In the fountain example of Fig. 3.9, water within the fountain would be at the pressure in the mains supply. This pressure is the source of the energy required to produce the jet of water in the fountain. Once water has emerged from the fountain into the jet above it, the pressure in the water rapidly equilibrates with the atmospheric pressure. Since this is essentially constant over the very modest height difference between levels 1 and 2 in Fig. 3.9, we neglected any difference in energy due to a pressure difference in the water in the jet. However, if we were to describe the motion of water at its source, entering the fountain at mains pressure, we could not neglect this pressure and the energy term associated with this pressure. This term is added to the potential- and kinetic-energy terms in Bernoulli's equation. This equation assumes

Figure 3.10 A stream tube of
flow defined by a series of
streamlines.

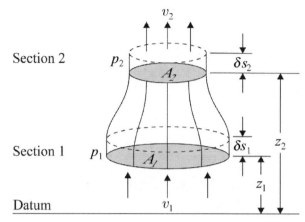

Figure 3.10 A stream tube of flow defined by a series of streamlines.

flow in a fluid to be steady and laminar, that the fluid is incompressible (a very good approximation for liquids such as water) and that the effects of viscosity can be ignored.

With these assumptions, Fig. 3.10 shows a stream tube of flow, which is a volume of fluid enclosed by streamlines. This stream tube could represent a stream tube of flow within the fountain of Fig. 3.9, where changes in water pressure, p, are important.

The bottom of the stream tube experiences a force due to the fluid pressure p_1, given by p_1A_1, where A_1 is the cross-sectional area of the stream tube at this lower section. Because the fluid is flowing at velocity v_1, this force moves a small distance δs_1 in a time δt given by $\delta s_1/v_1$. The energy given to the fluid will be equal to the work done by the force, and this is $p_1A_1\,\delta s_1 = p_1\,\delta\vartheta$, where $\delta\vartheta$ is the element of volume given by $A_1\,\delta s_1$.

The equation of continuity expresses the 'principle of mass conservation', where, as in the energy-conservation principle, the term conservation is used in the sense of constancy, mass being neither created or destroyed. Mass conservation requires that, since liquids are almost incompressible,

$$\rho A_1\,\delta s_1 = \rho\,A_2\,\delta s_2 = \rho\,\delta\vartheta \qquad (3.14)$$

Thus the energy of the fluid due to pressure at the higher section (level 2) will be $p_2A_2\,\delta s_2$ or $p_2\,\delta\vartheta$.

The potential energy per unit volume at a height h above some datum level (where this energy is arbitrarily taken to be zero) is given by Eq. (3.11), and the kinetic energy per unit volume is given by Eq. (3.12). If we now add to these two energies the energy due to pressure derived above, the total energy at section 1 in Fig. 3.10 will be given by

$p_1 \delta\vartheta + \rho g z_1 \delta\vartheta + \frac{1}{2}\rho v_1^2 \delta\vartheta$. It follows from energy conservation that the total energy given by the sum of component energy terms is constant. Expressing this energy per unit volume, then

$$p_1 + \rho g z_1 + \tfrac{1}{2}\rho v_1^2 = p_2 + \rho g z_2 + \tfrac{1}{2}\rho v_2^2 \qquad (3.15)$$

Equation (3.15) is called Bernoulli's equation, and, since the sections 1 and 2 in Fig. 3.9 can be generalised to any two points in fluid flow (not necessarily vertically related as in Fig. 3.9), Bernoulli's equation can also be expressed by saying that, under the assumptions stated earlier,

$$p + \rho g z + \tfrac{1}{2}\rho v^2 = \text{constant along a stream tube} \qquad (3.16)$$

The fluid pressure p is commonly called the 'static pressure', and the term $\frac{1}{2}\rho v^2$ the 'dynamic pressure' (being the pressure exerted when the flow is brought to rest, such as when the flow is impeded by a plate placed at right angles to the flow). The term $\rho g z$ is sometimes called the gravitational pressure due to the 'head' z.

Whilst viscosity was assumed negligible in deriving Bernoulli's equation, in practice there will be some continuous loss of this total mechanical energy through conversion to heat. This energy loss per unit volume, E_{loss}, which was neglected in Eq. (3.16), can be added to give the constant total energy per unit volume, E (J m^{-3}). Thus, more generally than Eq. (3.16), along a stream tube

$$E = p + \rho g z + \tfrac{1}{2}\rho v^2 + E_{\text{loss}} = \text{constant} \qquad (3.17)$$

Outside boundary layers (within which high viscous shear stresses can lead to significant conversion to heat), Bernoulli's equation is commonly found to give good agreement with experimental results. Bernoulli's equation can be applied to explain many natural as well as technologically important processes. For example, it provides the basis for understanding the flow of fluids around any reasonably streamlined object, natural or constructed, and describes the basis of flight.

Example 3.6

The figure shows the relative motion of a hump-back whale through the sea. (The flow pattern of fluid is the same irrespective of whether the whale is moving in a sea without a current, or the whale is stationary and there is a moving current.) The distance that sea water has to move over the top of the whale is 12% less on average than that over its underside. The length of the whale is 20 m, its average width is 4 m, and the speed of the whale relative to the sea water is 5 m s^{-1}. If the whale is swimming horizontally, calculate the magnitude of the vertical force due to the whale's motion through the water. Assume that, as water flows over the top of the whale, it is an average height of 3 m above the lower surface of the whale's body. The flow may be assumed to be laminar.

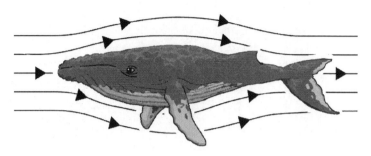

Solution

Consider a stream tube of sea water that, as it impinges on the snout of the whale, divides into two parts, one part moving over the upper (approximately horizontal) surface of the whale's body, the lower part moving over the bottom surface.

Since the flow is laminar, water in the upper half of the stream tube flowing over the upper part of the whale's body must arrive at the tail at the same time as water that has moved under the whale, since, if not, turbulence must be induced. Since sea water moving over the lower half of the whale travels 12% further than does water moving over the upper half, the velocity of water over the bottom of the whale must be 12% (or 1.12 times) greater (5.6 m s^{-1}) than the 5 m s^{-1} speed of flow over the upper half.

Let quantities such as static pressure over the lower half of the whale be denoted with a subscript 1 and those over the upper half with a subscript 2 (see the figure). The value of z for the lower half of the whale (z_1) can be arbitrarily taken as zero since it is only differences in height that are important. Then it follows from Eq. (3.14) that

$$p_1 + \rho g z_1 + \tfrac{1}{2}\rho V_1^2 = p_2 + \rho g z_2 + \tfrac{1}{2}\rho V_2^2$$

On substituting values, we have

$$p_1 + 0 + \tfrac{1}{2} \times 1000 \times (5.6)^2 = p_2 + 1000 \times 9.8 \times 3 + \tfrac{1}{2} \times 1000 \times (5)^2$$

Thus

$$p_1 - p_2 = 26\,220 \,\text{N}\,\text{m}^{-2}$$

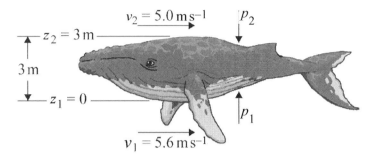

Thus the static pressure on the upper side of the whale is less than that on its lower side, and this difference in static pressure, multiplied by the plan area of the whale, will result in an upward force:

$$\text{upward force} = (p_1 - p_2) \times (\text{plan area in m}^2)$$
$$= 26\,220 \times 20 \times 4$$
$$= 2.51 \times 10^6\,\text{N}$$

This upward force is called a lift force because it lifts the whale against its immersed weight. This net lift force and the downward-acting immersed weight are only two of the forces acting on the whale during its motion relative to the ocean. The motion of the whale's tail in particular leads to a forward force or thrust, and this force is opposed by the resistance to motion, called a drag force. The origins of this drag force are discussed in the next subsection.

3.6 Flow around submerged solids

The forces acting on the swimming whale of Example 3.6 are shown in Fig. 3.11. That whales appear to float when stationary may indicate that the immersed weight, W, is small, or may be more than counterbalanced by the lift force.

The drag force D arises for two different reasons. Firstly, as illustrated in Fig. 3.5 and Eq. (3.3) (for laminar flow), or Eq. (3.5) (for turbulent flow), the fluid shearing stress τ is generated by the velocity gradient in the fluid boundary layer which surrounds all solids. The shearing stress acting on the surface of the body is commonly called 'skin friction'.

The second cause of drag resistance is due to the net result of the static pressure distribution over the surface of the solid body. This static pressure distribution depends on the characteristics of the boundary layer which develops, and these depend a great deal on the shape or form of the solid body and its speed. The net drag force due to this cause is hence commonly called the 'form drag'.

Figure 3.11 The various forces acting on a whale swimming in the sea. **L** = lift force, **W** = immersed weight, **T** = thrust and **D** = drag.

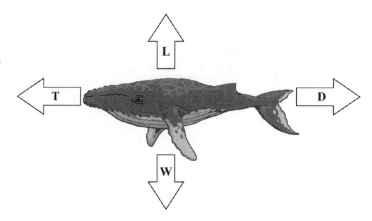

Figure 3.11 The various forces acting on a whale swimming in the sea. **L** = lift force, **W** = immersed weight, **T** = thrust and **D** = drag.

Form drag is the dominant drag component when the solid object is a bluff body, i.e. a body that is not streamlined. Conversely, skin friction can dominate for a streamlined body such as those of many mobile sea creatures (e.g. fish, seals and dolphins). A streamlined shape can greatly reduce form drag, but drag due to skin friction is not so readily avoided. The total drag is the sum of the form-drag and skin-friction components.

The total drag force, D, is commonly related to the dynamic pressure or kinetic energy per unit volume of the flow, $\frac{1}{2}\rho v^2$, and the cross-sectional area, A, of the object viewed in the direction of the flow around the object. The product of A and $\frac{1}{2}\rho v^2$ is a force, and this force is converted into the drag force D by introducing a non-dimensional term, C_d, called a 'drag coefficient'. Thus

$$D = C_d A \times \tfrac{1}{2}\rho v^2 \tag{3.18}$$

The magnitude of the drag coefficient depends on the shape of the body and the nature of the flow around the body, in particular whether it is laminar or turbulent, or in some transitional state. Thus the magnitude of C_d depends on the Reynolds number, Re, and, for a spherically shaped object, the nature of this relationship is given in log–log plotting form in Fig. 3.12.

At very low Reynolds numbers viscous forces dominate over inertial forces (Section 3.4), and with this constraint Stokes showed that the equations of fluid motion could be solved with the implication that $C_d = 24/Re$, yielding a linear relationship shown by the line A–B in Fig. 3.12, with slope of -1. Beyond the Stokes range, inertial forces lead to flow separating from the surface of the sphere, and the slope of the C_d versus Re relationship alters in the range B–C (Fig. 3.12). C_d becomes essentially constant in the range C–D, where eddies or vortices

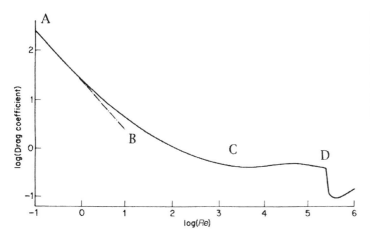

Figure 3.12 The nature of the relationship between the drag coefficient C_d of a spherical object immersed in a fluid and the Reynolds number, Re. (After Monteith and Unsworth (1990).)

are shed behind the sphere to form a 'wake', which rather suddenly (beyond D in Fig. 3.12) becomes fully turbulent in nature.

Figure 3.12 refers to flow of any fluid (liquid or gas) round a spherical solid object of any size, and covers enormous ranges of Re and C_d. In principle, being a non-dimensional plot, Fig. 3.12 summarises information on an indefinitely large number of realisations of fluid flow around a sphere, from a microscopic globule to a large hot-air balloon. Whilst the detail of Fig. 3.12 refers to flow around spherical objects, it applies well to natural objects that only approximate that shape. Similar relationships to that shown in Fig. 3.12, but differing in detail, have been determined for objects of a variety of shapes.

Example 3.7

Using Eq. (3.18), calculate the drag force D (in newtons) acting on a fish of effective diameter 0.1 m when it is swimming at the following two speeds relative to the water: (i) $2\,\mathrm{m\,s^{-1}}$ and (ii) $10\,\mathrm{m\,s^{-1}}$. Assume that $C_d = 0.035$ and 0.03, respectively, at these two speeds, and that $\rho = 1000\,\mathrm{kg\,m^{-3}}$. Also calculate the power being expended by the fish when it is swimming at these two speeds.

Solution

$$D = C_d(\pi/4)0.1^2(1/2)1000v^2$$

So

$$D = 0.55\,\mathrm{N} \quad \text{if } v = 2\,\mathrm{m\,s^{-1}}$$

and

$$D = 11.78\,\mathrm{N} \quad \text{if } v = 10\,\mathrm{m\,s^{-1}}$$

Power is the product of force and speed, Dv, so

$$\text{power} = 1.1\,\text{W} \quad \text{if}\,v = 2\,\text{m s}^{-1}$$

and

$$\text{power} = 117.8\,\text{W} \quad \text{if}\,v = 10\,\text{m s}^{-1}$$

3.7 Oceans and waves

Oceans, which cover 71% of the earth's surface, have major currents with some persistence, which, through interaction with the earth's atmosphere, play a substantial role in controlling the climate of the earth's land masses. Oceanic currents can be driven by the earth's rotation on its axis, by differences in density of sea water due to temperature, by attraction from the moon and sun and by persistent winds. Continents strongly affect the pattern of these currents.

Rivers bring enormous quantities of non-saline water, sediment and chemicals to the edge of oceans. More than half the world's population lives close to coastlines and river mouths, and this concentrated population along coastlines has tended in the past to use the sea as a convenient sink in which to dilute and disperse unwanted pollutants of human and industrial origin. There is now widespread recognition of the range of undesirable consequences of such releases by land and ocean dumping of waste, and most countries now seek to regulate such pollution of the oceans.

A major effect of oceans on the near-shore environment is via the action of waves. Waves continually shape coastlines, and are generated mostly by wind blowing over the sea surface. However, an undersea earthquake gives rise to a major sea wave, or 'tsunami', which can deliver enormous wave energy to coastal zones, with highly destructive consequences.

Thus, coasts are a quite dynamic interface of the land, the sea and the atmosphere, with tidal influences also being important.

Despite the vital importance of other features of oceans, here attention will be restricted to the general characteristics of waves and wave motion. Ocean waves are an important natural example of periodic motion, a feature of diverse physical phenomena. As wave forms move across oceans, the displacements of individual elements of water are dominantly at right angles to the direction of the wave motion. This is illustrated by the motion of an object floating on the sea; its motion is largely up and down as waves pass by. Despite this oscillatory motion, an object can remain almost in one place unless there is a general mass-transport current or drift. Water waves are an example of 'transverse' waves, where the motion of elements of the medium is dominantly trans-verse to the horizontal direction of movement of the wave. When the wave form moves, it is called a 'travelling' or 'progressive' wave to dis-tinguish it from the situation in which there is wave-like motion but no progression of the wave. Such a situation is described as a 'stationary' or 'standing' wave. An example of a stationary wave is a plucked guitar string; whilst the magnitude of the displacement of the string varies along the string between its two fixed ends, the time dependence of the trans-verse motion is the same everywhere along the string. The sound we hear of the plucked string is due to wave motion transmitted through the air in a kind of motion described as a 'longitudinal travelling' wave. The light we see by is a transverse electromagnetic form of wave motion. Thus the idea of wave motion is widespread in a physical description of the universe.

The motion involved in an ocean wave does have analogies on land. Wind is not usually regular in nature and commonly comes in turbulent bursts of packets. These exert fluctuations in wind speed and pressure on any surface. As such a burst moves across a wheat field, for example, an accompanying wave motion can be seen to progress across the field formed by the heads of wheat bowing in response to the travelling packet of air turbulence which we might sense as a gust of wind. Clearly each head of wheat has moved around in some restricted orbit as the gust moves over it, and this form of motion progresses or travels across the field with the turbulent burst. However, each head of wheat, despite its limited motion, stays basically fixed; it is the moving form of the bowing wheat heads which travels rapidly and extensively over the field.

Transferring this analogy to a long swelling ocean wave, individual water molecules will move up and down, and execute a limited oscillation as any wave passes by. However, identifying the water molecule with the head of wheat, the water molecule does not follow the extensive and more rapid movement of the wave form we can use to describe the shape of the ocean surface at any moment.

If, in addition to a progressive wave motion, there is a general lateral ocean current or drift in some direction, then this analogy is incomplete, since, unlike the tethered head of wheat, the water molecule is free to drift with the bulk of the water. One of the globally important examples of general directional currents in oceans is the flow of tides caused by the gravitational attractions of the moon and the sun. On a smaller scale, it is such a general current or near-shore rip that can carry an unwary swimmer out to sea despite their attempts to swim against it. It is these general 'mass-transporting' currents which determine the rate and direction of net transport of sand and other sediment by waves and tidal systems.

Figure 3.13 Motion of a floating ball on the sea surface as a wave passes it in the absence of a general current. The direction of movement of the ball at the different stages of the wave's passage is shown by the arrows. The ball (and thus water molecules also) executes an approximately circular orbit of diameter much smaller than the distance between wave crests. The wave motion is from left to right.

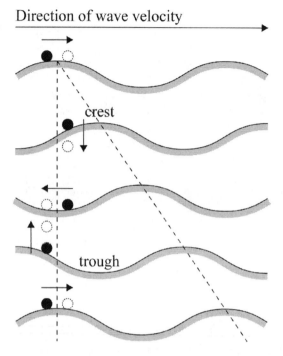

However, in the absence of such a general current the motion of a floating object, such as a ball, as a wave rolls by is illustrated in Fig. 3.13.

All waves have certain general characteristics that are constant for any particular regular periodic wave. A particular part of a wave, such as a crest, can be described as a particular 'phase' of the wave. These general characteristics, illustrated in Fig. 3.14, are the following.

(i) The 'wavelength' λ, which is the distance between corresponding parts or phases of adjacent wave forms, such as crest to crest, or trough to trough (unit: m).

(ii) The 'wave frequency', f, the number of wave crests (say) passing a given point per second (unit: cycles per second or hertz).

(iii) The period of the wave, T, the time for one complete wave to pass a given point (unit: s, being s/cycle if you wish).

Note that, from their definitions, it follows that

$$f = 1/T \tag{3.19}$$

(iv) Wave height, H (unit: m). For symmetrical waves, the 'amplitude' of the wave, A, is commonly used, where $A = H/2$ (Fig. 3.14).

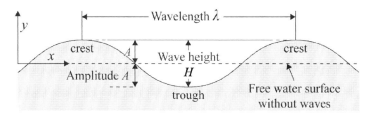

Figure 3.14
Characteristics used to
describe wave motion,
along with the period, T, of
the wave.

For progressive or travelling waves there is another general wave
characteristic, namely the velocity of movement, C, of any particular
'phase', such as a wave crest. C is thus also called the phase velocity,
since it is the velocity of any phase. Yet another name used for C is the
'celerity' of the wave. C is the distance moved by any phase in unit time.
By definition the distance λ is covered in time T. Thus the wave's phase
velocity, C, is given by

$$C = \lambda/T \qquad (3.20)$$

Also, from Eqs. (3.20) and (3.19),

$$C = f\lambda \qquad (3.21)$$

Equations (3.20) and (3.21) are general expressions for the velocity of
any progressive or non-stationary wave.

We will denote by y the water-surface displacement of the wave from
the still-water level that would be occupied if there were no waves. From
Fig. 3.14 it can be seen that the ordinate y can be positive or negative,
and can vary between the limits $+A$ and $-A$, where A is the amplitude
or half height of the wave.

Example 3.8

A person located on a rocky headland can observe waves moving by
the headland towards a beach. The observer times the passage of the
waves, and finds that 15 waves have passed by a fixed floating buoy in
one minute. A surfer among these waves on a surfboard estimates the
distance between wave crests to be 10 m. From this information calculate
the velocity and frequency of the waves.

Solution
The period is

$$T = (1/15)\,\text{min} = 4\,\text{s}$$

Thus

$$C = \lambda/T = 10/4 = 2.5\,\text{m s}^{-1}$$

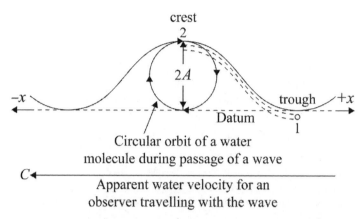

Circular orbit of a water
molecule during passage of a wave

$C \longleftarrow$

Apparent water velocity for an
observer travelling with the wave

Figure 3.15 The cross-section of a wave in deep water in which a molecule is seen to execute a circular orbit as the wave passes by. For convenience this orbit is shown under the wave crest. If we are travelling with the wave, it appears as if the whole body of water is travelling steadily in the $-x$ direction with velocity C. The apparent flow velocity is then enhanced by δC at position 1 in the trough, and reduced by δC in the crest (position 2).

The frequency is

$$f = 1/T = 0.25\,\text{s}^{-1}$$

(*Note*. The period of wind-driven sea waves is typically between 1 and 10 s.)

Whilst it can be shown that the depth of water is important for shallow-water waves, for 'deep-water' or 'short' waves, λ is much less than the depth of water, and then the depth is unimportant for the wave velocity. The phase velocity for deep-water waves, denoted C_{deep}, can be determined by considering the changes in flow velocity and elevation between a wave trough and its adjacent crest. As illustrated in Fig. 3.13, a molecule of water at the crest of a wave has a velocity a little greater than the phase velocity C, by an amount we will denote as δC. Similarly, a molecule in the trough of the wave has a velocity less than C due to a velocity in the opposite direction to C (Fig. 3.13). We will assume that the magnitude of this backward velocity is also δC. Hence, during the passage of a symmetrical wave, a particular molecule can be seen to execute a vertically oriented circular orbit of radius equal to the wave amplitude A, as shown in Fig. 3.15.

Whilst to a stationary observer waves move in the $+x$ direction with velocity C, to a surfboard rider moving with the same speed as the wave the water appears to be moving in the $-x$ direction with the same velocity C. This is illustrated in Fig. 3.15.

The big advantage of adopting this moving frame of reference is that the flow is then transformed from an unsteady to a steady flow, allowing Bernoulli's equation (which assumes steady flow) to be applied. Equation (3.15) will be applied to a stream tube of flow passing from position 1 in the trough, where the velocity is $C + \delta C$, to position 2 in the crest, where the velocity is $C - \delta C$ (Fig. 3.15). During this motion from positions 1 to 2, the potential energy of unit volume of water of mass equal to the density, ρ, has increased relative to the datum shown in Fig. 3.15 by $\rho g (2A)$. Since the static pressure p is the atmospheric pressure at both positions, application of Eq. (3.15) gives, for trough (position 1) and crest (position 2), respectively,

$$\tfrac{1}{2}\rho(C + \delta C)^2 = 2\rho g A + \tfrac{1}{2}\rho(C - \delta C)^2 \qquad (3.22)$$

From Eq. (3.22), we obtain

$$A = C\,\delta C / g \qquad (3.23)$$

During one wave period T, a molecule will completely traverse the circular orbital (shown dashed in Fig. 3.15) of circumference $2\pi A$. Thus

$$\delta C = 2\pi A / T$$

and, from Eq. (3.20), we obtain

$$\delta C = 2\pi A C / \lambda \qquad (3.24)$$

Eliminating δC between Eqs. (3.23) and (3.24) gives

$$C_{\text{deep}} = [\lambda g / (2\pi)]^{1/2} \qquad (3.25)$$

Thus, for deep-water waves, the depth of water is unimportant, but the wave's phase velocity increases with the square root of the wavelength. Thus longer waves in deep water travel faster than do waves of shorter wavelength.

Exercise 3.9

Calculate the values of C_{deep} and λ for an ocean wave of period $T = 7\,\text{s}$.

Solution
From Eq. (3.25),

$$\begin{aligned} C_{\text{deep}} &= 1.25\lambda^{1/2} \\ &= 1.25(CT)^{1/2} \end{aligned}$$

Thus

$$\begin{aligned} C_{\text{deep}} &= 1.56T \\ &= 10.9\,\text{m\,s}^{-1} \end{aligned}$$

Then

$$\lambda = 76.4\,\text{m}$$

3.8 An introduction to ocean-wave–coastline interactions

Since $\lambda = CT$ (see Eq. (3.20)), Eq. (3.25), which gives the phase velocity or celerity of waves in deep water, can alternatively be written as

$$C_{\text{deep}} = gT/(2\pi) \tag{3.26}$$

As deep-water waves approach a beach, passing through water of gradually decreasing depth, it is observed that the period or frequency of waves tends to remain almost constant. Thus the number of crests reaching the beach per minute is close to the number per minute approaching the coastline.

Since the wave frequency is $f = 1/T = C/\lambda$, a constant wave frequency implies a constant value of the ratio C/λ. As waves move towards a beach they pass over progressively shallower water, and it is observed that waves bunch closer together in this progression towards the beach. So, if λ progressively decreases, and C/λ is constant, then C must also decrease progressively as waves move towards the beach.

To the delight of people surfing the waves, it is a consequence of mass conservation that wave height increases at approximately the same rate as wave velocity decreases. Waves also become steeper as they approach the beach and, of course, can break or spill over.

Let us denote the depth of the body of water by h. Then, if h is not large relative to the wavelength λ (as it is in the deep ocean), the phase velocity C is affected by the depth of water h. In this situation, where λ/h is not small, waves are referred to as 'shallow-water' waves. It can be shown (e.g. Allen (1985)) that, for shallow-water waves, the wave phase velocity, here denoted C_{shallow}, is given by

$$C_{\text{shallow}} = (gh)^{1/2} \tag{3.27}$$

As the depth of water, h, decreases from deep-oceanic values, there is a gradual decrease in the wave's phase velocity, provided that other factors, such as wind speed over the ocean, remain constant.

Figure 3.16 illustrates how the wave speed C of waves of a constant period, T, of 7 s, decreases as the depth of water h gradually decreases, as is the case for ocean waves approaching a beach. Relationships for the limiting cases of deep- and shallow-water waves are shown (following Eqs. (3.26) and (3.27), respectively). The solid curve represents the result of more general theory, not given here, which is asymptotic to these two

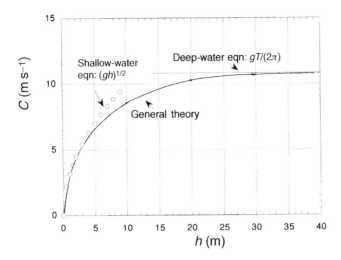

Figure 3.16 The variation in wave velocity or celerity, C, for waves of period 7 s, on water of gradually changing depth, h. Note the gradual transition, as h decreases, from the deep-water speed given by $gT/(2\pi)$ (shown by the horizontal line) to values (shown by open circular symbols) given by the shallow-water expression, $(gh)^{1/2}$.

limiting cases. With an appropriate change to C_{deep}, which is proportional to T (Eq. (3.26)), the general form of Fig. 3.16 is similar for values of T different from the adopted value of 7 s.

In deep water, the velocity of the wave of period 7 s is $10.9\,\mathrm{m\,s^{-1}}$, so $\lambda = 76.4\,\mathrm{m}$. With passage of the wave into progressively shallower water (decreasing h, Fig. 3.16), the wave speed C is increasingly reduced, since λ/C is constant as noted earlier, and as we assumed in Fig. 3.16. Near the beach, at a depth of (say) 1.5 m, C is reduced from $10.9\,\mathrm{m\,s^{-1}}$ offshore to $3.8\,\mathrm{m\,s^{-1}}$, and λ is reduced (in the same proportion) from 76.4 m to 26.8 m, nearly a threefold reduction.

You may well have noticed that, as waves approach a beach, the crests of waves are commonly approximately parallel to the beach line. Since, in the open ocean, wave crests are normally approximately perpendicular to the wave direction, and this direction can vary considerably, it is perhaps surprising that waves always end up so nearly parallel to the beach. However, we have seen that waves are slowed down as they move into shallower water and that the depth of water often decreases gradually as the shore is approached. Thus, any wave crests that approach the coastline obliquely, as in Fig. 3.17, will have those parts closest to the shore slowed down much more than those parts further out to sea. Those more distant waves will thus travel faster than those close to the shore, the effect being to swing the oblique offshore wave crests to a more parallel orientation with the beach.

The changing orientation of wave fronts due to change in velocity of wave propagation is referred to as 'wave refraction'. This phenomenon occurs in all wave motion, including sound and light waves. Just as light

Figure 3.17 Illustrating how wave crests approaching a beach at an oblique angle will, on meeting water of gradually decreasing depth, become more closely aligned with the direction of the shoreline. This is called wave refraction.

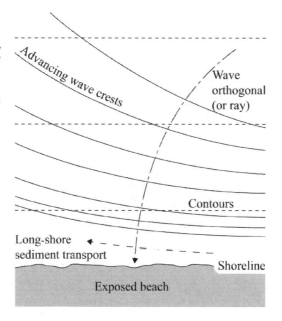

is reflected by most surfaces, at least a part of a water wave is reflected when it meets an obstacle, such as a cliff face. Any reflection of the wave motion tends to form standing waves whose amplitude can exceed that of the incoming wave, due to superposition of the incoming and reflected waves. Even if waves meet only a steeply sloping section of beach (perhaps at higher tides), they can be partly reflected rather than have their energy more gradually dissipated by spilling over a more gently sloping beach. Such differences in beach morphology also strongly affect the manner in which waves change shape as they approach the beach.

Figure 3.17 shows a dashed arrow indicating the direction of 'long-shore sediment transport', which, on beaches, may involve beach sand or pebbles. How is it that oblique waves (even if they are somewhat normalised by refraction) result in a current parallel to the beach? Any beach swimmer will have felt the effects of this long-shore current, and sometimes it is almost too strong to resist. The reason for this current is evident from Fig. 3.18 using the concept of mass conservation of water. The water transported across AB by the general on-shore flow that develops in the near-shore zone does not progress beyond the water's edge AC. Thus mass conservation requires that there be a long-shore current across BC.

At any particular coastline there is commonly a preferred average obliquity of offshore waves, associated with a preferred or common wind direction. In such cases, the long-shore current also has a preferred direction. This leads to long-shore transport of sediment that can be

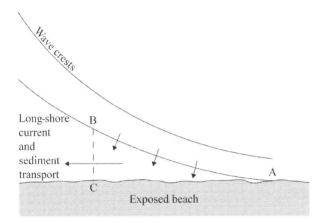

Figure 3.18 The origin of persistent long-shore currents and sediment transport.

geographically extensive. Also, any pollutant such as sewage entered at one point will be taken along with the long-shore current.

For a stable beach, the rate of removal of sediment from any stretch of beach is compensated by input from sediment transported into that stretch. Whilst net erosion can take place due to natural causes, such as cyclones and storm surges, human activity can also substantially affect beach stability. For example, jetties, piers, or river-training walls projecting beyond the breaker zone can be built to provide port facilities, or for navigational or recreational reasons. Such structures can interrupt any general long-shore sediment transport. As a consequence of this interruption, sediment accumulates on the upside of the structure, and the lack of supply of sediment on the downflow side of the structure leads to substantial net erosion (Fig. 3.19).

Near-shore sand dunes are a natural source of sediment that can be called on in severe storm-surge erosion. High winds lead to waves of high amplitude, A, and the wave energy involved in sediment transport is found to be proportional to A^2. Sand lost from dunes in extreme events can then be slowly restored by natural on-shore sediment-transport processes. However, if such dunes are built upon so that their erosional loss is a catastrophe for persons involved, huge expense in protective works with often uncertain outcomes can be involved.

This brief consideration of wave–coastline interactions provides an introduction to mathematical representation of the physical processes involved in such interactions, which have important environmental consequences. Such representations provide the basis of mathematical models that simulate near-shore processes. Martinez and Harbough (1993) describe a computer program named WAVE, which is used to simulate the behaviour of waves and the transport of sediment as they interact with coastal environments. This program has been linked to other programs

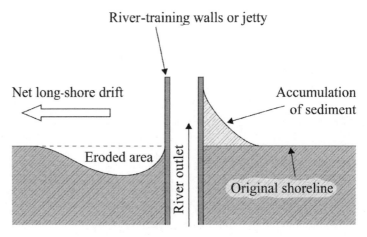

Figure 3.19 The sediment-accumulation and erosion effects of any structure that impedes the net long-shore drift of sediment.

to form a unified mathematical model that simulates sediment transport, not only by waves, but also by rivers.

Main symbols for Chapter 3

a Acceleration

A Area, or wave amplitude (Fig. 3.14)

C Phase velocity or celerity of a wave (Eq. (3.20))

C_d Drag coefficient (Eq. (3.17)).

d Size scale of an object

D Drag force

E_{loss} Energy loss (Eq. (3.17))

f Wave frequency (Eq. (3.19))

F Force

g Acceleration due to gravity

h Distance, such as depth beneath the water surface, or depth of water

K Kinematic eddy viscosity (Eq. (3.5))

m Mass

p Pressure (Eqs. (3.1) and (3.2))

r Radius

Re Reynolds number (Eq. (3.7))

s Displacement, or distance moved

v Velocity

t Time

T Wave period (Eq. (3.19))

W Weight, or work done by a force (Eq. (3.8))

z Distance

Δ	Finite difference in any quantity
η	Coefficient of dynamic viscosity (Eq. (3.3))
θ	Angle
λ	Wavelength
ν	Kinematic viscosity ($= \eta/\rho$)
ρ	Fluid density
τ	Shear stress

Exercises

3.1 Archimedes' principle states that the buoyant force on a body immersed in a fluid is equal to the weight of the fluid displaced by that object. Derive an expression for the buoyant weight or effective downward force on a solid of volume V and density σ immersed in a fluid of density ρ. Define any other symbol you may use in this derivation.

3.2 A viscometer can consist of two concentric cylinders, one fixed and the other free to rotate. In a particular viscometer the diameter of the outer cylinder is 11.60 cm, and that of the inner cylinder is 11.20 cm, the liquid whose viscosity is to be determined being poured into the annular space between the two cylinders, filling it to a depth of 10.0 cm. The outer cylinder is fixed, and a torque of 0.02 N m produces a steady rotational speed of 60 revolutions/minute in the inner cylinder. Calculate the viscosity of the liquid between the cylinders.

3.3 The velocity of flow of a liquid, v, through a tube of internal diameter d and length L, between the ends of which there is a static pressure difference of Δp, is given by

$$v = \frac{\Delta p \, d^2}{32 \eta L}$$

where η is the coefficient of viscosity.

(a) Using this equation, show that the volume flow rate, Q, of liquid through the pipe is given by

$$Q = \frac{\pi \Delta p \, d^4}{128 \eta L} \tag{3.26}$$

(b) Using Eq. (3.26), calculate the value of Q (in $m^3 \, s^{-1}$) for the flow of water up a single tubular xylem vessel of a tree, given that the internal diameter of the xylem vessel (assumed to be of circular cross-section) is 0.2 mm, the length of the xylem vessel (approximately equal to the height of the tree) is 40 m and the pressure difference between the ends of the xylem, Δp, is

30 atmospheres. Take one atmosphere to be 100 kPa, and assume that the dynamic viscosity of water, η, is 10^{-3} Pa s.

(c) The total transpiration stream for the tree with xylem vessels described in part (b) is contributed to by many such xylem vessels conducting water up the tree. The total transpiration rate of the tree results in a loss of 10 kg of water per hour. Assuming that the volumetric rate of water flow in all individual xylem vessels or tubes is given by the value of Q calculated in part (b), calculate the number, n, of xylem vessels that must be present in the trunk of the tree in order to supply water at the total rate of transpiration given above.

3.4 (a) Fluids can possess energy for various reasons. Describe three different types of contribution to the energy of a fluid, and express them symbolically in a form that shows how these energies depend on more fundamental quantities.

(b) Define and express symbolically the principles of conservation of mass and energy in forms suitable for application to a flowing fluid. If water is moving through a pipe of diameter 5 cm with a velocity of $4 \, \mathrm{m \, s^{-1}}$, what will the fluid velocity be if the same pipe is constricted to a diameter of 3 cm?

3.5 (a) Discuss and illustrate with examples the environmental significance of the two possible types of flow in fluids, namely laminar and turbulent.

(b) A grain of fine sand (of diameter 0.1 mm) is falling through water. Indicate why you might expect the flow of water around the grain to be laminar rather than turbulent.

3.6 It can be shown that the velocity of a wave in shallow water, denoted here as C_{shallow}, is given by

$$C_{\text{shallow}} = (gh)^{1/2} \qquad\qquad (3.27)$$

where h is the depth of water and g is the acceleration due to gravity $(9.8 \, \mathrm{m \, s^{-1}})$.

For shallow waves the ratio λ/h is not small. However, because the wavelength of oceanic tidal flows is thousands of kilometres and the world-average ocean depth h is about 4 km, the velocity of tidal flows is well given by Eq. (3.27), even though, in the ordinary sense of the word, the sea is not 'shallow'.

(a) Calculate the value of C_{shallow} for a tidal wave with an ocean of depth 4 km.

(b) Assuming that the wavelength of the oceanic tidal wave is 8000 km, calculate its period.

(c) Calculate C_{shallow} for water near a beach where the depth of water is $h = 3 \, \mathrm{m}$.

3.7 (a) By comparing Eq. (3.25), for waves in deep water, and Eq. (3.27), for waves in shallow water, discuss the factors affecting the velocity of water waves. (More general theory shows that there is a gradual transition between these two regimes.)

(b) Using Eq. (3.25), calculate the velocity of a deep-water wave whose period $T = 6\,\text{s}$. What is the depth of water at which the velocity of a shallow wave calculated using Eq. (3.27) is the same as that of the wave in deep water with $T = 6\,\text{s}$?

3.8 (a) Waves in water of depth h_i are approaching the beach, with wave crests inclined at an angle θ_i to the beach line as shown in the figure. There is a sand bar, which leads to a sudden decrease in depth of water from h_i to h_r. The figure shows the wave refraction caused by the sudden reduction in wave velocity from v_i to v_r as the wave crosses the bar. During time period t the incident wave front will have travelled a distance $BC = v_i t$; during the same time period, the refracted wave will have travelled a distance $AD = v_r t$.

Show that

$$\sin\theta_i / \sin\theta_r = v_i / v_r$$

(b) Assume that water depths h_i and h_r are both sufficiently shallow for the wave velocity to be given by Eq. (3.27). Given that $h_i = 6\,\text{m}$, $h_r = 3\,\text{m}$ and $\theta_i = 30^\circ$, calculate θ_r and $\theta_i - \theta_r$.

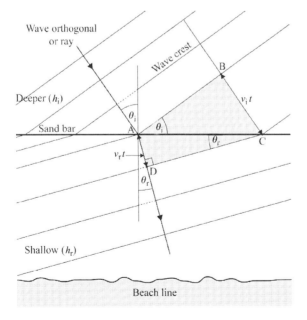

References and bibliography

Allen, J. R. L. (1985). *Principles of Physical Sedimentology*. London: George Allen and Unwin.

Giancoli, D. G. (1991). *Physics: Principles with Applications*, 3rd edn. London: Prentice-Hall International.

Martinez, P. A., and Harbaugh, J. W. (1993). *Simulating Nearshore Environments*. Oxford: Pergamon Press.

Monteith, J. L., and Unsworth, M. H. (1990). *Principles of Environmental Physics*, 2nd edn. London: Edward Arnold.

4
Soil, water and watersheds

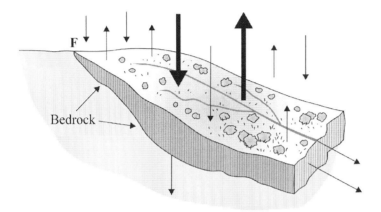

4.1 Introduction

In Chapter 3 we focused solely on the behaviour of liquids such as water. The physical behaviour and strength of soil were shown in Chapter 2 to depend strongly on the type of solid material in the soil and on its water content. In this chapter we develop further the recognition in Chapter 1 that water can enter into, and be stored in, the soil profile, but can also leave that store as vapour in evaporation to the atmosphere, or as liquid water in overland or subsurface flows.

Most environmental concerns involving soil and water require consideration at watershed scale before they can be adequately understood. Once this understanding has been achieved, the possible ways in which the environmental problem or issue may be addressed also typically involve action at this scale, though yet larger scales may need to be considered for some issues. These are the reasons why attention is drawn to the watershed scale early in this book, and this chapter introduces some of the terminology used to describe soil and water characteristics relevant to watersheds. How some of the important environmental variables can be measured is also illustrated.

Using the principle of mass conservation of water, the variety of inputs and outputs to a watershed is described and quantitatively related to changes in the amount of water stored within the soil profile of the watershed. Such considerations are also important at larger and even

125

national scales. For example, the island of Taiwan has had to import water from China, despite severely restricting the range of activities for which water can be used. It is feared that lack of water will increase the incidence of disease in that country. Also competition for above- and below-ground water resources is an issue in the difficult relationships between Israel and the Palestinian peoples.

A deeper physical interpretation of the factors controlling the rates of the various hydrological components outlined in this chapter is deferred until later chapters in this book. More explicit consideration of the environmental roles and implications of the basic information in this chapter also will be given in later chapters, where movements of water into and out of watersheds, and changes in water storage within the soil profile of watersheds, will be found to be directly linked to an important range of such concerns.

In Section 2.3 soil was described as a solid in which particles of various sizes and types of material are commonly stuck together into aggregates, leaving a pore space that can be filled with water or air (Figs. 2.2 and 2.4). A useful description of soil solids from a physical viewpoint is the proportions in various size ranges of the ultimate particles of the inorganic soil material. These size-range components were described in Table 2.1 as ranging from coarse sand, through finer silt, to clay as the finest fraction.

The fraction of the pore space filled with water affects not only the physical response of soil to trampling, cultivation, or the ability of plant roots to penetrate soil, but also the activity of the vast populations of biological organisms which inhabit and form a vital part of what we call soil. The ability of plants and trees to take up water from the soil through their roots is also strongly affected by soil water content, or more directly by soil water suction, a concept introduced in Section 2.7 and whose relation to soil water content depends on soil type. The higher the soil water suction (the pressure deficit below atmospheric pressure), the more energy vegetation has to expend on removing water from the soil. If soil in the root zone becomes very dry indeed, then plants might not be able to withdraw the only remaining water in the soil because it is held at very high suctions. In this situation a plant may visibly wilt, with its leaves hanging limply. A sustained water deficit can cause leaves to die and be shed from the plant in order that the plant itself may stay alive.

Water entering the soil surface is an important process called 'infiltration'. Except in very cold regions, with precipitation falling as snow or sleet, or where the soil profile is frozen, some fraction of precipitation enters the soil. Also, if the soil profile is saturated up to the soil surface, very little, if any, rainfall may infiltrate. With significant exceptions such as those just listed, some fraction of precipitation enters the soil surface.

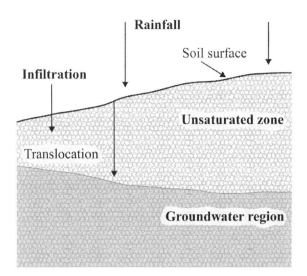

Figure 4.1 Distinguishing between the unsaturated zone and the region saturated with water – the groundwater zone. The processes of infiltration into the soil surface and subsequent translocation within the unsaturated zone are indicated.

This fraction is of great importance, replenishing the store of water in the soil profile which can be drawn upon by growing vegetation. The uptake of water by vegetation to form the transpiration stream is necessary for its growth. Water stored in the soil profile is required for the activity of soil organisms, and facilitates the wide array of chemical reactions of such great importance in soils.

Any precipitation not infiltrating or stored in depressions on the soil surface will run over the soil surface and can present an erosion hazard, especially if the soil surface is cultivated or not well covered by grass, leaf litter, or other forms of protection. Such surface runoff contributes to stream flow during its occurrence, and, in excess amounts, can lead to flooding on a local or more extensive scale.

In general there is a zone extending from the soil surface to some depth within the soil profile where the pore space contains air and gaseous water vapour. This region is called the 'unsaturated zone' to distinguish it from a 'saturated zone' where all the pore space is filled with liquid water (Fig. 4.1). The depth to the saturated zone can vary greatly from region to region, but spatial variation in this depth, even within a constrained region, can be considerable. This water-saturated volume is called the 'groundwater zone' (Fig. 4.1). Variation with time of depth to the groundwater zone at a given location is generally slow, substantial change possibly taking decades unless accelerated by groundwater pumping or extreme events.

Figure 4.1 illustrates both the infiltration of water into the soil surface and its subsequent translocation down through the soil profile. As discussed in association with Figs. 1.4 and 1.5, infiltrating water is intimately involved in developing the profile characteristics of soil. Processes discussed there included humification, translocation of solutes, weathering of parent material and other processes indicated in Figs. 1.3–1.5.

Infiltrating water can also bring with it pollutants that have been applied to, or disposed of on, the soil surface, and this occurs readily if they are water soluble, are in suspension, or have become attached to fine soil particles. Such pollutants may eventually, if not quickly, reach water stored in the groundwater region. Since water in this region is commonly used as a water supply for human consumption, there is widespread concern at the possible health consequences of the transport of natural, applied, or leaking contaminants to the groundwater.

Loss of water from the soil surface by evaporation may be quite small or even negligible during rainfall, and so is not represented in Fig. 4.1. Evaporation may also be small compared with infiltration for the duration of the infiltration process, although evaporation from water stored in depressions of the soil surface may be significant and contemporaneous with the infiltration of such stored water. However, in general there can be significant time periods of finer weather between rainfall–infiltration events, and during such periods it is the process of evaporation or evapotranspiration which dominates (these terms were discussed in Section 1.5). Water can change phase from liquid to vapour at many sites: at the soil surface, within the soil profile, from a moist litter layer, or from wet leaf surfaces. When evaporation occurs from within the stomatal cavities of leaves, this water has been taken up by the roots of the vegetation, and moved up through the stem of the vegetation as the transpiration flux (involving a different translocation process from that shown in Fig. 4.1, where movement occurs within the soil).

The total flux of water vapour from the land surface to the atmosphere, by whatever pathway, is commonly called the evapotranspiration flux, and is shown in Fig. 4.2. Evapotranspiration is a convenient term in the sense that it is quite difficult experimentally, and often unnecessary, to partition the total evaporative flux from a complex system of land cover into components that have, and components that have not, passed through the stems or trunks of vegetation.

The loss of water at or close to the soil surface in evapotranspiration causes the surface layer of the soil to dry out. If soil beneath this drying layer is adequately wet, then, for reasons explained in Chapter 11, water can move upwards in the soil profile from the wet to the upper dry layer of the soil. The reason for this upward movement of water, in the opposite direction to gravity, is somewhat akin to using a sponge to soak up water from a wet floor, and the process involves concepts of capillarity and suction described in Section 2.7.

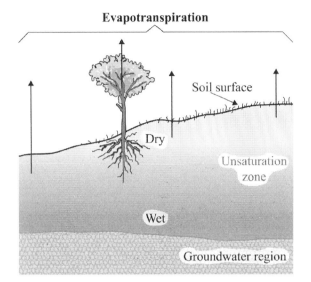

Evapotranspiration

Soil surface

Dry

Unsaturation zone

Wet

Groundwater region

Figure 4.2 The phase change of water from liquid to vapour can take place at many sites. Loss of water vapour by the land from whatever surfaces is commonly called evapotranspiration, but in some literature is simply called evaporation.

Adequate soil water within reach of the root system of vegetation is needed for its growth. Thus there is wide interest in understanding the factors affecting the plant-available water store. The magnitude and spatial extent of this store, and its typical fluctuations throughout a year, constitute a major constraining factor limiting the types of vegetation that can evolve in any particular region. These characteristics of the plant-available water store thus provide one key to understanding the spatial patterns in natural vegetation. Together with the energy balance at the earth's surface considered in Section 1.5, the characteristics of the water store also provide a constraining framework for all agricultural, pastoral and forestry activities. The climatic factors determining the water-store characteristics are also of vital significance regarding the availability of water for its many human uses in cities and urban areas, much of this water coming from surrounding regions.

For all these reasons there is wide interest in how the plant-available water store varies at a range of spatial scales, from field plots to 'catchments' or 'watersheds'. The terms watershed and catchment are alternative terms used to describe a landscape component whose water drains into a stream, or perhaps a number of connected streams or rivers.

In this chapter we will consider the ways in which water arriving at the earth's surface, either as rain or in frozen forms, is distributed within and beyond a catchment, or a component of a catchment, often called a sub-catchment. Quantifying the various ways in which water can be distributed is often done in terms of a 'water budget'. This term uses the analogy with a bank account into which there can be inputs, or deposits, and withdrawals. A bank statement of one's account allows one to check that the change in the balance of money residing in the account over the time period of the statement is correct. One can do this by checking that the difference between these two balances is consistent with the difference between financial inputs and withdrawals over that period. The assumption behind this checking is that the value of money in the various transactions is a 'conserved quantity'. A conserved quantity is one to which conservation principles, such as mass conservation, can be applied. Financial budgeting implies managing these gains and losses so that the balance remains positive.

Like all matter, water is also a conserved quantity. Thus water is subject to the principle of mass conservation, which is analogous to the principle of conservation of money referred to in the paragraph above. Thus a water budget uses the principle of 'mass conservation' of water to follow quantitatively the inputs and outputs of water from the catchment, and the balance of water stored within it. A water budget is therefore sometimes referred to as a 'water balance', again in an analogous manner to preparing a financial balance sheet. Hence there is an analogy between financial accounting and hydrological budgeting operations, and, indeed, in following the fate of any conserved quantity. In Chapter 12 we will use a similar approach in considering the important environmental land-management issue of 'salinity', since naturally occurring salt also can be regarded as a conserved quantity.

Finally, in this chapter we will illustrate a simple arithmetic system of water-balance accounting, showing how changes in the amount of water stored within the root zone can be followed. Just as financial accounting can show up an error if a balance is not achieved, so water-balance accounting has the important ability of being able to indicate errors in the measurement or estimation of the various component terms involved in the hydrological outcome for a watershed.

Alternatively, if all terms in the water balance can be measured except one, which may be more difficult than others to ascertain, then the water-balance or mass-conservation principle can be used to determine that unknown component. When this use is made of the water-balance principle then of course it is not possible to show the error involved in summing all the independently measured component terms such as rainfall and runoff. For example, evapotranspiration from a watershed over a significant time period is more difficult to measure than the delivery of water to the soil surface in precipitation, and so is often estimated as the only unknown component in the water-balance equation (which will be given later).

Let us now consider the characteristics and measurement of precipitation.

4.2 Precipitation and runoff measurement

Precipitation

Important characteristics of rainfall that affect the delivery of water to earth are its rate (or intensity) and its duration, or, more specifically, the time history of the rainfall rate. The terms rainfall rate and rainfall intensity are used interchangeably, and are defined as the volume of water received per unit horizontal area per unit time. Rainfall is measured by collection in a rain gauge of some kind, the collecting rim of the rain gauge being horizontal and located at some standard height above the surrounding surface. This height is made as low as is feasible in order to reduce errors from wind blowing around the rain gauge. To avoid splashing into the gauge, a rim height of 0.3 m above a turf surface is a common standard recommendation, achievable in a weather station, but other heights are also used for practical reasons, such as 0.75 m and 1.2 m.

Being given by the total volume of water collected per unit area, the total amount of rainfall in a given event is correctly expressed by the depth of water if it was ponded on collection, but is usually called the depth of rainfall or the rainfall amount, commonly expressed in mm. If rainfall is collected in a cylindrical container of the same diameter as the rim of the rain gauge, then the ponded depth gives the rainfall amount directly.

The rainfall rate will here be denoted by the symbol P, and, like the rainfall amount, rate is commonly expressed in non-SI units such as $mm\,h^{-1}$. The SI unit of rainfall rate is $m^3\,m^{-2}\,s^{-1}$ or $m\,s^{-1}$. A metre of water per second is a deluge that might be experienced under a waterfall, which is a reason why rainfall rate is commonly expressed in the much smaller unit of $mm\,h^{-1}$.

Example 4.1

What is a rainfall rate of $36\,mm\,h^{-1}$ expressed in SI units?

Solution

$$36\,\frac{mm}{h} = \frac{36\,mm}{1\,h} \times \frac{10^{-3}\,m}{mm} \times \frac{h}{3600\,s}$$
$$= 10^{-5}\,m\,s^{-1}$$

One way of measuring the rainfall rate is to pass the rainfall collected by the mouth of the gauge into one side of a quite small tipping device called a 'tipping bucket'. This small bucket is divided into two parts, and the bucket tips over when a given weight of collected rainfall has been received. (This given weight can vary somewhat due to dynamic effects

Figure 4.3 The principle of operation of a tipping-bucket rainfall-rate recorder. Water passing through the tipping bucket is not normally collected in the base of the instrument as shown for illustrative purposes in this figure. The half of the collecting cylinder towards the reader is removed (see dotted lines) to expose the mechanism of the tipping bucket. Whilst the drop size can be of some significance, for example in soil erosion, it is the time distribution of the rainfall rate which determines the rate of delivery of water to the land surface.

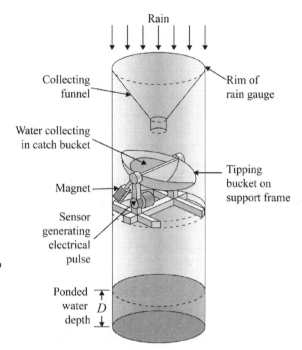

associated with the rate at which rainfall is received, and this requires recognition in calibration of the instrument.) Following a tip, the other side of the bucket is presented to the inflow (Fig. 4.3). The bucket tip moves a magnet past a device that generates an electrical pulse. The time of occurrence of each pulse can be recorded on an electronic data logger or a computer, or the number of pulses in a given time interval recorded, either indicator of the tipping rate yielding the rainfall rate at the recorder site. Thus the basic data are digital in nature.

Rainfall consists of raindrops of a variety of sizes, smaller drops being close to spherical in shape due to the dominance of surface-tension effects discussed in Section 2.7. (The sphere has the minimum surface area per unit volume of any shape, and surface tension tends to minimise the area of the liquid/air interface of the drop.) Larger raindrops deform from a spherical shape due to differences in the static air pressure over the surface of the drop. The mean size of raindrops tends to increase with rainfall rate, and the upper stability limit to drop size is about 5 mm, beyond which drops tend to break up if formed.

The amount of precipitation falling as snow is commonly recorded as a depth of snow accumulation, or accumulation from last snowfall, which is desirably measured on reasonably flat land as far away as possible from the effects of trees or buildings. However, the 'water equivalent' of a given depth of snow, defined as the equivalent ponded-water depth of melted snow, depends on the degree of packing or density of snow. For this reason it is recommended that samples of a snowfall be taken with a specially designed sampling tube, which is then weighed full of snow sampled down to the surface of the last snowfall. The weight of snowfall is obviously uninfluenced by packing effects.

Both rates of rainfall and snowfall, and accumulated amounts, vary spatially in any given precipitation event. Remote-sensing techniques such as weather radar are providing data on the spatial-distribution structure of rainfall amount. However, for fields up to several hundred square metres in size the spatial variation in total rain catch in a given event appears to be so small as to be of little consequence.

During rainfall, some water will infiltrate into the land surface, some will fill up depressions in it, and, if there is still rain left over, this excess will run off as overland flow, roughly in the direction of greatest downhill slope unless deviated by channels, for example.

If the rainfall rate P is assumed approximately constant for a short period of time, Δt, then the depth of rainfall received over that time period is $P\,\Delta t$. To obtain the total amount (or depth) of rainfall received during a rainfall event, D, the amount received during each time period Δt must be summed up for all periods in the event. This summation will be written as

$$D = \Sigma P\,\Delta t \qquad\qquad (4.1)$$

The summation sign, Σ, is taken to imply summation over the duration of the rainfall event being considered.

Example 4.2

The figure below shows the time history of the rainfall rate during a rainfall event. Calculate the amount, or depth, D, of rainfall received during the event.

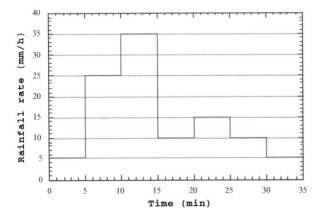

Figure 4.4 A runoff plot defined by boundaries across which there no flow except at its lower end, where runoff is collected and its rate measured. Measurement of flow rate could be by a tipping-bucket device of the type illustrated in Fig. 4.5. (After Rose (1993).)

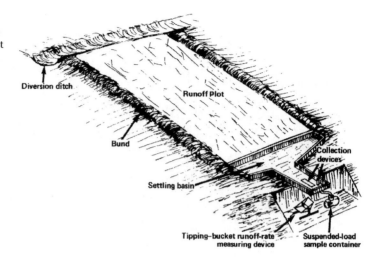

Solution

P (mm h^{-1})	Time (min)	Δt (min)	P (mm min^{-1})^a	Depth $D = \Sigma P \, \Delta t$ (mm)
5	0	5	0.083 333	0.4
25	5	5	0.416 67	2.1
35	10	5	0.583 33	2.9
10	15	5	0.166 67	0.8
15	20	5	0.250 00	1.3
10	25	5	0.166 67	0.8
5	30	5	0.083 333	0.4
Sum				8.8

^{a} Note that P (mm min^{-1}) $= P$ (mm h^{-1})/60.

Runoff

Measurement of runoff can be carried out on a hydrologically defined runoff plot of the type shown in Fig. 4.4. The plot is hydrologically defined by preventing entry into or loss of water from the plot boundaries except for runoff from the lower end of the plot.

Measurement of flow rate from runoff plots (such as that shown in Fig. 4.4) of area up to about 500 m^2 can be made conveniently by a tipping-bucket device having exactly the same principle of operation as that of the rainfall-rate recorder shown in Fig. 4.3. Figure 4.5 shows a drawing of such a device, where the collecting bucket is much larger than that in a rainfall-rate recorder. It is preferable that water falling into this device is previously freed of coarser sediment by deposition in a

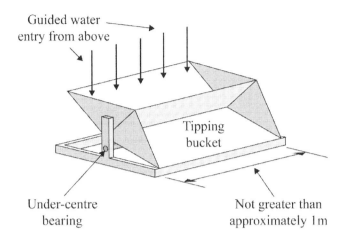

Guided water entry from above

Tipping bucket

Under-centre bearing

Not greater than approximately 1 m

Figure 4.5 A tipping-bucket flow-rate-measuring device suitable for measuring runoff rate from plots up to about 500 m² in area. Ciesiolka and Rose (1998) give design details for such devices.

low-slope collection trough (Fig. 4.4), because otherwise the calibration of bucket-tipping rate to flow rate for the device can be affected. For larger flow rates tipping-bucket devices become unwieldy, and flumes or weirs are then commonly used for flow-rate measurement as described in French (1985), for example.

4.3 Soil water content and its profile storage

Most water in a catchment is commonly stored within the soil profile. We will now briefly discuss how the amount of soil water accumulated in a profile is calculated and expressed.

In Section 2.3 we saw that soil water content could be expressed in two different ways. Firstly soil water content can be expressed as the ratio of mass of water to mass of solids (Eq. (2.7)). This ratio is called a 'gravimetric' expression of water content, w, because weighing is involved in its determination. However, in quantitatively following the changes in amount of water in a region or within the soil profile there are distinct advantages in using the expression for water content given by the ratio of the volume of water per unit volume of space, defined in Eq. (2.8). This method of expression gives what is called the 'volumetric water content', θ.

As shown in Eq. (2.9), there is of course a relationship between these two ways of expressing soil water content, w and θ, involving the bulk density of the soil (the relationship being defined in Eq. (2.4)).

The advantage in using the volumetric water content, θ, in following changes in movement and storage of water can be seen with reference to Fig. 4.6. Figure 4.6(a) shows θ constant to depth d, beyond which it falls to zero. The area enclosed within this simple profile form is given by θd,

Figure 4.6 Two profiles of volumetric water content θ as a function of depth z from the soil surface down into the soil.

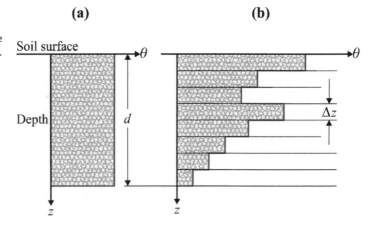

which has units $m^3\,m^{-3}$ m, or just m. Suppose that $\theta = 0.2\,m^3\,m^{-3}$ and $d = 3$ m in Fig. 4.6(a). Then, if all the water in a cylindrical soil column of $1\,m^2$ cross-sectional area were extracted from the column and ponded in a cylinder of the same ($1\,m^2$) cross section, it would have a ponded depth of $0.2 \times 3 = 0.6$ m. This figure of 0.6 m is referred to as the 'equivalent ponded depth' of water in the profile to that particular depth. Because of its computational convenience, this concept of equivalent ponded depth is commonly used when one is analysing the water budget of a particular watershed, or a given soil profile within it.

The concept of equivalent ponded depth corresponds to the ponded depth of rainfall collected in a rain gauge as discussed in Section 4.2 and illustrated in Fig. 4.3. If 100 mm (or 0.1 m) of rainfall all infiltrated into the soil column just discussed, then the equivalent ponded depth of water in the column would be $0.6 + 0.1 = 0.7$ m (assuming that no water was lost from the column during this input). This illustrates the advantage of simplicity in using volumetric water content to account for the components of water inputs and losses in a soil profile.

Figure 4.6(b) shows a somewhat more complex profile of θ. The profile is shown divided up into layers of equal thickness Δz, with Δz chosen so that the profile can be well represented with θ considered constant in each layer. The equivalent ponded depth of water in any layer is then given by $\theta\,\Delta z$. Hence, if the depths of water in all layers are summed up or added to depth d, this will give the equivalent ponded depth of water to this depth in the profile. Denote this summation by the Greek symbol capital sigma, which will be written Σ_d, the subscript d indicating summation over depth d. Then

$$\text{equivalent ponded depth of water to depth } d = \Sigma_d \theta\,\Delta z \qquad (4.2)$$

Methods and types of equipment used to measure the profile variation in water content are described in Marshall, Holmes and Rose (1996). Commercially available equipment allows $\Sigma_d \theta \, \Delta z$ to be directly and conveniently determined.

4.4 The water budget for a watershed using the principle of mass conservation of water

The idea of a water budget was introduced in Section 4.1 as describing the partitioning of water incoming to a system of defined volume over a given time period. The inputs may be either as precipitation (rainfall, snowfall, or dew) or as irrigation water.

A water budget applies the constraint of mass conservation to the water input. This application is a special case of the more general physical principle of conservation of mass, which states that mass is neither created nor destroyed (with the important exception of mass-to-energy transformations in nuclear transformations, which does not impinge on our considerations here).

Precipitation can either infiltrate into the soil or, if an excess of precipitation over infiltration occurs, generate overland flow. Some of this overland flow can find its way into rivers, though river flow is commonly dominantly dependent on subsurface flow.

A water budget or water balance is constructed over a particular volume of soil and for a selected time period. This chosen soil volume is commonly defined by an imaginary surface that would be generated by a vertical line moving around the land surface area selected for the budget, with the soil volume extending to some chosen depth. The surface area might enclose a farm unit, unit area of land, or a sub-catchment as in Fig. 4.7. The chosen depth might be the extent of the root zone of vegetation, the depth of a monitored drainage system, or the depth to bed rock or to a much less permeable soil layer beneath a catchment or sub-catchment, as in Fig. 4.7.

The width of the lines representing fluxes in Fig. 4.7 illustrates the possible relative magnitudes of the variety of water flows, both liquid and vapour, that cross the conceptual boundary surfaces of the watershed or catchment during the chosen time period of the budget. The surface boundaries FH and FG might be chosen as lines of steepest descent, thus minimising water fluxes across the surfaces beneath them. If there is a lower surface of much less permeable material, such as bed rock (Fig. 4.7), this would be chosen because of the difficulty otherwise of quantifying the vertical flow U_v beneath the volume selected for the budget. Likewise the vertical surface GHIJ might be chosen where the

Figure 4.7 Conceptual boundaries of a watershed underlain by bedrock that can be used in applying mass conservation or water-budget considerations expressed in Eq. (4.3).

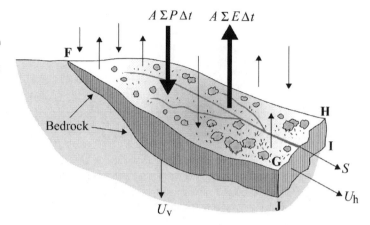

subsurface horizontal flow U_h could be small, or perhaps it might be chosen where the overland river flow, S, is measured (Fig. 4.7).

Now consider the fluxes shown in Fig. 4.7 to be volumetric amounts (in m^3) for the entire watershed of area A accumulated over the chosen budget time period (which could be a week, a month, or a year, for example). It will be assumed that the rainfall rate P, and the evaporation (or evapotranspiration) rate E are average values for the entire watershed. (As the size of the watershed increases, spatial variations in P and E will become increasingly important.)

The total rainfall amount or depth D given by Eq. (4.1) can be interpreted as the volume of water received per unit area. Hence the total volume of rainfall input for the rainfall event across the entire watershed of area A will be $A \Sigma P \Delta t$. Similar comments apply to E. Then the water-budget equation expressing conservation of mass of water for the entire watershed volume shown in Fig. 4.7 over the selected time period of the budget is given by

$$A \Sigma P \Delta t = S + \Delta D + \Delta M + U_v + U_h + A \Sigma E \Delta t \qquad (4.3)$$

where

$A \Sigma P \Delta t =$ total precipitation (or irrigation) received on the surface of the soil volume, summed or integrated over the selected duration of the budget;

$S =$ total volume of surface runoff;

$\Delta D =$ increase in volume of water (as liquid or as snow or ice) stored on the surface of the soil volume ('surface detention');

$\Delta M =$ increase in soil water stored in the selected volume;

U_v, U_h = vertical and horizontal volumetric flows of water from the soil volume; and

$A \Sigma E \Delta t$ = total evaporation (or evapotranspiration) from the surface of the volume.

On dividing all the volumetric terms in Eq. (4.3) by the watershed area, A, all quantities would be expressed in the alternative and convenient unit of the equivalent ponded depth of water (m^3/m^2, or m).

Assuming that all terms other than $A \Sigma E \Delta t$ in Eq. (4.3) can be measured, estimated, or, with some justification, assumed negligibly small, then evapotranspiration over the time period selected for the water-budget analysis can be calculated as the only unknown. This is one common application of Eq. (4.3), whose use in this way avoids the need to measure the temporal and spatial variations in E, which is not a straightforward task.

In hydrological studies of watersheds, the period of time sometimes chosen for the analysis is a 'hydrological year', at the commencement and end of which water-content profiles are judged to be identical. Thus, for a hydrological year the term $\Delta M = 0$. Also the budget period selected usually ensures that $\Delta D = 0$.

The principles of soil water movement described in Chapters 10 and 11 can be used to estimate the term U_h. Also, if the lower surface of the budget volume is not impermeable, then U_v may have to be estimated in a similar way. Assuming that soil water-content profiles have been monitored, ΔM can then be determined down to the surface of zero flux. Suppose that U_v, U_h and ΔD are all zero, and that runoff happens to be negligible over the time period chosen for the water budget – assumptions that may be satisfied only for a segment of a watershed. Then, on dividing terms in Eq. (4.3) to give terms expressed as equivalent ponded depth, it follows that

$$\Sigma E \Delta t = \Sigma P \Delta t - \Delta M \qquad (4.4)$$

where the soil water storage term M is now expressed in the alternative units of equivalent ponded-water depth (an expression commonly contracted simply to 'depth', or, with less definition, to 'amount').

Note that there are a number of assumptions to be satisfied in order for Eq. (4.4) to provide an accurate expression of the water budget. However, it is not uncommon to be able to make a judicious choice of time periods and parts of a watershed (or sub-catchment) for which these assumptions are satisfied, allowing this simpler equation, Eq. (4.4), to be used, as it often has been, to determine evapotranspiration.

If water falls on the watershed as snow or other frozen forms, then the increase in detention term, ΔD, in Eq. (4.3) could be considerable.

However, the equation applies equally whatever the form of water, provided that quantities are expressed in compatible terms.

The most commonly available measurement of evaporation is that from the surface of open water contained in a tank or pan of water that is usually 1–2 m in diameter sited in a meteorological enclosure (commonly grassed). There are several standard designs, the most widely used being the US Weather Bureau Class-A pan. This pan is 1.21 m in diameter and 0.25 m deep, located above ground on a slatted wooden supporting base. The advantage of such an instrument is the relative ease with which evaporation can be measured. Assuming that there is no rainfall, evaporation during the period of observation is indicated by the drop in water level in the pan over the measurement period, such as a day. Alternatively, the volume of water per unit pan area that must be added to bring the water surface back to a predetermined level below the rim can be used to give the evaporation. The evaporation rate from such a pan can be written E_{pan}.

Once evapotranspiration has been measured at some location, perhaps using Eq. (4.3) or Eq. (4.4), it is common practice to relate this rate of evapotranspiration to that from an evaporation pan located in the region as a reference value. This evaporation ratio varies with many factors, such as the climatic regime, the type of vegetation and its stage of growth, and can be reduced by water stress or senescence of the vegetation. However, a considerable body of experience has been built up over time on how these various factors affect the ratio E/E_{pan}. Thus, even with only descriptive information available, a useful, if approximate, estimate of this ratio can be made (Doorenbos and Pruitt, 1977). Information on this ratio E/E_{pan} is therefore widely used to estimate evapotranspiration in situations in which pan evaporation is the only available evaporation measurement.

In all but wet environments, evaporation from bare soil is considerably less than that from an evaporation pan. Thus, through time, following germination of an agricultural crop, for example, and as its leaf area develops, the increasing interception of radiation by the crop will increase evapotranspiration. Thus the ratio of evapotranspiration to pan evaporation will increase until, at full vegetative cover, in the absence of water stress, the ratio commonly reaches a stable value.

Figure 4.8 provides a particular example of this progression in the ratio E/E_{pan} with growth of a tropical pasture from germinating seed. Early in the growth of the pasture, evaporation is dominantly from the soil, from which evaporation is more restricted than that from vegetation. As the pasture grew, so the ratio of evapotranspiration in relation to evaporation from the open water surface of the pan increased (Fig. 4.8). The ratio was close to unity when the pasture completely shaded the soil.

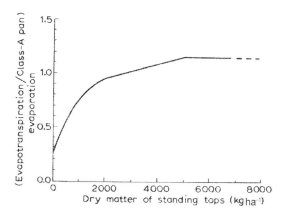

Figure 4.8 The
relationship between the
ratio evapotranspiration/
loss from a US Weather
Bureau Class-A pan, when
water is not limiting
transpiration, and the dried
mass per hectare of a
growing tropical legume
pasture crop. (After Rose
et al. (1972).)

The evapotranspiration rate, E, was measured by a weighing lysimeter, in which a volume of soil with its growing vegetation was isolated, though surrounded by similar vegetation. The lysimeter with its vegetation is continuously weighed, the rate of weight loss giving E.

4.5 Water-balance accounting

At the scale of a watershed, illustrated in Fig. 4.7, spatial variation in soil properties can be considerable. A property of particular importance is the rate at which rainfall, overland flow, or melt water can enter the soil surface. This rate, the infiltration rate, can be expressed in terms of the ponded depth of water entering during a given time period. The infiltration rate is defined as the volume of water entering unit area of the soil surface per unit time. Thus the SI unit is $m^2\,m^{-2}\,s^{-1}$ or $m\,s^{-1}$. Like the rainfall rate, the infiltration rate is usually expressed in units such as $mm\,h^{-1}$, even though the SI unit is $m\,s^{-1}$.

What factors affect the infiltration rate at the field and watershed scales will be considered in Chapter 6. When the necessary concepts have been further developed in Chapter 11, insight provided by soil physics into infiltration at a particular location will also be outlined.

At the watershed scale there is usually substantial spatial variability in all components of the water balance. The rainfall rate becomes increasingly variable in space as the scale increases, but this is also true of soil characteristics. For example, there is now substantial evidence that spatial variability in infiltration rate is typically high. However, despite there being substantial spatial variability in most soil properties, it is found that useful progress in following the general trends in soil water stored in the root zone of vegetation can be made by adopting suitable typical values for soil water characteristics. For example, assessing the likelihood that a given crop may be successfully grown in a new location can be done by calculating the expected changes in water storage at a potential root depth using historical data on rainfall variability (if such data are available). The methodology is also helpful in interpreting geographical variation in types of natural vegetation.

This type of investigation can make use of Eq. (4.3), and the methodology is called 'water-balance accounting'. The output of this procedure provides an estimate of water available in the root zone of vegetation during a chosen time interval within any period of interest for which data on precipitation and evapotranspiration are either available or can be estimated. The purpose of applying this procedure in agriculture can be to determine periods during which water stress or waterlogging are likely to occur, or to have occurred in the past, since both situations can reduce plant growth. The procedure can be employed also to improve the efficiency of water use in irrigation agriculture – an increasingly pressing issue as water resources become overextended, and the danger of salinity that can arise from over-irrigation is recognised. The procedure can also be useful as an aid in interpreting natural vegetation patterns, particularly in less-humid zones.

Water-balance accounting can be applied retrospectively, in real time, or predictively, using either long-term mean climatic data or historical or synthesised climatic data. Water-balance accounting is especially useful in areas with sparse meteorological or agronomic data, where it assists quantitative assessment of the effect of climatic variability on agronomic production. The methodology allows close simulation of general changes in the water content of the root zone over long time periods, provided that appropriate functions for evaporation are used for particular crops or fallow periods, as illustrated by Fitzpatrick and Nix (1969). This provides a convenient methodology for quantitative agroclimatic assessment, which is particularly useful in sub-humid-to-semi-arid climatic regions of the world where soil water availability is the major environmental factor controlling pasture growth, or the growth and yield of agricultural crops. Figure 4.9 illustrates the ability of water-balance accounting to provide long-term simulation of the soil water store available for plant growth. The ratio E_t/E_o in Fig. 4.9 describes a potential evapotranspiration function for a growing wheat crop; during the summer fallow period this ratio took varied values depending on the weekly rainfall received.

The methodology, whose results are illustrated in Fig. 4.9, has proved particularly useful in interpreting variations in crop yield due to seasonal variation in soil water availability. Hence long-term prediction of yield variation can be carried out given basic historical meteorological data, as illustrated by Nix and Fitzpatrick (1969). Figure 4.10 illustrates that a

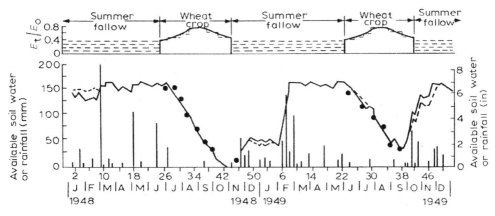

Figure 4.9 A comparison of estimated (——, - - -) and observed (· · ·) changes in availability of soil water under a wheat–fallow sequence (1948–1949) at Biloela, central Queensland, Australia. The solid curve (——) was obtained using actual weekly data to estimate evaporation; the broken curve (- - -) represents corresponding predictions using long-term values instead. (After Fitzpatrick and Nix (1969).)

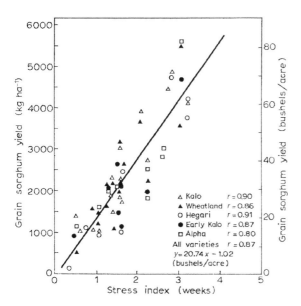

Figure 4.10 Relationships between the yield of grain sorghum and a computed water-stress index for five grain sorghum varieties grown over 16 years at Biloela, central Queensland, Australia. (After Nix and Fitzpatrick (1969).)

water-stress index for a growing crop, based on water-balance accounting, is most effective in interpreting large variations in yield of a number of grain sorghum varieties over a 16-year period.

Extensive long-term agroclimatic assessment of the type illustrated in Figs. 4.9 and 4.10 has yielded significant environmental benefits both in socio-economic and in biophysical terms for the following reasons. Such assessment delineates those regions in a country with limited water resources in which sustainable production systems are possible, and helps avoid the social hardship, and the potential degradation of soil and habitat, which can follow from clearing and tilling land that, in the long term, is climatically unsuitable for such land use. When it is suitably extended and adapted, the methodology can also be used to assess the potential for more sustainable primary production systems, as illustrated for the nation of Ghana by Rose and Adiku (2001).

To provide a simple illustration of the type of procedure used in the type of water-balance accounting whose results are illustrated in Figs. 4.9 and 4.10, let us now rewrite Eq. (4.3) in terms of equivalent ponded depth of water, essentially by dividing by the watershed area, A. After also expanding the term ΔM, we obtain

$$M_t = M_{t-1} + \Sigma P \Delta t - \Sigma E \, \Delta t - (S + U) \tag{4.5}$$

where

M_t = equivalent ponded depth of water stored in the root profile during time period t;

M_{t-1} = the value of M_t for the prior time period, $t - 1$; and

$S + U$ = combined losses from the water-balance volume by runoff and by drainage of water below the root zone. (S and U are shown combined, since, in non-instrumented situations, in which water-balance accounting is commonly carried out, it is not possible to partition this sum into its two components.)

Especially if soil in the root zone is either coarse textured (i.e. sandy) or has a good structure with pores through which water can drain readily, it is found that, a day or two after a soaking rainfall or thorough wetting, soil drains to an approximately reproducible water content, called the 'field capacity'. The advantages and limitations of this concept are discussed in Rose (1976). Expressed in terms of volumetric water content, field capacity will be denoted θ_{fc}. Because of the rapidity of drainage at water contents higher than θ_{fc}, at least for well-draining soils, θ_{fc} can be regarded as an effective upper limit to the amount of water that is available to vegetation for uptake by its root system. For any particular soil and plant species there is also an effective lower limit to soil water content below which the plant cannot extract water (though it is the pressure head or soil water suction that is directly involved, rather than water content). This lower limit to availability is commonly called the 'wilting point', θ_{wp}. (Since plants can wilt in a high-evaporation environment

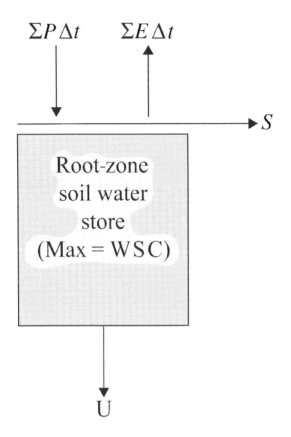

$$\Sigma P \Delta t \qquad \Sigma E \Delta t$$

Root-zone
soil water
store
$(\text{Max} = \text{WSC})$

S

U

Figure 4.11 The system used in the simple water-balance-accounting procedure described in the text.

even if they are well supplied with water, the wilting point is determined under low-evaporation conditions.)

The water-content difference $(\theta_{fc} - \theta_{wp})$ is called the 'available water content'. If the available water content, averaged over the rooting depth for the vegetation, is multiplied by the rooting depth, the resulting equivalent ponded depth of water is called the 'water-storage capacity', or WSC. The WSC is essentially the maximum effective value of M_t in Eq. (4.5). If water-budgeting calculations show the WSC to be exceeded, then this excess becomes the loss term $(S + U)$ in the same equation.

With these concepts, the system whose water budget is described by Eq. (4.5) is shown schematically in Fig. 4.11. The root zone is shown as a single water store with a maximum storage given by the WSC. Though losses by surface runoff (S) and drainage beyond the root zone (U) are shown separately, their separation is commonly not feasible in uninstrumented situations. Separate determination of S and U is essential for some purposes, for example in predicting soil erosion by overland flow and predicting loss of nitrate to groundwater. However, for much

Figure 4.12 A flow chart illustrating a simple water-balance-accounting procedure that depends on a measured or estimated value of the water-storage capacity (WSC). M_t is the equivalent ponded depth of water in the root zone at time period t, and $M_t(1)$ is the first estimate of of M_t, which may either be confirmed (if $M_t(1) <$ WSC), or altered (if $M_t(1) >$ WSC).

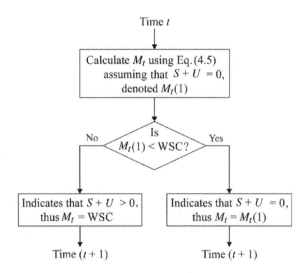

of the world's pastoral and agricultural lands, water is the nutrient most limiting plant growth. In this context of water available for transpiration and growth, determining whether water is lost to the soil root zone by runoff or by deep percolation is not the major objective, this being to interpret variations in plant growth or crop yield. In the context described, this interpretation can be made without requiring separation of S and U, and this will be illustrated in the following description of a simple water-balance-accounting procedure.

With the WSC experimentally measured or otherwise estimated for a particular crop–soil combination, a simple practical procedure for water-balance accounting can be set up knowing the amount of rainfall and having evapotranspiration estimated, perhaps as some fraction of pan evaporation as illustrated in Fig. 4.8. The logic of this accounting procedure, which has already been described, can be represented in flow-chart form as in Fig. 4.12.

A simple example of this water-balance-accounting procedure based on Eq. (4.5), and using the logic outlined in Fig. 4.12, is now given in Example 4.3.

Example 4.3

Using the methodology based on Eq. (4.5) and Fig. 4.12, carry out the water-balance-accounting procedure on a monthly basis for the months shown in Table 4.1. Data in this table are from the monsoonal tropics in the southern hemisphere, where rainfall decreases to very low values at the end of the wet season. Evaporation is from a tropical legume

Table 4.1 Data on monthly total rainfall and evapotranspiration for a tropical legume pasture in a monsoonal climate (all amounts in the table are equivalent ponded depths in mm, and the water-storage capacity (WSC) is 200 mm)

Month	Column 1 Precipitation $\Sigma P\,\Delta t$	Column 2 Evaporation $\Sigma E\,\Delta t$	Column 3 $\Sigma(P-E)\,\Delta t$	Column 4 M_{t-1}	Column 5 M_t (1)	Column 6 $S+U$ (if positive)	Column 7 M_t
Comment			C1 − C2[a]	150 mm or C7[b]	C4(t − 1) + C3t	C5 − 200 mm	$S+U>0$, $M_t=200$ mm; $S+U<0$, $M_t=$ C5
Initial (Dec)				150			
Jan	250	150	100	200	250	50	200
Feb	200	140	60	200	260	60	200
Mar	180	130	50	200	250	50	200
Apr	50	100	−50	150	150	(−50) 0	150
May	30	90	−60	90	90	(−110) 0	90
Jun	10	50	−40	50	50	(−150) 0	50
Jul	10	20	−10	40	40	(−160) 0	40
Aug	10	10	0		40	(−160) 0	40

[a] C1 refers to column 1, etc.
[b] C7t implies that, from January onwards, the value of M_{t-1} is obtained by transfer from column 7 in the same month (t).

pasture, measured by weighing lysimeter. Accounting is to begin in the month of January, the equivalent ponded depth of water available in the root zone in the previous month, denoted M_{t-1}, being 150 mm. The water-storage capacity (WSC) in the root zone for this pasture–soil combination is 200 mm. The logic of this procedure is illustrated in Fig. 4.12.

Solution

For ease of recognition, column numbers in Table 4.1 are given in bold type in the following text. Column **3** gives the net monthly balance between the rainfall input and loss by evapotranspiration. If it is positive, this net input can recharge the root profile with water up to the maximum limit set by the WSC (200 mm); if further water remains after this recharging, then this will be lost, this loss being given by $S + U$, which is the sum of surface runoff and drainage beneath the root zone. This combined loss term is given by the numbers in column **6** of Table 4.1, which are either positive or zero (as discussed later).

The initial value of M_{t-1} (150 mm) in the previous month (column **4** of Table 4.1) is used in calculating the first estimate, $M_t(1)$ of M_t for the month of January. $M_t(1)$ is calculated assuming that $S + U = 0$ (Fig. 4.11). Calculated values of $M_t(1)$ are given in column **5** of Table 4.1.

The sum $S + U$ is calculated as the difference between $M_t(1)$ and the WSC (200 mm), and is shown in column **6**. If the calculated value of $S + U$ is a positive quantity, it is accepted as a valid estimate of $S + U$. However, under these circumstances, $M_t(1)$ is not a valid estimate of M_t, since then the assumption invoked, namely that $S + U = 0$, in calculating $M_t(1)$ is not supported. Hence, under these circumstances, M_t is taken to have its maximum possible value of 200 mm. M_t is then shown on the left-hand side of column **7** under the appropriate comment heading (Table 4.1).

The calculated value of M_t in column **7** for January (200 mm) then becomes the value of M_{t-1} for the month of February; hence the value of M_t in column **7** is transferred back to column **4**, where it provides the value of M_{t-1} for February.

A different situation occurs in the month of April and later months (Table 4.1). In these months the calculated value of $M_t(1)$ is less than the WSC (200 mm). In such cases the assumption that $S + U = 0$ is valid, so the first estimate of M_t given by $M_t(1)$ is accepted as correct, and is shown as such on the right-hand side of column **7**. In these cases in which $M_t(1) <$ WSC, then, $S + U$ calculated as $M_t(1) - 200$ is negative, which has no physical meaning, so $S + U$ is then taken to be zero (see column **6** in Table 4.1).

The procedure illustrated in Fig. 4.9 can be expanded to deal with the situation in which the available water content varies with depth. Exercise 4.4 provides an opportunity to examine how this can be done.

How water-balance-accounting-type procedures can be used in the interpretation and management of salinity problems and in the analysis of contaminant transport in the soil profile will be discussed and illustrated in Chapter 12.

Main symbols for Chapter 4

A	Area
d	Depth or distance
D	Equivalent ponded depth of water (quite generally) or on the soil surface in particular
E	Evaporation rate
M	Water stored in a soil volume expressed as equivalent ponded depth; M can be subscripted with t or $t - 1$ to indicate the time of storage (Eq. (4.5))
P	Rainfall rate
S	Total volume of surface runoff
t	Time
U	Volumetric flow of water from a soil volume (U_h, horizontal flow; U_v, vertical flow)
WSC	Water-storage capacity
z	Distance beneath the soil surface
Δ	Finite difference of any quantity
θ	Volumetric water content
θ_{fc}	Volumetric water content at field capacity
θ_{wp}	Volumetric water content at wilting point
Σ	Summation symbol

Exercises

4.1 Following the first snowfall for the season, a cylindrical snow-sampling tube of diameter 0.3 m was inserted vertically into the snow and a sample of the snow retrieved. The tube of snow was then weighed and found to have a mass of 11.6 kg. If the mass of the sampling tube is 4.9 kg, calculate the magnitude of the snowfall expressed in terms of equivalent ponded depth of water in mm.

4.2 A short intense storm lasted for 20 minutes and had an average rainfall rate of 4.9 inches per hour. One half of this rainfall infiltrated into the ground, and the other half ran off. Calculate the infiltration and

Figure 4.13 In (a) a soil profile in which both water content, θ, and the available water content (AWC) vary with depth is shown. The resulting water-content profile following infiltration of water is shown in (b), in which the upper two soil layers have been recharged to their AWC value.

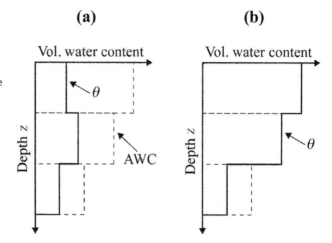

(a)

(b)

runoff amounts as equivalent ponded depths in mm. (Evaporation during the rainfall can be neglected.)

4.3 A ring infiltrometer ponds water on the soil surface in order to measure the infiltration rate for the ponded area of soil. Water is ponded within a cylindrical ring, of diameter 0.6 m in this instance. To maintain the water ponded in the infiltrometer at a constant level, it was found that 3 litres of water had to be added each 5 minutes. Calculate the infiltration rate at the site in $mm\,h^{-1}$.

4.4 Figure 4.13 illustrates a situation in which the available water content (AWC) varies with depth. In (a) a water-content profile in which θ is less than the AWC for each of the three soil layers is shown. The resulting water-content profile following infiltration of water is shown in (b).

Assuming that the values of θ and depths of measurement are known, explain how the change between (a) and (b) in the equivalent ponded depth of water stored in each of the three segments of the profile shown in Fig. 4.13 can be calculated. How then could the change between (a) and (b) in equivalent ponded depth of water stored in the complete profile shown be calculated?

4.5 (a) In relation to the procedure of water-balance accounting, briefly discuss
  • the purpose of the procedure,
  • the principle on which the procedure is based,
  • the nature of the component terms used in the procedure, and
  • how the component terms in the water-balance equation can be estimated or measured.

  (b) Table 4.1 gives the mean monthly rainfall (in inches) at the site of a major proposed land-development project at Pahang Tenggara

Table 4.2 *Monthly meteorological data*
for Pahang Tenggara, Malaysia

Month	Rainfall (inches)	Δe (in Hg)
January	7.3	0.49
February	7.6	0.51
March	8.9	0.60
April	8.7	0.59
May	8.1	0.54
June	5.1	0.53
July	5.8	0.53
August	6.8	0.53
September	7.7	0.54
October	12.6	0.51
November	13.5	0.40
December	13.2	0.38

in Malaysia. From meteorological data at the site a synthetic vapour pressure, Δe, was calculated, where Δe is used to estimate evaporation from an open water-filled pan in the following empirical equation due to Fitzpatrick (1963):

$$\Sigma E \, \Delta t = 1.0 + 10.0 \, \Delta e \quad \text{(inches of water per month)}$$

where the time summation extends over a one-month period, and the unit of Δe is inches of mercury (in Hg). Monthly data on Δe are also given in Table 4.2. The maximum equivalent ponded depth of water available to plant growth that can be stored in the soil at the Pahang Tenggara site is 5 inches (or 127 mm).

Assume that, prior to the initial calculation month of January, the available water stored in the soil profile is at its maximum value of 5 inches, and that pan evaporation provides an adequate approximation to evapotranspiration in this particular context. Then estimate the mean monthly amount of available water, M_t, stored in the profile at this site for each month of the year. Also estimate for each month the sum of surface runoff and through drainage $(S + U)$, and explain how you arrive at this estimate. Express all results both in inches and in mm of water, setting out your calculations in tabular form as in Table 4.1.

Also demonstrate that your monthly estimates lead to satisfaction of the principle of conservation of mass of water applied to the soil profile over a 12-month period.

List any assumptions you make, giving justifying comments, and comment on implications of the overall results of your calculations.

References and bibliography

Ciesiolka, C. A. A., and Rose, C. W. (1998). The measurement of soil erosion. In *Soil Erosion at Multiple Scales – Principles and Methods for Assessing Causes and Impacts*, eds. F. W. T. Penning de Vries, F. Argus and J. Ker. Wallingford: CABI Publishing in association with the International Board for Soil Research and Management (IBSRAM), pp. 287–301.

Doorenbos, J., and Pruitt, W. O. (1977). Crop water requirements. Irrigation Drainage Paper 24. Rome: UN Food and Agricultural Organisation.

Fitzpatrick, A. A. (1963). Estimates of pan evaporation from mean maximum temperature and vapour pressure. *J. Appl. Meteorol.* **2**, 780–792.

Fitzpatrick, E. A., and Nix, H. A. (1969). A model for simulating soil water regime in alternating fallow-crop systems. *Agric. Meteorol.* **6**, 303–319.

French, R. H. (1985). *Open-channel Hydraulics*. New York: McGraw Hill.

Nix, H. A., and Fitzpatrick, E. A. (1969). An index of crop water stress related to wheat and grain sorghum yields. *Agric. Meteorol.* **6**, 321–337.

Rose, C. W. (1976). *Agricultural Physics*. Oxford: Pergamon Press.

Rose, C. W. (1993). Erosion and sedimentation. Chapter 14 in *Hydrology and Water Management in the Humid Tropics – Hydrological Research Issues and Strategies for Water Management*, eds. M. Bonnell, M. M. Hufschmidt and J. S. Gladwell. Cambridge: Cambridge University Press, pp. 301–343.

Rose, C. W., and Adiku, S. (2001). Conceptual methodologies in agro-environmental systems. *Soil Tillage Res.* **58**, 141–149.

Rose, C. W., Begg, J. E., Byrne, G. F., Goncz, J. H., and Torssell, B. W. R. (1972). Energy exchanges between a pasture and the atmosphere under steady and nonsteady state conditions. *Agric. Meteorol.* **9**, 385–403.

5

Evapotranspiration and exchange of energy at the earth's surface

5.1 An introduction to vegetation-based ecosystems

The origin of energy at the earth's surface is radiation from the sun of our solar system, and this energy provides the ultimate physical basis for life on earth. It is a consequence of energy laws that this 'high-quality', short-wave solar energy is conserved, but it is also ultimately or quickly converted into other forms of energy described as of 'low quality' because of the difficulty of converting it into useful forms of energy. This is a simple statement of the first and second laws of thermodynamics.

As discussed in Section 1.5 and illustrated in Fig. 1.8, it is this solar radiation, and its partitioned forms in the atmosphere and at the earth's surface, that dominates the life-suppporting environment close to the earth. Figure 1.8 also shows that evaporation commonly consumes a significant fraction of the energy available at the earth's surface, so that energy systems and hydrological systems are closely linked. This chapter considers evaporation both from water surfaces and from land surfaces, including the transpiration stream through living plants or trees. The heating of soil and air will also receive some attention.

Life on earth depends specifically on that small fraction of the global radiation income that is absorbed in the process of photosynthesis (Fig. 1.8). From a strictly chemical point of view all life consists of exquisitely organised chemical systems. Maintaining the order in these systems requires a virtually continuous input of energy in order to overcome the processes of energy degradation. Green-leaved trees and plants are now the dominant organisms possessing the ability to convert the raw products of water (H_2O) and carbon dioxide (CO_2) into much more complex molecules, carbon being the basic building block of life-sustaining large organic molecules. These complex molecules, initiated as some form of carbohydrate, are created using solar radiant energy in the visible waveband. In chemical symbolism, writing carbohydrate as $(CH_2O)_n$, the complex photosynthetic process can be symbolised as

$$CO_2 + H_2O + \text{light energy} = (CH_2O)_n + O_2 \tag{5.1}$$

This remarkable transformation depends also on the presence in the photosynthesising vegetation of essential mineral nutrients such as nitrogen (N), phosphorus (P) and sulphur (S). Such essential elements are obtained from the soil, which also provides the water and, importantly, houses the multitude of soil-living organisms that process these essential elements into a form that can be taken up by the roots of plants and trees. Thus, as well as the obvious above-ground leaves and stems, land vegetation requires below-ground parts, the root system.

For plant growth, the nutrients, water and, ultimately, even carbon all come only from the earth, so their continued availability requires that they must be recycled following uptake. All these mineral cycles, including the water or hydrological cycle, are driven by solar energy (Fig. 5.1). Note that solar energy is not recycled, although it is continuously available at time-varying intensities and with location-varying annual energy totals.

Some of the products of photosynthesis, for which Eq. (5.1) indicates only a starting point, serve the growth needs of the photosynthesising vegetation itself. Human populations harvest increasing quantities of some of the products of the photosynthesisers such as grain and fruit, but more generally they are consumed by the large group of organisms collectively referred to as herbivores (Fig. 5.1). Herbivores in turn can be the food source for carnivores (Fig. 5.1). Humans participate as partners in this carnivorous dimension of the 'food chain' when they consume animals such as cattle and chickens.

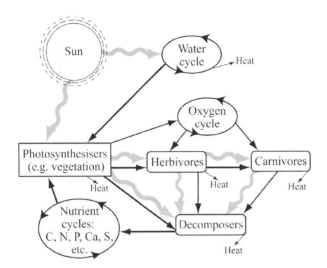

Figure 5.1 Illustrating the one-way conversion of solar energy, which drives the cycles of water and nutrients required for growth of vegetation. Nutrient cycles are both gaseous and sedimentary in nature and depend on the decomposers. Energy flows are shown as wavy lines, and dominantly material flows as straight lines.

The photosynthesisers also provide the oxygen for the oxygen cycle (Fig. 5.1), without which all oxygen-dependent organisms would not exist. Living organisms consume oxygen in their metabolic respiratory life processes and also in physical activity, thus effectively reversing the direction of product formation of Eq. (5.1), the direction then being from right to left, instead of from left to right. The currently oxygen-rich atmosphere on earth exists because, over evolutionary time, the conversion of carbon dioxide to oxygen in photosynthesis has exceeded the re-conversion of oxygen to carbon dioxide in respiration.

Net production by photosynthesisers cannot continue indefinitely without renewal of the essential minerals and water which are taken up from the soil (Fig. 5.1). This continual renewal is achieved dominantly by return to the earth of dead, shed or excreted products, which are the food source for another group of organisms essential to all sustainable ecosystems – the 'decomposers' (Fig. 5.1). The decomposers consist chiefly of bacteria and fungi, and these organisms break down and absorb the range of dead organic material returned to earth. However, as well as meeting their own needs, decomposers release minerals in forms that can be taken up by the photosynthesisers, thus providing the essential link in the mineral nutrient cycle (Fig. 5.1).

The food requirements of rapidly increasing human populations, amongst other factors, have led farming practice in many countries to supplement these natural mineral nutrient cycles with industrially produced chemical fertilisers, often using fossil-fuel energy to bring about the required chemical conversion. 'Waste products', such as ground bones and blood from animal slaughter, are also used to supplement mineral cycling. The major decrease and alteration in the world's natural vegetation, dominantly due to its replacement by food crops over the last century of rapidly rising human population, is causing corresponding changes in the available land area and in the effectiveness of the mineral cycling which supports such photosynthesisers. Much of this decrease is due to common agricultural practices, in which food or fibre photosynthesisers are substituted for those removed, but, unlike in natural systems, the products extracted from these crops are often not accompanied by appreciable mineral recycling. The long-term energy issues involved in this modification of natural mineral cycles include the continuing increase in concentration of greenhouse gases in the atmosphere discussed in Section 1.5. These issues must be carefully considered, since it is the photosynthesisers that provide the base of the food chain which extends from photosynthesisers to carnivores, as shown in Fig. 5.1.

As we know from human experience, all organisms lose some of their energy as heat (Fig. 5.1). For decomposer organisms, the heat produced in active composting illustrates this fact. Thus even the high-quality solar energy temporarily captured by the photosynthesisers eventually ends up as low-grade heat energy. On absorption by bare soil, this energy conversion from high- to low-grade heat occurs immediately. Even when it is absorbed by vegetation, most of the energy is used in evaporating water (Fig. 1.8).

Figure 5.1 does not indicate the major processes of extraction, concentration and chemical synthesis of materials now carried out by industry. One dimension of environmental concern is the increasing concentrations of substances from such activity that are not recycled, which can have deleterious effects on the natural cycles on which life depends. Examples of such concerns are with the over-concentration of materials naturally occurring in the earth's crust, such as cadmium, lead, mercury and uranium, as well as synthesised products. Such products come from very varied activities, such as production of nuclear energy, and from the vast variety of chemical industries – even supposedly benign new compounds can cause environmental problems, such as the chlorofluorocarbons found to be dominantly responsible for depleting the earth's ozone layer.

This brief introduction to the material cycles of terrestrial ecosystems is amplified in texts such as Kormondy (1969) and Miller (1979). This introduction has illustrated the central dependence of all life on the photosynthesisers such as trees and plants. The physics of some of the processes involved in the growth of vegetation is considered in Rose (1979) and Marshall, Holmes and Rose (1996).

The rate of plant and tree growth is intimately related to the rate of water flow through the vegetation in the process of transpiration. This is because the small pores on leaf surfaces, or 'stomata', through which most of the transpiration stream is lost as water vapour, are also the port of entry of carbon dioxide for photosynthesis (Eq. (5.1)).

The decrease in the magnitude of evapotranspiration caused by human activity is a dominant factor leading to the development of dryland salinity in susceptible landscapes. Thus changes in evapotranspiration rate can have important environmental consequences.

Because of the significance of evapotranspiration, this chapter considers the factors which affect its rate and some of the ways in which this rate can be measured. In Section 4.4 the principle of mass conservation of water was applied to the variety of water inputs to, and losses from, a watershed shown in Fig. 4.7. The water budget so constructed for a watershed is given by Eq. (4.3), one of the terms in that equation being the total evaporation (or evapotranspiration) from the surface area A of the watershed, namely $A \Sigma E \Delta t$. The summation sign, Σ, recognises that the total amount of evaporation during any time period must be obtained by summing up the amounts of evaporation during smaller periods Δt in which the average evaporation rate is E.

If all component terms other than total evaporation in the water-conservation equation (4.3) can be measured or estimated, then this equation allows the determination of $A \Sigma E \Delta t$.

After considering atmospheric humidity as one factor affecting the evaporation rate (Section 5.2), the chapter considers the evaporation process and a variety of ways of measuring it (Sections 5.3–5.6). Brutsaert (1982) provides a general, but more advanced, reference text on evaporation. The effects associated with absorbed solar radiation warming the lower atmosphere and the soil are also considered (Section 5.7).

5.2 Atmospheric humidity

Evaporation of water depends not only on the solar radiation received, but also on other factors, including the humidity of the air above the evaporating surface. Water is chiefly present in the atmosphere as an invisible gas called 'water vapour'. When a high-flying aircraft is said to be shedding vapour trails, what is visible is not really vapour, but vapour condensed into small droplets, such as we see in clouds. The humidity of the atmosphere refers to the relative concentration of water vapour in its mixture with dry air. Provided that no condensation of water vapour takes place, as occurs in rainfall, the behaviour of water vapour is similar to that of any other gas. Thus the pressure exerted by water vapour, e, is quite well described by the general equation of an 'ideal gas', often called the 'ideal-gas law'. For water vapour this equation or law can be represented by

$$e = \frac{\rho_v}{M_w} R_u T \tag{5.2}$$

where ρ_v is the density of water vapour of relative molecular mass M_w, R_u is a universal constant, called the 'universal gas constant', and T is the absolute temperature in degrees Kelvin (K). (Note that K $=$ $^\circ$C $+$ 273.) The magnitude of R_u is 8.315 J mol^{-1} K^{-1}. A 'mole' (abbreviated mol) is the number of grams of a substance which is numerically equal to its relative molecular mass, which is 18 for water (H_2O). The density ρ_v is also called the 'absolute humidity' of the atmosphere.

From Eq. (5.2) the density of water vapour ρ_v at any temperature is proportional to eM_w. Denoting the total atmospheric pressure by p_a, the pressure of dry air is $p_a - e$. Thus the density of dry air will be proportional to $(p_a - e)M_a$, where M_a is the relative molecular mass of dry air. Under normal atmospheric conditions e is 5% or less of p_a, so, to an accuracy adequate for most purposes, $p_a - e$ can be approximated by p_a, with the density of moist air being taken as proportional to $p_a M_a$. A commonly employed measure of humidity, called the 'specific humidity', q, is defined as the mass of water vapour per unit mass of moist air, and is represented by

$$q = \frac{eM_w}{p_a M_a}$$

or

$$q = 0.622e/p_a \tag{5.3}$$

since $M_w/M_a = 0.622$.

A common method of measuring e (or q or ρ_v) is to use a wet-and-dry-bulb type of hygrometer, also called a 'psychrometer'. The 'bulb' referred to is the bulb of a liquid-in-glass thermometer, though any form of temperature sensor can be used, such as a thermocouple or electrical-resistance thermometer, or a thermistor, as in Fig. 5.2. As the name implies, one bulb or temperature sensor is dry and the other wet. The degree of cooling induced by evaporation of water from the wet bulb relative to the dry-bulb temperature provides a measure of the humidity of the air flowing over the sensors. Maintaining a wet bulb with an adequate but not excessive water supply requires careful design, a traditional solution being to cover the wet bulb with a close-fitting muslin sheath, kept moist by a wick connected to a reservoir of distilled water (Fig. 5.2).

Fluid-in-glass thermometers are still used for standard daily observations in a Stevenson screen, a freely ventilated box used to house

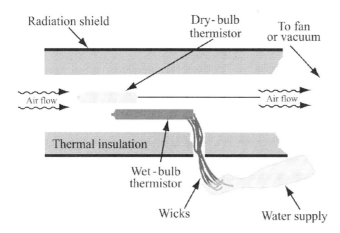

Figure 5.2 The principle of construction of a ventilated wet-and-dry-bulb hygrometer.

air-temperature- and humidity-measuring equipment in standard meteorological enclosures. However, in order to allow automatic electronic recording, such thermometers can be replaced by some type of thermometer with electrical output.

Consider a closed container partly filled with pure water. When equilibrium has been achieved at some constant temperature, the numbers of water molecules leaving and returning to the liquid surface per second will be the same, and the air is said to be 'saturated' with water vapour. The vapour pressure is then the maximum possible for that temperature, and is known as the 'saturation vapour pressure' (SVP). The SVP is a unique function of temperature, increasing somewhat rapidly with temperature, as shown in Fig. 5.3.

Denote the dry-bulb air temperature by T, and the wet-bulb temperature by T_w. The vapour pressure, e_w, achieved by air flowing over the wet bulb at the wet-bulb temperature is assumed to reach the SVP at that temperature; e_w is therefore a known quantity, as shown by Fig. 5.3. Theory reproduced in Penman (1955) and in Rose (1979), for example, shows that, with some assumptions,

$$e = e_w - \gamma(T - T_w) \tag{5.4}$$

where e_w is the SVP at temperature T_w and γ is a 'psychrometric constant' defined by

$$\gamma = \frac{c_p}{\lambda} \frac{p_a}{0.622} \tag{5.5}$$

(Note that the symbol γ is also used in some literature to represent c_p/λ.) In Eq. (5.5), c_p is the 'specific heat' of air measured under constant-pressure conditions. (The specific heat of any substance is the heat required to increase the temperature of unit mass of the substance by

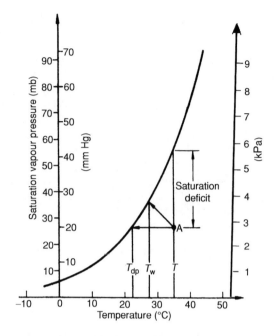

Figure 5.3 The saturation vapour pressure of pure water as a function of temperature. T is the air temperature T_w is the wet-bulb temperature, and T_{dp} the dew-point temperature. (After Rose (1976).)

1 °C.) Also in Eq. (5.5), λ is the 'latent heat of vaporization', which is discussed in Section 5.4.

When wind speeds are low, as can occur with unaspirated wet-and-dry-bulb thermometers in a Stevenson screen, the appropriate value of γ is higher, and appropriate psychrometric tables should be consulted (e.g. Smithsonian Institution (1951)).

For reasons of history and psychrometric-table usage, the use of non-SI units for expressing vapour pressure is still commonly found in the literature. Thus in Fig. 5.3, as well as the SI unit of kPa (kilopascas) for SVP, other units are also given, namely

(a) 1 millibar (mb) $= 10^{-3}$ bar $= 0.1$ kPa and
(b) the pressure exerted by a column of mercury (commonly used in the past to measure atmospheric pressure), 1 mm Hg $= 0.133$ kPa approximately.

Suppose that point A in Fig. 5.3 represents the vapour pressure of air entering the psychrometer in Fig. 5.2. As this air flows over the wet bulb it will become saturated and also cools due to the extraction from it of the latent heat required to evaporate water from the water-coated wet bulb. The line from point A sloping up to the left to intersect the SVP curve at the wet-bulb temperature traces the history of this change in condition of the air. As indicated by Eq. (5.4), the slope of this line is $-\gamma$.

Figure 5.3 also indicates the 'dew-point temperature', T_{dp}, corresponding to air initially described by point A. The horizontal state path from point A represents cooling at a constant vapour pressure. This can occur during radiative cooling of air on a calm clear night. On falling to temperature T_{dp} the air is saturated with vapour, and dew would form on any surface in contact with the vapour if it were cooled to T_{dp} or below. This forms the basis for a 'dew-point' hygrometer, an alternative instrument for measuring vapour pressure.

Finally, Fig. 5.3 also shows the difference between the SVP and the actual vapour pressure at any temperature, which is called the 'saturation deficit'.

5.3 Evaporation from an open-water surface

In Section 4.4, and as illustrated in Fig. 4.8, evapotranspiration is commonly expressed as a ratio to evaporation from an open water surface, which is commonly measured on a daily basis in an evaporation pan (E_{pan}). E_{pan} is an effective and experimentally simple integrator of climatic factors affecting evaporation. Hence use of E_{pan} as a reference evaporation rate allows more effective indication of the role of other non-climatic factors in controlling evapotranspiration (cf. Fig. 4.8).

Over a century ago, a scientist called Dalton showed experimentally that evaporation from a water surface such as a lake depended on at least two factors. The first of these was the difference in the pressure of water vapour between the evaporating surface and the bulk air above, this difference being the driving force behind the upward evaporative flux of water vapour. The second factor was the wind speed, which directly affects the efficiency of removal of the water vapour from the surface to the air above.

Expressed alternatively, Dalton found that the vapour flow occurred along a gradient in concentration of atmospheric moisture, with the efficiency of that transfer depending mainly on the intensity of turbulent mixing, which can be expressed by an empirical relationship with the wind speed measured near the site of interest.

Right at the water surface the pressure of the water vapour is a maximum, the SVP (e_w), which increases quite rapidly with water temperature (Fig. 5.3).

Outside the boundary layer over the water surface, gradients in vapour pressure are small, so the exact height in the air at which the vapour pressure e is measured is not of major importance in defining the driving difference in vapour pressure, $e_w - e$. In any case, the minor effect of measurement height for e is in practice incorporated into the form of the experimentally based function of mean wind speed, $\bar{u}$, used in an open-water-evaporation equation of the type developed by Dalton. Denoting this function by $f(\bar{u})$ and the SVP of the evaporating water surface by e_o, the Dalton-type evaporation equation is written as

$$\lambda E = f(\bar{u})(e_o - e) \tag{5.6}$$

where the vapour-pressure difference $(e_o - e)$ is the saturation deficit. A form commonly adopted for $f(\bar{u})$ is $f(\bar{u}) = a(1 + b\bar{u})$, where the coefficients a and b require experimental determination and can depend on factors such as water-surface roughness.

Although the form of $f(\bar{u})$ requires calibration, Eq. (5.6) has been applied quite successfully to evaporation from open water surfaces such as evaporation pans, lakes and oceans, where the direct measurement of water-surface temperature by thermometry is not difficult.

Evaporation from an evaporation pan (such as the US Weather Bureau Class-A pan referred to in Section 4.4) is not regarded as a standard meteorological observation, partly because of the difficulty of ensuring representative and reproducible siting of the pan, on which the measurement obtained depends. Nevertheless, evaporation-pan equipment is quite commonly located at meteorological recording stations. This has allowed the development of empirical relationships between the rate of pan evaporation and standard meteorological observations made at the site, such as measurements of atmospheric humidity and daily wind run (or average wind speed). These empirical relationships can then be used, with some caution, to predict 'open-water-surface evaporation'.

Open-water-surface evaporation, either measured or predicted, has also been related to evapotranspiration measured by methods such as those outlined in Sections 4.4 and 5.4. For example, Penman (1948), working in England, found a somewhat regular seasonal variation in the ratio of evapotranspiration rates from pastures to that from an evaporation pan.

5.4 Measurement of evapotranspiration using the principle of conservation of energy

It was shown in Section 4.4 that one method of measuring evapotranspiration from a watershed was by using the principle of mass conservation of water. The accuracy achievable with this methodology is improved if the vertical flux beneath the watershed, denoted U_v in Fig. 4.7, is negligible. As discussed in Section 4.4, this mass-conservation methodology is particularly suitable when measurement of evapotranspiration over longer time scales, such as a season or a hydrological year, is required. At least at watershed scale, this methodology is quite inappropriate for use if measurement of evapotranspiration at daily time scales is required.

It is especially at daily or sub-daily time scales that the energy-conservation principle provides a suitable basis of feasible and reasonably accurate methodologies for the measurement of the evapotranspiration rate. However, as commonly applied, this methodology provides information on the evapotranspiration rate over a limited area surrounding the location of measurement. Replication of the measurement methodology in space to provide watershed-wide measurement is possible but rarely attempted.

A general statement of the energy-conservation principle or law is that energy cannot be created or destroyed, despite the many transformations in forms of energy and exchanges of energy that continuously take place. In his famous equation $e = mc^2$, Albert Einstein quantified the prodigious amount of energy that can come from a transform of mass into energy in the nuclear reactions which drive the sun's energy output, for example. So, although we should strictly conceptually expand the energy-conservation principle to include mass as well as energy, this can be ignored when we are considering the fate of solar energy absorbed at the earth's surface.

From the point of view of energy used, there is no significant difference in the energy consumed in evaporating a unit mass of water from a water surface, from moist soil, or from vegetation. This is why in some literature the term 'evaporation' is used inclusively to incorporate the transformation of liquid water into its vapour, regardless of what the source of that vapour may be. Used in this way, the term evaporation would be applied even if some of the flux of water vapour came from transport of water through vegetation (i.e. transpiration).

In other literature the term 'evaporation' is used in a more restricted sense to exclude transpiration. However, a flux into the atmosphere of water vapour generally will have multiple sources (as discussed in Section 4.1). These sources can include transpiration, evaporation from non-plant sources such as the soil and evaporation from water intercepted on the exterior of plant surfaces, such water not being involved in the transpiration stream passing through the plant's interior vascular system. Because of the restricted meaning ascribed to the term evaporation in some literature, the term 'evapotranspiration' was coined to describe evaporation in the common multiple-source situation. This term, evapotranspiration, is now widely adopted, and is used in this text. However, the term evaporation is also used in a synonymous way to the term evapotranspiration, the meaning of evaporation not being restricted as referred to earlier in this paragraph.

Heat is commonly defined as the energy transferred from one body to another because of a difference in temperature between them. Such a flow of energy can be picked up by the senses, such as when heat flows from your bare feet to a cold tiled floor. Such heat is sometimes called 'sensible' heat to distinguish it from latent heat, which is involved solely in creating the change in phase from liquid to gas or vapour, rather than in raising temperature. For evaporation to occur, the heat energy required to separate liquid water molecules into the gaseous form of water vapour must be supplied. The heat energy required to convert unit mass of water into vapour is called the 'latent heat of vaporisation', λ. Thus the rate at which energy must be supplied to sustain an evaporation rate of $E \, \mathrm{kg \, m^{-2} \, s^{-1}}$ is λE where λ has units of $\mathrm{J \, kg^{-1}}$. Thus λE has SI units of $\mathrm{J \, m^{-2} \, s^{-1}}$ or $\mathrm{W \, m^{-2}}$. For water, the magnitude of λ is $2.45 \times 10^6 \, \mathrm{J \, kg^{-1}}$ or $2450 \, \mathrm{kJ \, kg^{-1}}$ at $20 \, ^\circ \mathrm{C}$, falling to $2260 \, \mathrm{kJ \, kg^{-1}}$ at $100 \, ^\circ \mathrm{C}$.

As described in Section 1.5, and illustrated in Figs. 1.8 and 1.9, the atmosphere is traversed by a variety of type of short- and long-wave radiation. When they are absorbed in the atmosphere, these radiation streams affect meteorological conditions directly, and, when they are absorbed at the land surface, they can support evaporation or generate sensible heat fluxes. Representing the relative magnitudes of these streams by the widths of arrows representing energy fluxes, Fig. 1.8 illustrates possible outcomes by day (when most evaporation occurs), and Fig. 1.9 shows possible outcomes by night.

Figure 1.8 shows that some of the incoming short-wave (or global) radiation, R_s, is reflected by the land surface (with a reflection coefficient r), with the remainder of the radiation being absorbed at the land surface. Some of this absorbed solar energy is re-radiated with the longer wavelengths characteristic of emission at terrestrial temperatures (Figs. 1.7 and 1.8), the net flux density of long-wave radiation being denoted R_L. However, the remainder of the absorbed short-wave radiation can heat either the soil (G), or the atmosphere in contact with the surface (H), or be used to evaporate any water that may be available from any soil, plant, or litter surfaces (λE).

The principle of energy conservation applied to this energy exchange at the earth's surface can be expressed as a balance between all incoming and outgoing energy flux densities:

$$R_s(1 - r) = R_L + G + \lambda E \quad (\mathrm{W \, m^{-2}}) \qquad (5.7)$$

where

$\quad R_s$ = flux density of global short-wave radiation received by the land
$\qquad$ surface from the sun and sky;
$\quad r$ = reflection coefficient (or albedo) of the land surface;

R_L = net flux density of long-wave radiation emitted by the land
surface, the difference between that emitted and that absorbed
from incoming long-wave radiation;

G = sensible heat flux density directed into the earth's surface;

H = sensible heat flux density directed from the ground surface up
into the atmosphere;

λ = latent heat of vaporisation of water; and

E = evapotranspiration rate (mass flux density). (Condensation
from the atmosphere could be treated as a negative evaporation
rate.)

The way Eq. (5.7) is written assumes that the terms on the right-
hand side of this equation are positive if energy is removed from the
land surface – the common situation by day. In this equation the energy
used in photosynthesis is neglected, usually being no more than about
1% of R_s. Equation (5.7) is also assumed to apply to a surface, meaning
that any change in heat storage in vegetation is neglected. Alternatively,
the term G in Eq. (5.7) can be regarded as the sum of the heat fluxes into
the vegetation and into the soil. If fluxes H and λE are measured above a
forest canopy (as is commonly done with forests), change in the storage
of heat in the timber stand could be a significant component term when
changes in mean temperature are rapid.

If the soil surface is moist or carrying transpiring vegetation, then,
as suggested in Fig. 1.8, a large fraction of incoming solar energy is
used to evaporate water. In a dry arid environment, however, most of
the absorbed solar energy is used to heat the soil or the atmosphere,
so that terms G and H are dominant over λE. Whatever the nature of
the environment, the various terms of Eq. (5.7) must be adjusted so that
Eq. (5.7) is satisfied (provided that the assumptions made in the equation
are also satisfied).

The sensible (H) and latent (λE) heat fluxes are usually measured
at some modest height above the surface, vegetated or not. The one-
dimensional energy-conservation equation (5.7) assumes that there is no
net horizontal or 'advective' flux of heat energy across the region where
evaporation is being determined. An oasis in a desert is an extreme example where this assumption could be significantly violated. A significant
advective flux is quite unlikely if evapotranspiration is measured from a
spatially extensive and somewhat uniform vegetation canopy. Despite its
general significance, in what follows in this section advection of energy
will be assumed negligible.

The difference between incoming solar radiation absorbed at the land
surface and net outgoing long-wave radiation is called the 'net radiation
flux density', or simply the 'net radiation', R_n. From this definition,

$$R_n = R_s(1 - r) - R_L \quad (\text{W m}^{-2}) \tag{5.8}$$

Figure 5.4 The sensor head of a net radiometer for measuring R_n. (After Szeicz (1975).)

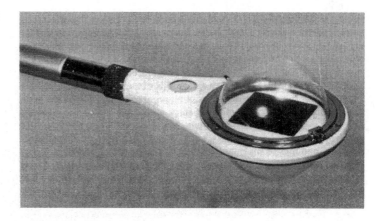

On writing Eq. (5.7) in terms of the energy involved in the evaporation flux and using Eq. (5.8), we obtain

$$\lambda E = R_n - G - H \quad (W\ m^{-2}) \qquad (5.9)$$

The net radiation, R_n, can be measured with a 'net radiometer' such as that illustrated in Fig. 5.4. The sensor consists of upward- and downward-facing blackened surfaces specially designed to absorb both long- and short-wave radiation. Radiation in both wavebands is well transmitted by the twin inflated thin polythene hemispherical covers which protect the upward- and downward-facing black absorbing surfaces from wind and dust. A bank of thermocouples converts the difference in temperature between the upper and lower radiation-absorbing surfaces into a small electrical voltage that can be recorded and converted into net radiation using a calibration factor determined for the instrument.

If R_n is not measured, then, by using Eq. (5.8), its magnitude can be inferred with useful accuracy if the flux density of incoming global short-wave radiation, R_s, is measured. If R_s is not measured at the site of interest, then it is possible to fall back on the results of 'solarimeter' networks set up in some countries to record R_s continuously. Even in countries with such a network, measurement sites may be sparse, and spatial interpolation of R_s for the site of interest usually can give good accuracy only if data are averaged over significant time periods, such as a month or a year. Also required in order to calculate R_n using Eq. (5.8) is the reflection coefficient or 'albedo', r, which can be measured, for example, by inverting a solarimeter. Generalised tables of typical values of r for a variety of described surfaces can be found in texts and other sources. For natural vegetated surfaces the value of r typically lies between 0.1 for forests and 0.25 for agricultural crops.

The total outgoing emitted terrestrial radiation from the earth's surface, L_{oe}, whose temperature is T_s (K) can be calculated from a relationship known as the Stefan–Boltzmann law. From this law,

$$L_{oe} = \varepsilon\sigma T_s^4 \tag{5.10}$$

where ε is the 'emissivity' of the surface, whose value is usually in the range 0.95–0.99 for natural surfaces, and σ is the Stefan–Boltzman constant (5.67×10^{-8} W m^{-2} K^{-4}). Especially with a vegetation-covered surface, the commonly measured air temperature, T_a, provides an approximate estimate of T_s.

The long-wave radiation incoming to the earth's surface from the atmosphere, L_i, also can be estimated from Eq. (5.10) provided that T_s is replaced by an effective atmospheric radiative temperature, T_e. In the absence of clouds T_e is commonly some 20 °C cooler than air temperature at ground level. Empirical equations for L_i based on air temperature at ground level, T_a, also give useful accuracy in prediction (see Eq. (5.12)).

With L_{oe} and L_i thus estimated, the net long-wave radiation can be calculated from

$$R_L = L_{oe} - L_i \tag{5.11}$$

As illustrated in Fig. 1.10, the atmosphere is a good absorber of the earth's long-wave radiation except for a narrow-waveband window. Since absorption and emission of thermal radiation are related to the same processes at the molecular or atomic level, a good absorber is also a good emitter of radiation. It is the strong emission back to earth from the atmosphere (i.e. L_i) that provides the greenhouse effect (Fig. 1.9). This is illustrated in the following example.

Example 5.1

(a) Calculate the magnitude of the outgoing emitted long-wave radiation (L_{oe}) from a vegetated surface at the same temperature as the air, T_a, namely 20 °C or 293 K. The emissivity of the vegetation is 0.97.
(b) Estimate the long-wave incident irradiance, L_i, using an empirical equation due to Swinbank (1963), which allows L_i under clear skies to be calculated from the absolute air temperature, T_a (K):

$$L_i = 0.92 \times 10^{-5}\sigma T_a^6 \quad \text{(W m}^{-2}) \tag{5.12}$$

(c) Using Eq. (5.12) to estimate L_i, calculate the net outgoing long-wave radiation, R_L.
(d) Using Eq. (5.12), calculate the effective or black-body radiative temperature of the cloud-free atmosphere. (For a black body $\varepsilon = 1$.)

Solution

(a) From Eq. (5.10),

$$L_{oe} = 0.97 \times 5.67 \times 10^{-8} \times 293^4$$
$$= 405.4 \text{ W m}^{-2}$$

(b) $L_i = 0.92 \times 10^{-5} \times 5.67 \times 10^{-8} \times 293^6$
$$= 330.1 \text{ W m}^{-2}$$

(c) Thus

$$R_L = L_{oe} - L_i$$
$$= 75.3 \text{ W m}^{-2}$$

(d) Denoting the effective radiative temperature of the atmosphere by T_e, then, from Eqs. (5.10) and (5.12);

$$\sigma T_e^4 = 0.92 \times 10^{-5} \sigma T_a^6$$

Thus

$$T_e = \left(0.92 \times 10^{-5} T_a^6\right)^{1/4} \quad \text{with } T_a = 293 \text{ K}$$
$$= 276 \text{ K}$$
$$= 3°C$$

which is 17 °C cooler than the air temperature (20 °C).

The presence of cloud can increase T_e, especially if it is low and complete in cover. Atmospheric humidity can also affect T_e, since water vapour is a strong absorber of long-wave radiation. Nevertheless, T_e is usually less than T_a, resulting in the net outgoing long-wave radiation, R_L, being positive, as shown in Fig. 1.9.

With R_n evaluated in one or other of the ways described earlier, the remaining terms in the energy-balance equation that require measurement in order to yield the evaporation rate are G and H. The value of the soil heat-flux component of G can be measured with an instrument buried just below the soil surface called a 'heat-flux plate'. The other component of G, the heat flux into the vegetative canopy, is sometimes considered as a separate term, V. The flux V is not frequently measured because of the need to measure mass and temperature variations through the canopy profile. Thus, if this flux component is likely to be important, as in a forest, methods other than energy balance are usually adopted to measure E above the canopy. The contribution to G of short vegetation is so small that it is neglected in this context, as is the component of energy used in photosynthesis.

The irregular eddying flow of turbulent liquid motion was described in Section 3.4 and Fig. 3.3. This was contrasted with orderly laminar flow.

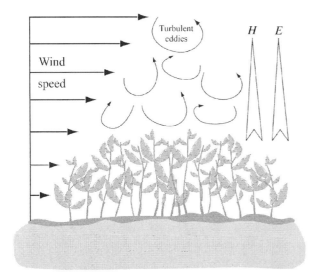

Figure 5.5 Mechanically generated turbulence due to air flowing over a rough vegetative surface.

Air flowing over vegetation can be laminar under very calm conditions, usually at night, but by day air flow is typically turbulent, which is consistent with our experience of blowing leaves or dust if wind speeds are sufficiently high. Especially over dry surfaces, this turbulence can partly be due to convective air flow caused by differences in density between cooler air aloft and less-dense air heated by contact with the hot dry surface. However, for air flowing over growing vegetation, turbulence is dominantly generated by the mechanical disruption it experiences on flowing over the aerodynamically rough vegetative surface or through vegetation (Fig. 5.5).

If the vegetative surface is warmer than the air above it, the turbulent mixing due to the structure of eddies sketched in Fig. 5.5 leads to an upward heat flux density, H, from the vegetation to the air. Since the vapour pressure will be higher within the vegetation than in the air above, exactly the same turbulent transport mechanism as that resulting in the sensible heat flux H will also be responsible for the evaporation flux density E (Fig. 5.5).

Since the size of the eddies expands with height above the vegetative surface, the effectiveness of mixing due to the eddies also increases with height. However, the gradients in vapour pressure and temperature decrease with height in such a manner that the resulting flux is constant with height in the absence of advective effects.

This intimate connection between the fluxes H and E indicates that there should be a close connection in the 'micrometeorological' theory describing these two processes. The material in Box 5.1 indicates that this is indeed the case. Details of micrometeorological theory are given more fully in texts such as Monteith and Unsworth (1990), Oke (1988), Sharma (1984), Rose (1979) and Campbell (1977).

Box 5.1 Measurement of flux densities of heat and water between vegetation and the air

As noted in discussing Fig. 5.5, the mechanically generated turbulent transport mechanisms result in the transport both of sensible heat and of water vapour. The profile in wind speed shown above the vegetation in Fig. 5.5 reflects the shear stress, τ, exerted on the vegetation by the air flow. For liquid density ρ, Eq. (3.5) shows that the magnitude of the shear stress is given by

$$\tau = \rho K \frac{\partial v}{\partial z} \qquad (5.13)$$

where the eddy diffusivity K is a measure of the effectiveness of eddies in transporting the momentum ρv of the fluid down to the surface along the velocity gradient $\partial v / \partial z$.

Exactly the same equation, Eq. (5.13), holds for the shear stress and momentum transport in air if ρ is replaced by the air density ρ_a, and K is the eddy diffusivity for air. It is a fascinating fact of turbulent flow that, although K increases with height above the vegetation, there is a corresponding decrease in the velocity gradient $\partial v / \partial z$ so that the product $K \partial v / \partial z$ is constant, as is the shear stress or momentum flux. With ρ replaced by the density of air, ρ_a, exactly the same type of 'micrometeorological' theory shows that the evaporation flux density, E, is given by

$$E = -\rho_a K \frac{\partial q}{\partial z} \qquad (5.14)$$

where the eddy diffusivity in Eq. (5.14) is often almost the same as K in Eq. (5.13), and q is the specific humidity of the air. If there are no horizontal advection effects (i.e. no spatial change in horizontal fluxes) the flux E will be constant with height and again the variations with height in K and $\partial q / \partial z$ will be such that the product of the two is constant. The negative sign in Eq. (5.14) indicates that the flux E is in the direction of decreasing q, as it must be.

The same theory shows that the sensible heat flux density, H, is given by

$$H = -\rho_a c_p K \frac{\partial T}{\partial z} \qquad (5.15)$$

where K is often very closely of the same value as in Eq. (5.13) and (5.14), since transport of the various air characteristics is driven by the same turbulent eddy system (Fig. 5.5). c_p is the specific heat of air under conditions of constant pressure. The negative sign in the equation indicates that H would be upwardly directed into the air if temperature decreased with height above the vegetation surface, and vice versa.

Thus H could be calculated if the temperature profile were measured and K determined. A way of removing the need to determine K will be illustrated in the text using the Bowen-ratio method.

In practice the terms v, q and T would be averaged over a time period because of their rapidly fluctuating nature.

It is useful to partition the net energy flux $(R_n - G)$ between H and λE by considering their ratio, known as the 'Bowen ratio'. The heat flux H is related to the gradient of the temperature profile in the air, and the evaporative flux E to the gradient in the specific humidity, q, defined as the mass of water vapour per unit mass of moist air. If the differences ΔT in temperature and Δq in specific humidity are measured over some height interval (say 1–2 m) as illustrated in Fig. 5.6, then it follows from Eqs. (5.14) and (5.15) that the Bowen ratio, β, is given by

$$\beta = \frac{H}{\lambda E} = \frac{c_p \, \Delta T}{\lambda \, \Delta q} \tag{5.16}$$

In practice, data on ΔT and Δq would be averaged over time periods no longer than an hour for daily values of E to be reliable.

With β determined experimentally with equipment as illustrated in Fig. 5.6, it follows from Eqs. (5.16) and (5.9) that E can be calculated from

$$E = \frac{R_n - G}{\lambda(1 + \beta)} \tag{5.17}$$

For reasonably moist surfaces, or vegetation not stressed for water, β is commonly between zero and 0.2. However, if vegetation is unable to take up water from the soil through its root system at a rate close to the rate of loss of water by transpiration, a 'water deficit' develops in the vegetation, which is usually accompanied by at least a partial closure of stomata and thus a reduction in the rate of loss of water by transpiration. A reduction in transpiration rate reduces the energy flux λE used in evaporating water, so that more of the net radiation must be partitioned into heating the air, the vegetation, or the soil. Thus, for a given R_n, the terms G and H must increase if λE is reduced (Eq. (5.9)). Hence the Bowen ratio $H/(\lambda E)$ will increase (Eq. (5.16)).

This procedure using the Bowen ratio is a combination of energy-budget and micrometeorological methods of flux measurement. The

Figure 5.6 Experimental determination of the Bowen ratio β, using micrometeorological techniques. ΔT could be measured by thermocouples or other temperature-sensing devices, Δq by hygrometers.

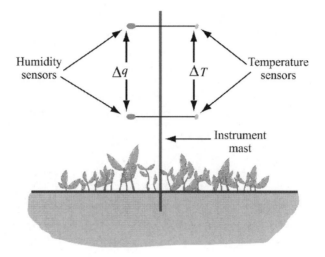

Bowen-ratio procedure is particularly suitable where interest focuses on following dynamic changes in fluxes in response to climatic input and vegetation response. Figure 5.7 illustrates how flux changes within a single day can be monitored on quite a fine time scale using such techniques. These data are for a pasture in a tropical monsoonal climate at a time of high radiation input. Because of the good availability of water, most of the net radiation is used in evapotranspiration, with the heat fluxes to the air (H) and soil (G) being relatively small. The very small heat flux into the vegetation was determined and shown as a separate term, V, rather than being incorporated as a component of G. For this pasture crop, V can clearly be neglected with negligible error. Incoming global radiation, R_s, is temporarily reduced by cloud between hours 14 and 16. At about 14 hours H changes direction from crop to air (H positive) to air to crop (H negative).

With drier surfaces, where β may be considerably greater than unity, the value of β can be affected by the buoyant movement of surface-heated air. If so the values of K in Eqs. (5.13)–(5.15) will differ.

Example 5.2

On an average June day of duration 12 hours, the global solar irradiance on a horizontal surface at Raleigh, North Carolina, is $22.5 \, \mathrm{MJ \, m^{-2} \, d^{-1}}$.

(a) Calculate the daily average irradiance in $\mathrm{W \, m^{-2}}$.
(b) Then calculate the daily average rate of evaporation, $\overline{E}$, from actively transpiring vegetation in units of $\mathrm{mm \, h^{-1}}$ given the following information.

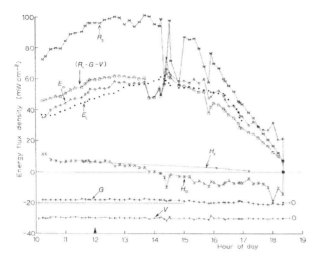

Figure 5.7 The history of variation on an almost cloud-free day of the components of the energy-budget equation (5.7) for a growing pasture of Townsville stylo (*Stylosanthes humilis* H.B.K) on 29 March 1967, at Katherine, Northern Territory, Australia. E_m indicates λE determined using micrometeorological techniques described earlier; E_L is also λE, but measured by a weighing lysimeter, showing that there is approximate agreement between these two independent estimates of evapotranspiration. Zeros are displaced by -20 mW cm^{-2} for G and -30 mW cm^{-2} for V, the heat flux to the vegetation. Other symbols are as in the text. Note that 1 mW cm^{-2} is equivalent to 10 W m^{-2}. (After Rose *et al.* (1972).)

- The vegetation's reflection coefficient for solar radiation, r, is 0.1.
- The average Bowen ratio for the day is 0.08.
- The temperature of the vegetation is the average daily air temperature of 28 °C.
- The emissivity of the vegetation is 0.98.
- On average $G = 0.1R_n$.

Use Eq. (5.12) to calculate the amount of incoming long-wave radiation, L_i, and take $\lambda = 24.5 \times 10^5$ J kg^{-1}.

(c) What would the daily evapotranspiration expressed as an equivalent ponded depth of water be?

Solution

(a) Denote the daily average irradiance by $\overline{R}_s$. Then

$$\overline{R}_s = \frac{22.5 \times 10^6 \text{ J}}{12 \times 60 \times 60 \text{ m}^2 \text{ s}}$$

$$= 520.8 \text{ W m}^{-2}$$

(b) From Eqs. (5.10)–(5.12), the daily average is

$$R_L = \varepsilon \sigma T_a^4 - 0.92 \times 10^{-5} \sigma T_a^6$$
$$= 5.67 \times 10^{-8} \times [(0.98 \times 301^4) - (0.92 \times 10^{-5} \times 301^6)]$$
$$= 58.9\,\text{W m}^{-2}$$

Now

$$R_n = R_s(1 - r) - R_L \quad (\text{Eq. (5.8)})$$

Thus

$$\overline{R}_n = \overline{R}_s(1 - r) - R_L$$
$$= 520.8(1 - 0.1) - 58.9$$
$$= 409.8\,\text{W m}^{-2}$$

Then, from Eq. (5.17),

$$\overline{E} = \frac{409.8(1 - 0.1)}{24.5 \times 10^5(1 + 0.08)}\,\text{kg m}^{-2}\,\text{s}^{-1}$$
$$- 1.394 \times 10^{-4}\,\text{kg m}^{-2}\,\text{s}^{-1}$$

The density of water is $1000\,\text{kg m}^{-3}$. Thus 1 kg of water spread over $1\,\text{m}^2$ would have a ponded depth of 10^{-3} m or 1 mm, or $1\,\text{kg m}^{-2}$ is equivalent to 1 mm depth of ponded water.

Thus

$$\overline{E} = 1.394 \times 10^{-4}\,\text{kg m}^{-2}\,\text{s}^{-1}$$
$$= 1.394 \times 10^{-4}\,\text{mm s}^{-1}$$
$$= 0.502\,\text{mm h}^{-1}$$

(c) So, over the 12-hour day, the evapotranspiration expressed as an equivalent ponded depth of water is given by

$$\Sigma \overline{E}\,\Delta t = 0.502 \times 12\,\text{mm}$$
$$= 6.0\,\text{mm}$$

The combined energy-budget-and-micrometeorological type of measurement method discussed earlier in this section is often used for investigating evapotranspiration on a daily or hourly time scale. Measurement at such relatively fine time scales is useful for recognising processes that affect transpiration, such as partial closure of stomata due to water stress. Such information is needed in order to predict transpiration.

If, however, the objective is to determine the evapotranspiration from a catchment area or watershed, where interest might centre on the longer-term effects of significant changes in vegetation, an energy-balance approach has practical disadvantages compared with use of the principle of conservation of mass of water. This principle is applied by drawing up a water budget for the watershed, as described in Sections 4.4 and 4.5. Though it is often suitable for larger-scale applications, this methodology is not constrained to the watershed scale. At all scales there is a strong interdependence between the energy balance and the water balance at the earth's surface; this is because the amount and energy status of water present in the root profile of vegetation strongly control the partitioning of net radiation between evapotranspiration and heating the air, soil and vegetation.

5.5 Determination of evapotranspiration from vegetated surfaces using standard meteorological data

Sections 5.4 and 4.4 outlined two physically based methods of general application for measuring evapotranspiration. Both methods require some special measurements, which are not available from conventional sources of relevant data, such as observations kept at standard meteorological recording sites. Standard meteorological data have a wide geographical spread, regularity and reliability of measurement and are publicly available. For these reasons it has long been a desire to have a method of estimating evapotranspiration in which most, if not all, the data required are those which are available from standard meteorological records.

Various related methods seeking to fulfil this desire have been reported. These commonly consist of a combination of the surface-energy-balance and Dalton-type equations. The best-known equation of this type is that due to Penman (1948). However, the approach will be illustrated using the method of McIlroy (reported in Slatyer and McIlroy (1961)). McIlroy's method is chosen since it very clearly indicates the general influence of surface factors on evapotranspiration.

It is an assumption common to such methods that the vegetated surface is considered to be uniform and sufficiently extensive for advective effects to be negligible. Air-buoyancy effects due to temperature gradients will also be assumed negligible, and the sources (or sinks) of sensible and latent heat will be assumed to be located at a common flat vegetative surface. The detailed morphology of individual plant-leaf arrays is neglected, so the surface can be imagined to be a 'large green leaf' at temperature T_o. For the present let us ignore the technical difficulty of accurately measuring the specific humidity at the leaf surface and denote it by q_o (Fig. 5.8).

The measurements made at a reference height Z above the surface are air temperature, T, and specific humidity, q (Fig. 5.8). From Eq. (5.15), the sensible heat flux density is given by

$$H = \rho_a c_p K \left(\frac{T_o - T}{Z} \right) \tag{5.18}$$

Figure 5.8 Temperature and specific humidity at a vegetative surface (T_o and q_o) and measured at reference height Z in a Stevenson screen (T, q).

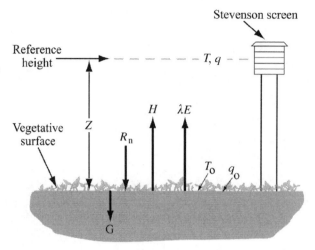

Figure 5.9 Possible relationships between air and wet-bulb temperatures at a surface and at a reference height above the surface. T and T_o are air temperatures at the reference height and the surface, respectively. T_w and T_{wo} are the corresponding wet-bulb temperatures.

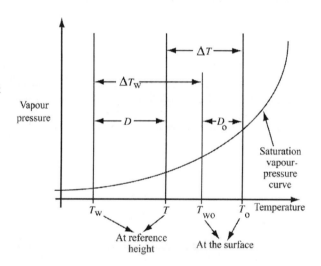

The ratio K/Z is a measure of the effective conductance of the entire air layer of height Z with respect to heat transport (Fig. 5.8). We will denote the term $\rho_a c_p K/Z$ as an effective conductance h, so that

$$h = \rho_a c_p K/Z \qquad (5.19)$$

Also let

$$\Delta T = T_o - T \qquad (5.20)$$

where ΔT is illustrated in Fig. 5.9. Then

$$H = h\,\Delta T \qquad (5.21)$$

Denote by T_w the wet-bulb temperature at the reference height (i.e. in the Stevenson screen of Fig. 5.8).

Let us write Eq. (5.4) as $e = e_w - \gamma D$, where $D = T - T_w$ is the 'wet-bulb depression', defined as the difference in temperature between the dry and wet bulbs of a wet-and-dry-bulb hygrometer as illustrated in Fig. 5.9, and, as in Eq. (5.5),

$$\gamma = \frac{c_p}{\lambda} \frac{p_a}{0.622}$$

Two further quantities are involved in developing the theory of McIlroy. The first is D_o, shown in Fig. 5.9 as the difference between the surface temperature, T_o, and the wet-bulb temperature at the surface, T_{wo}. The second quantity, s, arises from the need to represent a relationship between the two wet-bulb temperatures, T_w at the reference height and T_{wo} at the surface, and their corresponding saturation vapour pressures. Figure 5.10 shows s defined as the average slope given by the ratio of the differences Δe_w, between the two wet-bulb vapour pressures, and ΔT_w, between these two temperatures. That is,

$$s = \frac{\Delta e_w}{\Delta T_w} \qquad (5.22)$$

As shown in Slatyer and McIlroy (1961), McIlroy's evaporation equation is then given by

$$\lambda E = \left(\frac{s}{s + \gamma} \right)(R_n - G) + h(D - D_o) \qquad (5.23)$$

where all terms in this equation have been defined earlier.

The coefficient $(s/(s + \gamma))$ of $R_n - G$ in Eq. (5.23) is dependent on temperature as shown in Fig. 5.11.

The accuracy of coefficients such as $s/(s + \gamma)$ can be enhanced by evaluation at the average wet-bulb temperature, T_{wA}, shown in Fig. 5.10, and defined by

$$T_{wA} = T_w + \Delta T_w/2 \qquad (5.24)$$

where T_w in this equation is the measured wet-bulb temperature at the reference height.

An expression for ΔT_w is given by Slatyer and McIlroy (1961) as

$$\Delta T_w = \left(\frac{R_n - G}{h} \right)\left(\frac{1}{1 + s/\gamma} \right) \qquad (5.25)$$

In calculating ΔT_w from Eq. (5.25), adequate accuracy is provided by evaluating s/γ at the known reference-height wet-bulb temperature, T_w. (s/γ is given in Fig. 5.11.)

Use of these evapotranspiration-prediction equations will be illustrated in Example 5.3.

Figure 5.10 The relationship between the difference Δe_w in the wet-bulb saturation vapour pressure and ΔT_w, the difference in wet-bulb temperature between the surface and reference heights of Fig. 5.8. The average slope of the relationship over the temperature range is denoted by s.

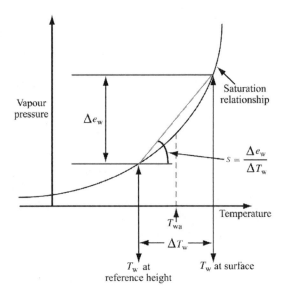

Figure 5.11 The saturation vapour pressure, e_w (kPa), together with the non-dimensional ratios s/γ and $s/(s + \gamma)$, as functions of temperature (°C). The ordinate for $s/(s + \gamma)$ is given on the right-hand side of the figure. The symbol key is given above the figure.

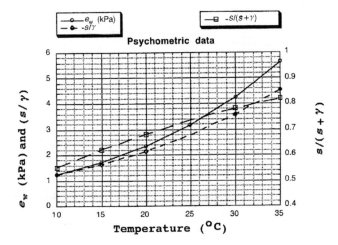

For a well-watered surface it can be assumed that the wet-bulb depression D_o is zero. If also the ground-heat-flux term G is neglected then Eq. (5.23) becomes

$$\lambda E = \left(\frac{s}{s + \gamma} \right) R_n + h(T - T_w) \qquad (5.26)$$

which is one of the forms of the widely used equation due to Penman (1948). Slatyer and McIlroy (1961) showed that Penman's equation contains more approximations than does that of McIlroy, and the latter clarifies the nature of these approximations.

In use of Penman's equation (Monteith and Unsworth, 1990) the term $s/(s + \gamma)$ is evaluated at the air temperature T, since it contains no theory corresponding to Eqs. (5.24) and (5.25) with which to calculate T_{wA}. Especially if ΔT_w is substantial, this can lead to significant error in $s/(s + \gamma)$, since T will typically be greater than T_{wA} (see Figs. 5.9 and 5.10).

Particularly because of the simplifying assumption that $D_o = 0$ in Penman's equation, Eq. (5.26), this equation has been used extensively for predicting evapotranspiration rates in well-watered situations – the context for which it was developed. Under these circumstances this rate is often referred to as the 'potential evapotranspiration rate', or the amount evaporated is described as 'potential evapotranspiration'. The neglect of the term G in Eq. (5.26) is justified when one is averaging over periods of a day or longer, and, with well-watered vegetation in temperate climates, it will be small compared with R_n at all times.

Equation (5.26) is a little different from the original form of Penman's equation (Penman, 1948). The original form resulted from combining the energy-balance equation (5.9) with a Dalton-type equation, Eq. (5.6), as shown in Rose (1979). Various forms of Penman's equation have been used and tested widely (Campbell, 1977).

The theories represented by Eqs. (5.23) (that of McIlroy) and (5.26) (that of Penman) have a general advantage over the Dalton method (Eq. (5.6)). This advantage is that the necessarily oversimplified empirical transfer coefficient, $f(\overline{u})$, is of dominant importance in Dalton's equation, Eq. (5.6), whereas the corresponding transfer or conductance coefficient h in the McIlroy and Penman equations occurs only in one term, and this is usually the smaller of the two terms. Noting that this smaller of the two terms was often some 20%–30% of the radiation-driven term, Priestley and Taylor (1972) suggested the following further-simplified equation for estimating evapotranspiration:

$$\lambda E = \alpha \left(\frac{s}{s + \gamma} \right) (R_n - G) \qquad (5.27)$$

where α is an empirical parameter that is often of order 1.2–1.3 in non-arid environments. Equation (5.27) is called the Priestley–Taylor equation.

McIlroy's form of equation also has the conceptual advantage that the value of the wet-bulb depression at the surface, D_o, has the potential to be estimated, or even measured. Aridity will ensure that D_o is not zero, as can be assumed in well-watered contexts, and its value is likely to depend on a range of soil moisture and plant factors. However, the term $D - D_o$ in

Eq. (5.23) can be very small under moderately dry conditions, and with surfaces of dry soil or desiccated vegetation D_o can exceed D. If $D < D_o$ the second term in these equations is negative, and evaporation is reduced below $[s/(s + \gamma)](R_n - G)$. Under such conditions G is likely to be a substantial fraction of R_n, and then soil heating, rather than evaporation, will consume a considerable fraction of R_n. This is the typical situation of arid-zone environments.

The term h in the McIlroy and Penman equations is the effective aerodynamic conductance for transport of water vapour or heat energy from the surface to the reference measuring height (Fig. 5.8). The magnitude of h will depend at least on wind speed and surface roughness. Empirical equations used to describe h will be illustrated in Example 5.3.

From the gas law (given in Eq. (5.2) for the gas water vapour), it follows that, at a given pressure, the air density will decrease as temperature rises. Thus, especially under somewhat arid conditions, it is possible for air heating to be sufficient to cause air-buoyancy effects induced by differences in density. This results in a kind of turbulent motion of the air that is often described as 'convective turbulence' as distinct from mechanical turbulence, which is generated by the flow of air over the rough earth surface (Fig. 5.5). When convective turbulence is significant, the conductance term h can have different values for sensible and latent heat fluxes for reasons indicated in Section 5.6. The theory is then more complex (Priestley, 1955).

Example 5.3

Table 5.1 contains a selection of evaporation and meteorological data collected by the (then-named) CSIRO Division of Meteorological Physics located close to Melbourne, Victoria, Australia. Measurements were made above a sward of pasture grass. Data were collected using standard meteorological equipment except for the heat-flux term, G. G was measured either with heat-flux plates buried at a shallow depth in the soil, or from soil-temperature-profile observations. R_n was measured with equipment shown in Fig. 5.4. On a daily basis McIlroy's equation, Eq. (5.23), was found to give good agreement with evapotranspiration measured independently with a weighing lysimeter, which continuously weighs an isolated soil column bearing vegetation. A purpose of the data set given in Table 5.1 was to test the accuracy of Eq. (5.23) for a period as short as an hour. A second purpose was to assess the error in particular contexts of assuming that $D_o = 0$, since the measurement of D_o is not a standard meteorological measurement. Results from calculation of evapotranspiration using these equations were compared with independent measurements made by a weighing lysimeter. The data are from

Table 5.1 Examples of hourly calculation of evapotranspiration rate using Eq. (5.23) compared with the same rate measured with a weighing lysimeter (data from Slatyer and McIlroy (1961))

Date	2/4/61			17/4/61			1/7/61			25/9/61		
Time (h)	07–08	08–09	09–10	08–09	09–10	10–11	12–13	13–14	14–15	11–12	12–13	13–14
R_n (mW cm^{-2})[a]	3.9	17.0	29.2	11.5	19.6	21.6	13.8	12.7	9.4	24.6	25.5	36.0
G (mW cm^{-2})[a]	1.0	4.3	7.3	2.9	4.9	5.4	0.8	1.9	1.4	1.7	1.3	2.4
$(R_n - G)/\lambda$ (mm water d^{-1})[b]	1.0	4.5	7.7	3.0	5.2	5.7	4.6	3.8	2.8	8.0	8.5	11.8
T (°C)	15.0	16.0	16.0	14.0	14.5	15.0	12.5	12.5	13.0	14.0	14.0	14.0
T_w (°C)	12.0	12.0	12.0	11.0	11.5	12.5	10.5	10.5	11.0	10.5	10.5	10.0
D (°C) $= T - T_w$	3.0	4.0	4.0	3.0	3.0	2.5	2.0	2.0	2.0	3.5	3.5	4.0
D_o (°C) (assumed = 0)	0	0	0	0	0	0	0	0	0	0	0	0
T_{wa} (°C) (Eqs. (5.24) and (5.25))	12.5	13.5	15.0	12.0	13.0	14.0	11.5	11.5	11.5	12.0	12.0	12.0
$s/(s + \gamma)$ (Fig. 5.11)	0.58	0.59	0.61	0.57	0.59	0.60	0.56	0.56	0.56	0.57	0.57	0.57
$[s/(s + \gamma)][(R_n - G)/\lambda]$ (mm water d^{-1})	0.6	2.7	4.7	1.7	3.1	3.4	2.6	2.1	1.6	4.6	4.8	6.7
$\bar{u}$ (m s^{-1})	0.5	1.1	1.2	3.2	3.2	3.0	4.3	4.6	5.6	4.1	4.4	4.7
$h/\lambda = 0.18(2 + \bar{u})$ (mm water d^{-1} °C^{-1})	0.4	0.6	0.6	0.9	0.9	0.9	1.1	1.2	1.4	1.1	1.2	1.2
$(h/\lambda)(D - D_o)$ (mm water d^{-1})	1.2	2.4	2.4	2.7	2.7	2.2	2.2	2.4	2.8	3.8	4.2	4.8
E (mm water d^{-1}) (Eq. (5.23))	1.8	5.1	7.1	4.4	5.8	5.6	4.8	4.5	4.4	8.4	9.0	11.5
Compare the above calculated E with												
E (by lysimeter) (mm water d^{-1})	2.1	5.5	7.9	3.7	4.0	4.6	3.7	3.7	4.3	7.6	7.0	10.1

[a] 1 mW cm^{-2} = 10 W m^{-2}.

[b] See the text for an example showing how to convert units.

Slatyer and McIlroy (1961), which can be consulted for more experimental detail.

Use of a locally derived relationship between conductance h and mean wind speed $\bar{u}$ (m s^{-1}) is made in Table 5.1. This relationship is

$$h/\lambda = 0.18(2 + \bar{u}) \quad (\text{mm H}_2\text{O d}^{-1}{}^{\circ}\text{C}^{-1})$$

(The numbers in such a relationship vary with site and type of vegetative surface.)

The degree of agreement between the hourly calculated and lysimeter-measured evapotranspiration rates in Table 5.1 is perhaps typical of that obtained when calculations are based on the use of standard meteorological measuring equipment. The degree of agreement was better when comparison was made on a daily basis. As measurement moved into a drier southern-hemisphere spring, the data for 1/7/61 and 25/9/61 show the lysimeter-measured evapotranspiration rates falling below the calculated rates, possibly due to the assumption $D_0 = 0$ not being satisfied as well as it was for the earlier two measurement periods shown.

Table 5.1 indicates that Eq. (5.23) provides a practical basis for calculation of evapotranspiration rates using standard meteorological measuring equipment, the only non-standard measurement being that of soil heat flux G. As shown in Table 5.1, G can be a significant term in hourly calculations, but is much less important on a daily-calculation basis. As well as varying with the period over which the evapotranspiration estimate is required, the relative importance of G declines with the density of vegetation and degree of wetness of the surface.

Here is an example of conversion of units for 25/9/61 hour 11–12 data for Table 5.1:

$$R_\text{n} - G = 22.9\,\text{mW cm}^{-2} = 229\,\text{W m}^{-2} = 229\,\text{J s}^{-1}\,\text{m}^{-2}$$

λ (at 14 °C) $= 2467$ kJ kg^{-1}, so

$$\frac{R_\text{n} - G}{\lambda} = 229\frac{\text{J}}{\text{s m}^2}\frac{\text{kg}}{2.467 \times 10^6\,\text{J}}\frac{10^{-3}\,\text{m}^3}{\text{kg}}\frac{1000\,\text{mm}}{\text{m}}\frac{24 \times 60 \times 60\,\text{s}}{\text{s d}}$$

$$= 8.0\,\text{mm d}^{-1}$$

5.6 Non-radiative sensible heat exchange between the land surface and the lower atmosphere

Mechanisms by which sensible heat is exchanged between the land surface and the atmosphere above it are discussed in association with Fig. 5.5, which illustrates the mechanical generation of turbulent eddies

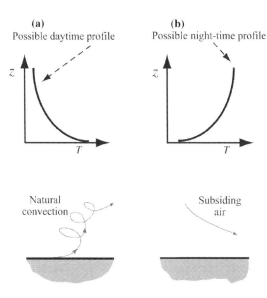

Figure 5.12 Illustrating possible air-temperature profiles above the land surface by day and night. T is the air temperature; z represents height above the land surface.

in the wind blowing over a vegetated surface. The transport of sensible heat from warmer to cooler layers by this type of turbulence is described as 'forced convection'.

When forced convection is the dominant mechanism of sensible heat transport, then the sensible heat flux, H, can be described well by Eq. (5.15). The eddy diffusivity, K, in this equation is almost identical in value with the diffusivity involved in the evaporative flux (Eq. (5.14)). Hence measuring the Bowen ratio described in association with Fig. 5.6 is one appropriate way of determining H. With E given by Eq. (5.17), H is given by $\beta\lambda E$ (Eq. (5.16)).

In addition to forced convection, another mechanism can cause mixing of volumes of air, thus resulting in convective heat transfer. When the land surface is warmer than the air above it, and an air temperature profile exists as illustrated in Fig. 5.12(a), air-buoyancy effects result in convective mixing. Temperature profiles as in Fig. 5.12(a) can occur by day, especially if the land surface is rather dry and the wind speed low, so that the lower atmosphere is not well stirred by mechanical turbulence, which acts to moderate temperature gradients. (It may be noted that air temperature quite generally decreases slowly with altitude (Rose, 1979). However, close to the ground, which is of concern here, neglect of this general small decrease causes negligible error.)

When air is warmed by the land surface, its density decreases, causing it to rise up into the cooler and denser air above it, much like a released balloon that has been inflated with a gas less dense than air (Fig. 5.12(a)).

This type of convective mixing, which arises from temperature-induced differences in density, is referred to as 'free convection' or 'natural convection'. An area of land that is warmer than its surroundings, perhaps due to clearing of vegetation or cultivation, can give rise to an ascending plume of warmed air, which, with calm winds, can extend many hundreds of metres into the atmosphere. (Figure 5.12(a) is intended only to be schematic. There will be areas of ascending air and adjacent areas where cooler air must descend.) The rising plumes of warm air, commonly referred to as 'thermals', are used by large birds and glider pilots alike to maintain or increase their altitude.

Work has to be done in stirring volumes of air and in raising an air mass against the downward gravitational force acting on it. Priestley (1955) and Oke (1988), for example, explain general criteria determining whether convective transport of heat in the atmosphere near the ground is likely to be fully forced or dominantly free, or natural, in character. One guide to answering this question is the 'Richardson number', Ri. This non-dimensional quantity describes the ratio of the rate of working of buoyancy forces to that of eddy shear forces. Buoyancy forces are closely related to the spatial change in air temperature with height, $\Delta T / \Delta z$. Eddy shear forces are related to the spatial change (or gradient) in mean wind speed $\bar{u}$, or $\Delta \bar{u} / \Delta z$. The Richardson number is approximated well by

$$Ri = \frac{g(\Delta T/\Delta z)}{T(\Delta \bar{u}/\Delta z)^2} \tag{5.28}$$

Whether Ri is positive or negative depends on the sign of $\Delta T / \Delta z$, which is the slope of the air-temperature profile (Fig. 5.12). In Fig. 5.12(a) the slope $\Delta T / \Delta z$ is negative, conditions favouring natural convection. However, this slope can also be positive, for example on a calm clear night, when the temperature of the air can exceed that of a cold land surface (Fig. 5.12(b)). The type of temperature profile illustrated in Fig. 5.12(b) is called an 'inversion', as distinct from that of Fig. 5.12(a), which is referred to as a 'temperature lapse'.

Priestley (1955) showed that the convective transfer of heat in the lower atmosphere is effectively fully forced only with in a rather narrow band of values of Ri. As shown in Fig. 5.13, convective transport is fully forced when Ri lies between zero and a small negative number of about -0.02 or -0.13. As Ri decreases below such values, free, or natural, convection increasingly dominates sensible heat transport.

Except in relatively arid climates, the value of Ri quite commonly falls within the narrow band within which convection is fully forced. Positive values of Ri are associated with the temperature-inversion situation of Fig. 5.12(b). Under these conditions initially turbulent air tends

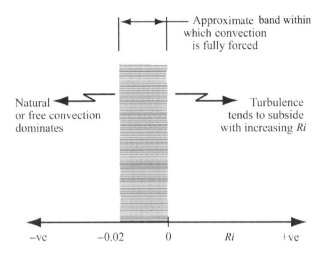

Approximate band within
which convection
is fully forced

Natural
or free convection
dominates

Turbulence
tends to subside
with increasing *Ri*

−ve   −0.02   0   *Ri*   +ve

Figure 5.13 Illustrating that convection is fully forced only within a rather narrow but important band of negative values of the Richardson number, *Ri*.

to lose its turbulent nature, subsiding into flow characteristics typical of laminar flow.

The theory which is able to predict the value of H outside the fully forced convective band of Ri shown in Fig. 5.13 is more complex than that of Eq. (5.15), which is not appropriate for use when the value of Ri is outside that narrow band (Priestley, 1955).

5.7 Ground heat flux and soil temperature

The magnitude of the ground heat flux (G in Eq. (5.9)), has direct and significant relationships with the variation in soil temperature throughout the soil profile and with time. The soil temperature directly affects the rates of all biochemical processes fundamental to the life of plants, insects and soil organisms. The soil temperature and the air temperature close to the ground are tightly coupled (Section 5.6), and many physiological processes in plants and trees are temperature-dependent. The rate of respiration, the process which utilises carbohydrate to make available chemical energy for the operation of plant cells, increases with temperature, at least up to some critical value. However, the rate of photosynthesis of carbohydrates is not strongly temperature-dependent, at least over the range 10–25 °C. It follows that, at sufficiently high temperatures, depending on the type of vegetation, the carbohydrate supply, and hence plant growth, will become limited.

From Eq. (5.9), the ground heat flux, G, is one of the ways in which the net radiation input is dissipated. Alternatively, if there is a loss of net radiation from the soil surface, as can occur at night, then part of that loss can be supplied by the flux G directed upwards in the soil towards the cooler soil surface, instead of downwards as is typical by day. The relative magnitude and significance of G increase as the degree of vegetation cover declines, since vegetation can intercept much of the radiation.

The mechanism of sensible heat-energy transfer in soil is largely 'thermal conduction'. In the kinetic-theory picture of thermal conduction in a solid such as soil, temperature is an expression of the energy of vibration of its constituent atoms and molecules. Thus heat energy is transmitted from hotter to cooler regions in thermal conduction by the flow of vibrational energy from hotter, more-energetic molecules to their adjacent cooler and less-energetic neighbours. Thus thermal conduction takes place in response to any difference in temperature.

Thermal conduction is the dominant overall process of heat-energy transfer in soil, the energy-transfer mechanisms of radiation and convection being negligible in soil, despite their being so dominant in air. Since soil usually has some water content, a spatial difference in temperature implies a corresponding difference in vapour pressure (Fig. 5.3). Thus, when the pore space of soil has a gaseous component (Fig. 2.4), water vapour will move from warmer to cooler regions. Because of its high latent heat of vaporisation (λ), the amount of sensible heat required to provide this latent heat in evaporation is substantial. Evaporation can take place within the soil profile, thus modifying the conductive or sensible heat flux. Water evaporated from the upper warmer layers of soil by day will not all be lost to the atmosphere; some can move downwards to cooler regions where the vapour pressure is lower, partially condensing back to liquid water.

Thus the movement of water in its vapour phase is not only a diurnally fluctuating mechanism of water movement, but also a process that modifies the flux of sensible heat in soil by thermal conduction (Rose, 1979).

A 'temperature gradient' is the difference in temperature, ΔT, between two sites, divided by the distance between the two sites, Δz, or $\Delta T/\Delta z$. The symbols ΔT and Δz are taken to imply positive quantities.

Figure 5.14 shows a temperature profile in soil typical of just after midday, when the maximum temperature of the profile is at the soil surface. The depth z is taken as downwards positive from the soil surface (Fig. 5.14). Note that, as the depth increases in the positive direction from z to $z + \Delta z$, the temperature decreases, so the temperature difference is written $-\Delta T$ (Fig. 5.14). Since heat flow is directed from warmer to cooler sites (i.e. downwards in Fig. 5.14), the direction and magnitude of G are proportional to $-\Delta T/\Delta z$. This is written

$$G \propto -\Delta T/\Delta z$$

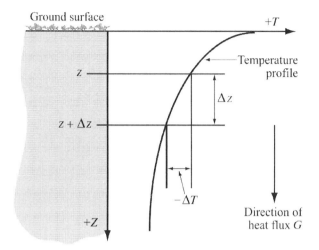

Ground surface

z

$z + \Delta z$

$+Z$

$+T$

Temperature profile

Δz

$-\Delta T$

Direction of heat flux G

Figure 5.14 Illustrating a temperature profile in the surface soil by day when the soil temperature decreases with depth into the soil profile.

or

$$G = -k(\Delta T / \Delta z) \qquad (5.29)$$

where k is the 'thermal conductivity' of the soil. Equation (5.29) applies to any medium in which the heat flux is due solely to thermal conduction. The negative sign in Eq. (5.29) is necessary in order to provide agreement with the experimental observation that heat energy flows in the direction of decreasing temperature.

The thermal conductivity of soil increases non-linearly with the water content. However, this does not imply that a cool wet soil (which can follow a wet winter, for example) will warm up more quickly than a drier cool soil – rather the reverse, as farmers in temperate climates are well aware, as they await an adequate rise in soil temperature for crop planting. This farmers' experience interprets the fact that the rate of rise in soil temperature, whilst depending directly on the thermal conductivity, k, does not depend on k alone. Rather, as shown, for example, by Rose (1979), the soil factor on which the rate of rise in soil temperature depends is the ratio $k/(\rho_b c)$, where ρ_b is the bulk density of soil, and c is its specific heat (the heat energy required to raise the temperature of unit mass of soil by $1\,^{\circ}$C). The product $\rho_b c$ is called the soil's 'thermal capacity'. The thermal capacity governs the change in temperature which will result from a given addition of heat, the latter quantity depending on k. Thus, the soil factor determining the changing heat flux and temperature in soil is a ratio, κ, defined by

$$\kappa = k/(\rho_b c) \qquad (5.30)$$

Figure 5.15 A schematic
illustration of the diurnal
variation of temperature, T,
with time, t, at the soil surface,
and at a depth of say 0.15 m
for cloud-free conditions.

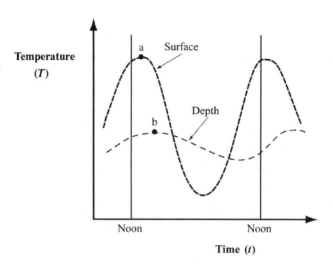

Figure 5.15 A schematic illustration of the diurnal variation of temperature, T, with time, t, at the soil surface, and at a depth of say 0.15 m for cloud-free conditions.

where κ is called the 'thermal diffusivity' or 'thermometric conductivity' of the soil. A typical value of κ in dry soil is of order 5×10^{-7} m^2 s^{-1}.

Both k and $\rho_b c$ increase with water content, but not at the same rate (as shown, for example, in Rose (1979)). As a result, the ratio κ can vary in a complex way with water content, but, under relatively wet conditions, it continues to decrease with increasing water content, because $\rho_b c$ then increases more rapidly than does k (Eq. (5.30)).

Under clear skies the diurnal variation in soil temperature is similar to the type of variation shown in Fig. 5.15 for the soil surface and a depth beneath the surface. The temperature variation at depth is of similar form to that at the surface, but the range of swing in temperature is typically smaller. There is also a delay in time between the surface-temperature maximum at 'a' and the maximum at depth, 'b' (Fig. 5.15). There is a similar delay in arrival of the temperature minimum or any other 'phase' of the temperature wave, so this effect is called a 'phase delay'. This phase delay is an indication of the fact that any fluctuation of temperature at the soil surface is propagated or transferred down into the soil profile with a certain speed, v. It can be shown that this speed is given by

$$v = (4\kappa\pi/\tau)^{1/2} \tag{5.31}$$

where τ is the 'period' of the temperature fluctuation (a day, in the case of a diurnal fluctuation in Fig. 5.15). Thus v is the speed at which any particular phase of a temperature wave is transmitted down into the soil profile.

If the diurnal variation in temperature at any depth in the soil is averaged over a period such as a month and then plotted against time on an annual basis, a figure with features very similar to those of Fig. 5.15 is obtained, except that, of course, the time difference between successive maxima (or any other phase) is a year, rather than a day. Since the period, τ, of the annual temperature wave is 365 times greater than that for a daily wave, Eq. (5.31) indicates that the speed of propagation will be much slower for the annual than for the daily wave, as will be illustrated in Example 5.4. However, the maximum swing or excursion in temperature is generally considerably greater on an annual than it is on a daily basis. For both these time scales the magnitude of the temperature swing decreases with depth beneath the soil surface. However, the depth of penetration into the earth of the temperature wave is much greater for the annual than for the daily wave. The temperature swing can be very small at a depth of some fraction of a metre for the daily wave; it can be several metres before the swing of the annual wave is negligible and the temperature becomes essentially constant.

Example 5.4

Using Eq. (5.31), estimate the speed of propagation into soil both of a daily and of an annual temperature wave. Assume that a constant value of $\kappa = 5 \times 10^{-7} \, \text{m}^2 \, \text{s}^{-1}$ applies.

Solution
For a daily wave,

$$\tau = 24 \times 3600 \, \text{s} = 8.64 \times 10^4 \, \text{s}$$

Thus

$$v = (4 \times 5 \times 10^{-7} \times \pi / 8.64 \times 10^4)^{1/2}$$
$$= 8.53 \times 10^{-6} \, \text{m s}^{-1}$$
$$= 30.7 \, \text{mm h}^{-1}$$

For an annual wave,

$$\kappa = 365 \times 8.64 \times 10^4 \, \text{s}$$

Thus

$$v = 1.6 \, \text{mm h}^{-1}$$

which is considerably slower than for the daily wave.

Main symbols for Chapter 5

A Area
c Specific heat of soil

c_p	Specific heat of air at constant pressure
D	Temperature difference $T - T_w$ (Fig. 5.9)
D_o	Temperature difference $T_o - T_{wo}$ (Fig. 5.9)
e	Vapour pressure of water (Eq. (5.2))
e_o	Saturation vapour pressure at water-surface temperature
e_w	Saturation vapour pressure of water
E	Evaporation rate
G	Sensible heat flux density directed into the earth's surface
h	Effective conductance of air (Eq. (5.19))
H	Sensible heat flux density directed from the earth's surface up into the atmosphere
k	Thermal conductivity of soil (Eq. (5.29))
K	Eddy diffusivity
L_i	Incoming long-wave radiation from the atmosphere to the earth's surface
L_{oe}	Total outgoing radiation emitted from the earth's surface (Eq. (5.10))
M_a	Relative molecular mass of dry air
M_w	Relative molecular mass of water
p_a	Total atmospheric pressure
q	Specific humidity (Eq. (5.3))
r	Reflection coefficient (or albedo) of the earth's surface
Ri	Richardson number (Eq. (5.28))
R_L	Net flux density of long-wave radiation emitted by the land surface (Eq. (5.11))
R_n	Net radiation received by the earth's surface (Eq. (5.8))
R_s	Flux density of global short-wave radiation received at the earth's surface
R_u	Universal gas constant
s	A slope defined in Eq. (5.22) and Fig. 5.10
t	Time
T	Temperature in degrees Kelvin ($K = °C + 273$)
T_a	Absolute air temperature
T_{dp}	Dew-point temperature
T_o	Ground-surface temperature
T_s	Earth-surface temperature
T_w	Wet-bulb temperature
v	Speed of penetration of temperature wave in soil (Eq. (5.31))
Z	Distance from ground surface to reference height (Fig. 5.8)
α	An empirical parameter defined in Eq. (5.27)
β	Bowen ratio (Eq. (5.16))
γ	Psychrometric constant (Eq. (5.5))

Δ	Finite difference in any quantity
ε	Emissivity of a surface
κ	Thermal diffusivity of soil (Eq. (5.30))
λ	Latent heat of vaporization of water
ρ_b	Bulk density of soil
ρ_v	Density of water vapour
σ	Stefan–Boltzman constant
Σ	Summation symbol

Exercises

5.1 (a) Define the following terms relating to moisture in the atmosphere:
   • absolute humidity,
   • vapour pressure,
   • saturation vapour pressure,
   • relative humidity,
   • dew-point temperature,
   • saturation deficit and
   • wet-bulb depression.
(b) It follows from the ideal-gas law (written for water vapour in
Eq. (5.2)) that the vapour density, ρ_v (in units of $g\,m^{-3}$) and
vapour pressure, e (Pa), are related by

$$\rho_v = 2.17\frac{e}{T} \quad (g\,m^{-3})$$

Calculate the vapour pressure if the vapour density is $10\,g\,m^{-3}$
at $30\,^\circ C$.

5.2 Figure 5.16 is from Campbell (1977). Two different conditions of
humidity and temperature are shown by the points on this figure
marked with solid dots. For parcels of air characterized by the two
points, determine
   • the relative humidity,
   • the saturation deficit,
   • the dew-point temperature and
   • the absolute humidity or vapour density (equivalent terms).
Notes to help use of Fig. 5.16 follow.
(i) In Fig. 5.16 relative humidity (expressed as a fraction rather than
   as a percentage) is given on the right-hand ordinate. Lines curving
   up to the right are lines of constant relative humidity. When
   the relative humidity is 1.0 (or 100%) this curve represents the
   variation in saturation vapour density with temperature.

Figure 5.16 The relationships between (dry-bulb) temperature and vapour density and relative humidity at sea level at atmospheric pressure. The inset figure is for temperatures below 0 °C. The straight lines sloping diagonally up to the left are lines of constant wet-bulb temperature at 2 °C intervals, the wet-bulb temperatures being shown on the saturation-vapour-pressure line (with relative humidity = 1.0 or 100%). (After Campbell (1977).)

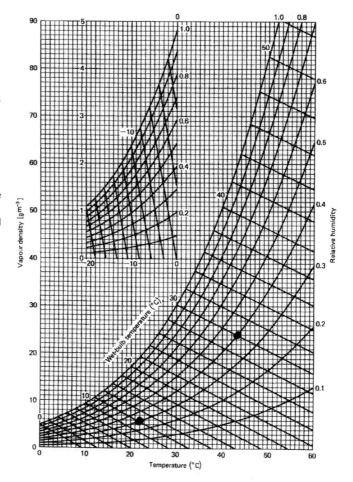

(ii) As air flows over a water surface it both cools and has its humidity increased. The straight lines sloping up towards the left in Fig. 5.16 represent the change in state during such a process. This describes the change in state of air as it flows over the saturated wick of a wet-bulb thermometer. At sufficient ventilation or wind speed over the wet bulb air becomes saturated, and where any upward sloping line meets the saturation line this represents the wet-bulb temperature for that parcel of air.

5.3 (a) Assuming that the temperature, T_s, of a vegetated surface is equal to the air temperature, T_a, what is an expression for the outgoing emitted long-wave radiation, L_{oe}? (See Eq. (5.10).)

(b) Using the empirical equation (5.12) for the incoming long-wave irradiance, calculate the net outgoing long-wave-radiation flux density ($L_{oe} - L_i$) assuming that the emissivity of the vegetated surface, ε, is 0.97 and that $T_a = 293$ K (or 20 °C).

(c) Using Swinbank's equation, Eq. (5.12), for L_i, calculate and plot L_i over the air-temperature range from 0 °C to 40 °C.

(d) Over the same temperature range calculate and plot on the same diagram as in part (c) the outgoing emittance L_{oe} assuming that the emissivity is $\varepsilon = 1$ (the value for what is called a 'black body').

Hence show that $L_{oe} - L_i$ declines approximately linearly with T_a under clear skies. How might this relationship be affected by the presence of clouds?

5.4 Let the magnitude of incoming solar flux density arriving at the earth's atmosphere be represented by 100 solar units (SU). The earth's 'albedo' is the fraction of incoming solar radiation reflected back to outer space.

(a) Calculate the number of solar units of energy at the earth's surface, given the following information on the fate, for a cloud-free atmosphere, of the 100 SU of radiation incoming to the earth's atmosphere:

- 2 SU are absorbed in the stratosphere, mainly by ozone;
- 10 SU are scattered back into space by the cloud-free atmosphere;
- 10 SU are scattered towards the earth's surface;
- 20 SU are absorbed by dust and other atmospheric components; and
- the average albedo of the earth's surface is 0.1.

(b) Calculate the overall albedo of the earth and its cloud-free atmosphere.

5.5 (a) Describe two requirements for (net) evaporation of water from a surface to take place.

(b) The rate of evapotranspiration from a land surface (E, $\text{kg m}^{-2}\,\text{s}^{-1}$) can be approximately calculated using the Priestley–Taylor equation, Eq. (5.27). Assume that the latent heat of vaporization of water is $\lambda = 2.4 \times 10^6\ \text{J kg}^{-1}$, that the location-specific parameter $\alpha = 1.2$ and that $s/(s + \gamma) = 0.75$. Then calculate the magnitude of E if $R_n = 400\ \text{W m}^{-2}$ and $G = 0.2R_n$.

(c) Using the same equation and assumed values of parameters as in part (b), but taking $G = 0$, calculate the evaporation during a 10-hour period of sunshine from an irrigation ring tank of surface area 2.5 ha if the average value of R_n during the 10-hour period is 300 W m^{-2}. Express this amount of evaporation in terms of

(i) an equivalent ponded depth of water (in mm) and

(ii) a mass of water in tonnes, where 1 tonne = 1000 kg.

(*Note.* The density of water $= 1000 \, \mathrm{kg \, m^{-3}}$ and 1 ha $= 10^4 \, \mathrm{m^2}$.)

5.6 (a) A form of the Penman equation for calculating the rate of evaporation from a shallow open-water surface is

$$\lambda E = \frac{s}{s+\gamma} R_n + \frac{\gamma}{s+\gamma}(\rho_w - \rho_v)f(\bar{u})$$

Using the data in Fig. 5.11, obtain values for the ratios $s/(s + \gamma)$ and $\gamma/(s + \gamma)$ for a temperature of 23 °C, the mean daily temperature at a particular location where the mean daily difference $(\rho_w - \rho_v)$ between saturation and actual vapour densities is $8 \, \mathrm{g \, m^{-3}}$. With the vapour-density difference in these units, the empirical function $f(\bar{u})$ is given by

$$f(\bar{u}) = 5.3(1 + \bar{u})$$

where $\bar{u}$ is the mean daily wind speed, which is $2 \, \mathrm{m \, s^{-1}}$ at this location.

The energy equivalent of the daily evaporation rate, λE, from an evaporation pan at this location was $300 \, \mathrm{W \, m^{-2}}$. Using the above equation, calculate what the average daily net radiation, R_n, required to support this rate of evaporation must have been.

(b) Convert the energy-equivalent value of the mean evaporation rate, $\lambda E = 300 \, \mathrm{W \, m^{-2}}$, into a rate E, given that $\lambda = 2.45 \, \mathrm{MJ \, kg^{-1}}$. Then calculate the equivalent ponded depth of water evaporated (in mm) if the length of day was 9 hours. (The density of water is $1000 \, \mathrm{kg \, m^{-3}}$.)

5.7 A form of the Dalton equation, Eq. (5.6), used by Penman (1948) to predict evaporation from an evaporation pan is

$$E = 0.4(1 + 0.17\bar{u})(e_w - e) \quad (\mathrm{mm \, d^{-1}})$$

where $\bar{u}$ is the mean wind speed measured in miles per hour (m.p.h.) at a height of 2 m, with e_w the saturation vapour pressure at the surface water temperature and e the vapour pressure at screen height (in units of mm Hg). Calculate E, given that $\bar{u} = 2 \, \mathrm{m \, s^{-1}}$, the temperature both of the water surface and of air at screen height is 12 °C and the relative humidity was 55%.

5.8 (a) Critically compare Eq. (5.23) (combined with Eqs. (5.24) and (5.25)) of McIlroy with the equation of Penman (Eq. (5.26).)

(b) Under dry-soil conditions the wet-bulb depression D_o can be similar to D. Examine the effect on the calculated evaporation rate of assuming that $D_o = D$ for the data in Table 5.1. (With this assumption approximate agreement with the evapotranspiration rate measured by lysimeter might no longer be expected.)

5.9 Discuss the significance of the Richardson number (Ri, Eq. (5.28)) for determining the likelihood that convective heat transfer in the lower atmosphere will be forced or free.

5.10 It may be shown that the amplitude of a simple temperature wave in soil will have fallen to 5% of its value at the soil surface at a depth z, when $z[\omega/(2\kappa)]^{1/2} = 3$. In this expression, ω, called the 'angular frequency' of the wave, is given by $\omega = 2\pi/T$, where T is the wave period. Assuming that the thermal diffusivity of the soil, κ, is constant with depth and of value $5 \times 10^{-7}\,\mathrm{m^2\,s^{-1}}$, calculate the value of the depth z at which the amplitude has fallen to 5% of its surface value for

(a) the daily temperature wave and

(b) the annual wave.

References and bibliography

Brutsaert, W. (1982). *Evaporation into the Atmosphere, Theory, History, and Applications*. Dordrecht: Kluwer Academic Publishers.

Campbell, G. S. (1977). *An Introduction to Environmental Biophysics*. New York: Springer-Verlag.

Kormondy, E. J. (1969). *Concepts of Ecology*. Englewood Cliffs, New Jersey: Prentice-Hall International, Inc.

Miller, G. T. Jr. (1979). *Living in the Environment,* 2nd edn. Belmont, California: Wadsworth Publishing Company.

Monteith, J. L., and Unsworth, M. H. (1990). *Principles of Environmental Physics*, 2nd edn. London: Edward Arnold.

Oke, T. R. (1988). *Boundary Layer Climates*, 2nd edn. New York: Halsted Press.

Penman, H. L. (1948). Natural evaporation from open water, bare soil and grass. *Proc. Roy. Soc. Lond.* A **193**, 120–146.

Penman, H. L. (1955). *Humidity*. London: Chapman and Hall, on behalf of the Institute of Physics, UK.

Priestley, C. H. B. (1995). Free and forced convection in the atmosphere near the ground. *Quart. J. Roy. Meteorol. Soc.* **81**, 139–145.

Priestley, C. H. B., and Taylor, R. J. (1972). On the assessment of surface heat flux and evaporation using large-scale parameters. *Mon. Weather Rev.* **100**, 81–92.

Rose, C. W. (1979). *Agricultural Physics*. Oxford: Pergamon Press.

Rose, C. W., Begg, J. E., Byrne, G. F., Goncz, J. H., and Torssell, B. W. R. (1972). Energy exchanges between a pasture and the atmosphere under steady and nonsteady state conditions. *Agric. Meteorol.* **9**, 385–403.

Szeicz, G. (1975). Instruments and their exposure. In *Vegetation and the Atmosphere, Volume 1, Principles*, ed. J. L. Monteith. London: Academic Press, pp. 229–273.

Sharma, M. L. (1984). *Evapotranspiration from Plant Communities*. Amsterdam: Elsevier Science Publishers B. V.

Slatyer, R. O., and McIlroy, I. C. (1961). *Practical Microclimatology*. Canberra, ACT: Commonwealth Scientific and Industrial Research Organisation.

Smithsonian Institution (1951). *Smithsonian Meteorological Tables*, 6th revised edn. Washington: The Smithsonian Institution.

Swinbank, W. C. (1963). Long-wave radiation from clear skies. *Quart. J. Roy. Meteorol. Soc.* **89**, 339–348.

6
Infiltration at the field scale

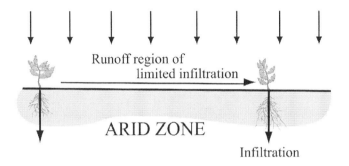

ARID ZONE

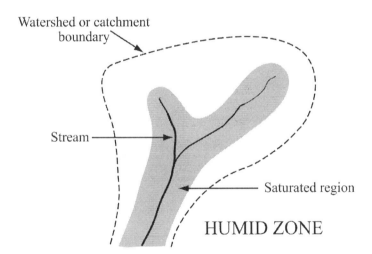

HUMID ZONE

6.1 Introduction: from rainfall to the sea

The ways in which precipitation falling on the land surface can be partitioned were introduced in Chapter 4. General relationships between the components of this partition in a watershed, using the principle of mass conservation of water, were also considered. These processes of precipitation falling onto land, of water infiltrating and moving through the soil profile and of water flowing over the the land surface are of vital importance to all life on earth. Much of the water that flows through the

Figure 6.1 The hydrological
cycle. Transpiration is
evaporation from vegetation.

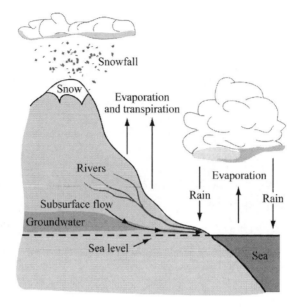

Figure 6.1 The hydrological cycle. Transpiration is evaporation from vegetation.

soil is collected into streams and rivers at the scale of watersheds, with overland flow periodically adding to this stream flow. With those exceptions where rivers feed inland lakes or evaporation sites, or when rivers are themselves saline, rivers deliver fresh water to the vastly greater body of sea water.

Evaporation of water from the land and the sea returns water back into the atmosphere, where cloud formation may lead again to the possibility of precipitation as rainfall, or, in colder regions, as snow or ice. Evaporation from the oceans leaves salt behind, although salt can be driven inland great distances by wind whipping up fine droplets of sea water and delivering the salt back to the land, either with rainfall or by deposition of the dried salt from the atmosphere. Some water is stored in the atmosphere, dominantly as the commonly invisible gas called water vapour, which we can experience as humidity, or else in visible forms as clouds or mist.

Thus there is an enormous global-scale cycling of the water delivered by precipitation to the land and the sea. This process, referred to as the 'hydrological cycle', is illustrated in Fig. 6.1.

The rain that falls can bring with it salt and dust and also the vast range of chemicals that are emitted into the atmosphere, either by natural processes or as the result of human activities. In general the concentrations of such chemicals in precipitation are quite low. However, their consequences can be serious, as in rainfall made slightly acidic by emitted chemicals.

The fate of precipitation, whether by rain, snow, or sleet, is of vital concern to all life on earth. This is because the infiltration of rain or snow melt is by far the the dominant source of water for the growth of all vegetation on the surface of the earth. Infiltration is the only extensive source of the water required to recharge the root zone so that it can be withdrawn by root systems and subsequently transported up through trees or plants as the transpiration stream. Water around the root system is also necessary for the uptake of plant nutrients essential for growth. The water that enters the soil recharges the root profile of plants and trees, and that which moves below the root zone can take chemicals with it into the groundwater.

In this chapter we describe how the rate at which rainfall arrives at the land surface interacts with the rate of uptake of water by the land surface, or infiltration rate, defined in Section 4.5. It is when the rainfall rate exceeds the infiltration rate that the excess water on the soil surface accumulates and runs downslope, some eventually ending up in streams, rivers, or lakes. Human activities of all kinds, rural, urban, industrial and recreational, impinge in particular ways on the infiltration of water. Therefore this process will receive particular attention in this chapter.

The effect of human activity modifying the infiltration rate is usually to reduce it, and this has consequent environmental impacts. One consequence is that a decrease in infiltration implies increases in volumes and rates of overland flow. This has possible consequences including the flooding of low-lying areas, damage to infrastructure such as road and rail systems and excessive rates of soil erosion, especially on relatively steep land and from river banks.

All these effects are related to dynamic processes at the landscape scale. A major priority in this chapter is to seek a description of infiltration at a range of scales, paying particular attention to the adequacy of such descriptions for predictive purposes. The utility of one such model will be tested by comparison with rainfall and hydrological data collected at the field scale. This infiltration model will be built on in Chapter 7, where we consider how rainfall that has not infiltrated collects on the soil surface and runs downslope towards the water-collecting systems of streams and rivers.

6.2 Infiltration into small areas of soil

It is quite possible that all rainfall received by the land surface can infiltrate into it. This is more likely to occur if rainfall rates and amounts are modest, the soil surface is protected from rainfall impact by litter, mulch, or vegetation and the soil is initially dry. An example of this combination of requirements could be a temperate forest. However, if

Figure 6.2 Idealised water-content profiles in a uniform soil column (initially at a uniform volumetric water content), after periods during which (a) water was applied at a steady flux rate such that ponding did not occur and (b) with ponded water maintained on the soil surface. The volumetric water content with the soil saturated is denoted θ_s.

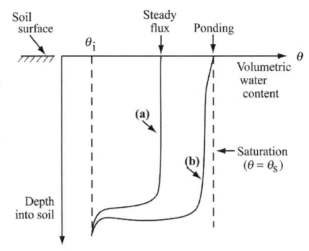

the soil profile is fully charged with water, little, if any, infiltration may occur.

With complete infiltration of all rainfall, no overland flow is generated, and the infiltration rate, I, is equal to the rate of delivery of water to the soil surface. If the storage of rainfall by interception on vegetation, surface litter or mulch is negligible, or if these interceptors are fully saturated, then $I = P$, where P is the rainfall rate. Interception of precipitation by vegetation can be quite significant under some circumstances, but, even when this is so, the rate of infiltration is controlled by the rate of delivery of water to the soil surface. When all water delivered to the soil surface enters it, the infiltration rate is said to be 'flux-controlled'. Infiltration of water delivered in frozen forms requires different considerations that are not discussed here.

Experiments have been carried out in which a constant steady rate of water has been applied to a specially prepared column of soil of uniform properties and uniform initial water content, θ_i. If this steady rate of application is continued for some time with no surface ponding, investigation of the profile of water content in the soil at termination of the experiment shows it to be as illustrated by profile (a) of Fig. 6.2. As the steady flux proceeds, water moves through a 'transmission zone' of an approximately constant water content such that the rate of flow through the soil is equal to the application rate (Fig. 6.2).

Notice that, at the lower end of this transmission zone in Fig. 6.2, the water content decreases rather suddenly, falling rather abruptly to the initial or prior water content θ_i. This zone where water content decreases rapidly with depth is called a 'wetting front'. The extent of the wetting front is more compressed in soils of light texture (e.g. sandy soils). This

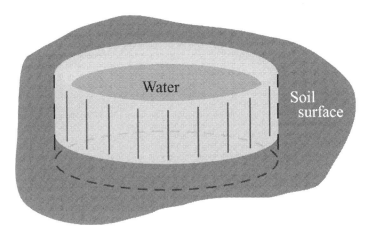

Figure 6.3 A ring infiltrometer consisting of a cylindrical ring driven into the ground surface. The rate of fall of the surface of the water ponded within the ring is a measure of the infiltration rate for soil within the ring.

finding, illustrated in Fig. 6.2, provides some justification for the approximations made in the water-balance-accounting procedure described in Section 4.5.

If water is continuously ponded on the surface of the soil which has a uniform initial water content of θ_i, then after a period of time the water-content profile can be as shown by profile (b) in Fig. 6.2. This profile displays similar features to profile (a) of Fig. 6.2, which was formed without ponding, the main difference being that the water content in the transmission zone for profile (b) is close to the saturation value.

As discussed in Section 4.5, the rate at which water enters the soil surface is called the infiltration rate. A measure of the infiltration rate at a particular location in the landscape is often sought by ponding water within a cylindrical ring of some convenient arbitrary diameter that has been driven into the soil (Fig. 6.3). The rate of fall of the surface of water ponded over the soil within the ring gives a measure of the infiltration rate for the particular enclosed area. In practice this is commonly determined by the amount of water required to be added per measurement period in order to keep the level of ponded water in the infiltrometer constant. A refinement of this technique is to surround the infiltrometer ring with a guard ring in which water is maintained at the same level as in the measurement ring. The reason for this refinement is that water infiltrating from the outer annulus formed by the guard ring helps ensure that infiltration from the inner measuring ring occurs vertically, rather than spreading out laterally, which would complicate interpretation of the measurement data.

There are several uncertainties in infiltrometer measurement of the infiltration rate. The rate obtained at a given location can vary with the diameter of the infiltrometer ring used, perhaps depending on whether or not the ring encloses a crack in the soil, a worm hole, or some other pore produced by some other type of soil fauna. Spatial variability in infiltration rate is also revealed by the common experience that, even using the same size of infiltrometer ring, the indicated infiltration rate can vary quite substantially over distances of the order of metres. As might be expected, the infiltration rate also depends on how dry or wet the enclosed soil is prior to ponding.

From these comments, it is not surprising that the scale of measurement of infiltration plays an important role in determining the nature of infiltration characteristics. There can also be some definite spatial pattern in the local infiltration rate, with some location or locations possibly having infiltration rates so high that water can rapidly infiltrate and move deeply to contribute rather quickly to the groundwater.

Another technique used to investigate the infiltration rate is applying water to the soil surface as simulated rainfall produced by a sprinkler system, for example. A rainfall simulator is a device in which a spray of droplets is produced to simulate rainfall. Irrigation by some form of sprinkler system can have similarities to, as well as differences from, the more uniform application of water by rainfall simulation. If the rate of application of water in simulated or actual rainfall is not too high, there is a period of time following commencement of application during which the surface water content increases, but is less than saturation. Whether the surface water content reaches its saturation value depends on the rate of application of water. At saturation, ponding of water on the soil surface becomes evident, and overland flow may subsequently commence.

Since the soil is unsaturated during the pre-ponded period of application of water or rainfall, water is not free to enter larger channels or 'macropores'. The term macropore is commonly used to describe a void or channel in the soil of any origin whose size is greater than about 1 mm. Such voids can result from soil shrinkage, or from the common pedal structure of soil, whereby soil is somewhat separated into distinct blocks, between which there exist passages through which water can flow. Both flora and fauna can also result in pores or channels being formed, in which case such voids are referred to as 'biopores'. In field soils macropores are the rule rather than the exception.

Ponded infiltration into macropores leads to much greater spatial heterogeneity in soil water content compared with that resulting from a rate of application sufficiently low that no ponding results (Clothier and Heiler, 1983). Indeed, field observations by Clothier and Heiler of the water-content distribution in the soil profile following contrasting non-ponding and ponding irrigation regimes illustrated the decisive role played by macropores in determining the depth of penetration into the soil profile of free-flowing water prior to its complete absorption by the soil matrix. Such penetration by water can be well beneath the root zone of an irrigated crop. On the basis of such observations, Clothier and Heiler (1983) recognised such excessive penetration by water as a cause of inefficiency of irrigation and possible pollution of groundwater. Hence they offered the following practical advice for the design of sprinkler or rotating irrigation equipment. It is desirable, in terms of greater efficiency of water use, to limit the rate of sprinkler application so that ponding does not occur. This insight into the significance of macropore flow also indicates that gains in efficiency of water use can be achieved through the use of sprinkler or rotating irrigation systems rather than application of water by surface flooding.

The control over rainfall rate provided by a rainfall simulator makes it a useful technique for investigating infiltration prior to the ponding of water on the soil surface. However, by using higher rates of simulated rainfall, it can also be used to investigate the infiltration rate after ponding. The values of the post-ponding infiltration rate obtained using rainfall simulators are usually smaller than those yielded by use of a ring infiltrometer at the same location. This may be due to effects associated with the breakdown in structural aggregates in soil which is produced by rainfall impact, or to the blockage of pores by the infiltration of sediment-laden water. Preferential intake, and hence blockage of larger pores, may also be a reason for the lower variability in infiltration rate usually obtained with this technique.

For soil in which the damage to structure in the soil surface by rainfall is what controls the infiltration rate (rather than in-profile soil characteristics), the use of simulated rainfall can provide more realistic data than using ponded water as in the ring infiltrometer of Fig. 6.3. Whilst the area of soil receiving simulated rainfall is usually larger than that for a ring infiltrometer, the area is still limited by practical considerations.

If a ring infiltrometer such as in Fig. 6.3 is inserted into somewhat dry soil, and a constant ponded depth of water maintained, it is found that the infiltration rate decreases with time, but ultimately reaches an approximately constant value as indicated in Fig. 6.4. The exact form of this relationship depends substantially on the initial water content of the surface profile of the soil prior to ponding (Fig. 6.4) and to a lesser extent on the depth of water ponded in the infiltrometer ring.

A considerable body of theory seeking to describe the infiltration of water into soil has been developed assuming the soil to be spatially homogeneous (see, for example, Marshall, Holmes and Rose (1996) and Hillel (1998)). Such work provides many useful insights and depends on concepts from soil physics that will be introduced in Chapter 11, when necessary prior concepts have been developed further.

Figure 6.4 The variation with time in the rate of infiltration measured with an infiltrometer.

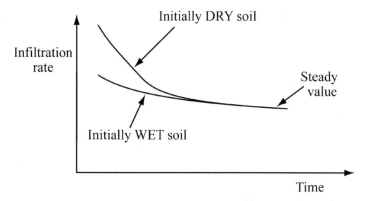

Two scientists, Green and Ampt, in 1911 developed an infiltration equation based on such concepts, but making simplifying assumptions. Their equation can be written in the simple form

$$I = I_c + b/\Sigma I \qquad (6.1)$$

where I_c is the long-term steady-state value towards which the value of I declines with time and with the cumulative depth of water infiltrated since the commencement of infiltration denoted ΣI. The two characteristic parameters in Eq. (6.1), b and I_c, are site-dependent. One of the limitations of Eq. (6.1) is that initially $\Sigma I = 0$, so that an infinite infiltration rate is predicted at the commencement of infiltration. Since I falls rapidly with time (Fig. 6.4), in practice this unrealistic feature of Eq. (6.1) doesn't present a major problem in fitting the equation to experimental data.

Example 6.1

The infiltration rate was measured in a ring infiltrometer. After a total depth of 80 mm of water had infiltrated, the infiltration rate was 15 mm h^{-1}. After some hours of infiltration, the rate appeared to be approximately steady at 4 mm h^{-1}. Using the Green–Ampt infiltration equation as a model, predict what equivalent ponded depth of water would have infiltrated when the infiltration rate is predicted to be 8 mm h^{-1}.

Solution

Using Eq. (6.1) and taking I_c to be 4 mm h^{-1},

$$15 = 4 + b/80 \,\mathrm{mm\,h^{-1}}$$

Thus

$$b = 880 \,\mathrm{mm^2\,h^{-1}}$$

Using this value of b with Eq. (6.1),

$$8 = 4 + 880/\Sigma I$$

Thus the cumulative infiltration required would be expected to be 220 mm.

If the soil profile is not uniform, or the depth of ponded water varies with time, then the form of infiltration with time will not be the simple form shown in Fig. 6.4. The depth of ponded water used in ring-infiltrometer measurement does not have a major influence on the measured infiltration rate. Much more important is whether or not placement of the infiltrometer ring has included voids or passages such as biopores formed by worms or other burrowing soil fauna. Channels formed by decaying plant roots and shrinkage cracks formed following drying in some clay soils are other examples of macropores that can have a very dramatic effect on the infiltration rate subsequent to their filling with water. Some appreciation of the potential great importance to infiltration of macropores can be obtained using an equation developed by the French scientist Poiseuille (1799–1869). 'Poiseuille's equation' for the laminar flow of a liquid of dynamic viscosity η through a cylindrical tube of inside radius r is given by

$$G = \frac{\pi r^4}{8\eta} \frac{\Delta p}{\Delta x} \qquad (6.2)$$

where G is the volume of liquid per second flowing through the tube in which there is a spatial gradient $\Delta p/\Delta x$ in fluid pressure p.

If the pressure gradient $\Delta p/\Delta x$ in Eq. (6.2) were the same for water filling approximately cylindrical passages of different sizes in soil, the dependence of G on r^4 in Eq. (6.2) would indicate that there is an enormous variability in G. A worm hole in soil could have a radius of 2 mm, whereas passages in a soil without worm holes could easily be a factor of ten smaller. Equation (6.2) then indicates that, if $\Delta p/\Delta x$ were the same for flow in both these cases, entry of water into the smaller hole would be less than that for the worm hole by the enormous factor of 10^4!

Since macroscopic passages exist in all field soils, entry of water through them is a dominant factor affecting the infiltration rate, at least following ponding (Clothier, 1988).

Such field data and the implication of Poiseuille's equation provide one reason for the spatial variability always found in ring-infiltrometer measurements, even for soil of an apparently uniform type and identical prior water content. This variability can be up to one or two orders of magnitude over quite short distances.

Figure 6.5 illustrates the consequence of either a biopore or an interpedal or shrinkage crack in soil on the distribution of water in the soil

Figure 6.5 Illustrating the effect of worm holes or soil cracks on water distribution in the soil profile following ponded infiltration.

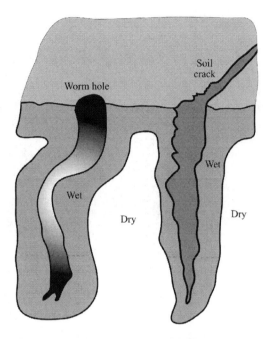

profile, from which their enhancing effect on infiltration can be understood. Figure 6.6 indicates that another common source of preferential pathways for entry of ponded water into soil is provided by its 'pedological' organisation (in which soil units are clumped together in somewhat discrete units) and, on a smaller scale, by its structural aggregation. These two types of soil organisation are by no means always as pronounced as suggested by Fig. 6.6, or of the type shown in this figure. However, some degree of aggregation and some pedological organisation are common characteristics of field soils.

The implication of Figs. 6.5 and 6.6 is often described as a dichotomy between macropores and the 'soil matrix' (defined as the soil fabric apart from the macropores). Addressing the consequences of this dichotomy for the movement of water and chemicals in field soils has been a major focus of soil-science research during the past couple of decades.

Entry of water into macropores as illustrated in Figs. 6.5 and 6.6 requires the displacement of air, which tends to be trapped and slightly compressed. Such entrapped air would restrict further entry of water if it could not escape. However, there is generally sufficient connection to surface-vented macropores for this compressed air to escape quite readily unless such vents are restricted in efficacy or destroyed by vehicular traffic, or blocked by surface-sealing processes. (Clothier (1988) reviews some of the literature on this issue.) The entry of sea water over a flat sand

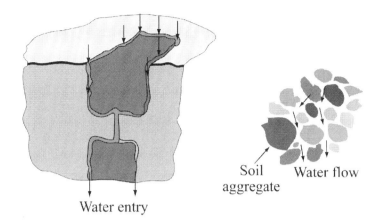

Figure 6.6 Preferential flow pathways for water provided by pedological organisation (left) and aggregate structure (right). Water in pedal units or aggregates is much less free to move than is water in the macropores, and in models of flow is often regarded as 'immobile' compared with water moving freely through inter-pedal or inter-aggregate passages (called 'mobile water').

beach can often be seen to be accompanied by a multitude of bubbles produced by displaced air, indicating the necessary escape of entrapped air which accompanies infiltration of ponded water.

The implication of such research at the scale of watersheds is by no means clear, since, at this scale, much larger and more effective passages for entry of water can be provided by burrowing fauna, decayed tree stumps etc. Other causes of spatial variability in infiltration rate will be discussed in the following section.

6.3 Spatial variability in infiltration rate

As reported in Section 6.2, infiltration-rate measurement at points in the landscape using ring infiltrometers invariably exhibits very considerable spatial variation. Such variability is an important general characteristic of infiltration into the small areas selected by this measurement technique. Such high spatial variability led soil scientists to investigate how large the number of measurement sites might need to be in order to obtain a representative value. This raises sampling problems dealt with in mathematical statistics, and addressed particularly in an application called 'geostatistics'. Geostatistical techniques, which originated in investigating the extent of ore bodies in mining, have been applied extensively to investigating the spatial variability of the infiltration rate (Nielsen, Biggar and Ehr, 1973).

Geostatistical analysis of the spatial variation in infiltration rate efficiently describes spatial correlation in this type of data. However, such studies obviously depend on making a large number of infiltrometer measurements, which is labour intensive. Whilst investigations of this kind have been most useful in documenting examples of the ubiquitous characteristic of spatial variation in infiltration rate, this type of

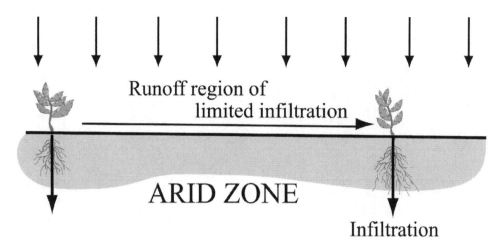

Figure 6.7 A natural arid-zone example of extreme periodic spatial variation in infiltration rate in which bands of vegetation effectively harvest and concentrate limited rainfall.

documentation does not appear to provide a ready basis for hydrological prediction at field or small-watershed scale.

Evidence for spatial variability in infiltration rate is by no means limited to infiltrometer studies. For example, there are natural arid-zone ecosystems whose structures and functioning provide an example of a somewhat regular contrasting variation in infiltration rate (Fig. 6.7). In such ecosystems, vegetation is not uniformly or regularly distributed, but is dominantly restricted to bands of trees or shrubs, separated by almost vegetation-free areas of bare soil. The infiltration rate in these bare areas is so low that, on the infrequent occasions when rainfall does occur, the rainfall rate considerably exceeds the very low infiltration rate, thus generating overland flow. When this overland flow meets the band of vegetation with high infiltration rate, due in part to biological activity, infiltration provides a store of water on which the roots of the vegetation can draw during the usually long periods without rain. This banded system, in which vegetation obtains extra water harvested from the runoff region in addition to rain falling directly on it, is illustrated in Fig. 6.7.

Because of the origin of overland flow in the inter-vegetation bare-soil areas of Fig. 6.7, it is referred to as 'infiltration excess overland flow', and is also called 'Hortonian flow', after the US hydrologist, Horton. Such flow is common with cultivated soil, or where surface soil structure has degraded or consolidated to form a 'seal'. However, this form of overland flow can occur on unsealed soil surfaces, especially with high rates or amounts of rainfall.

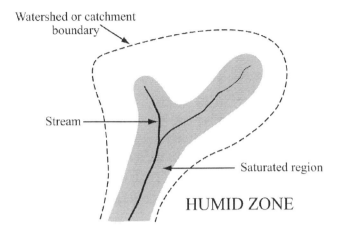

Watershed or catchment
boundary

Stream

Saturated region

HUMID ZONE

Figure 6.8 Illustrating a humid-zone watershed drained by streams surrounded by a variable saturated zone. Rain falling onto the saturated zone does not infiltrate but runs off to the streams as saturation overland flow.

Humid environments also provide common examples of spatially variable infiltration, though for very different reasons from those in the arid-zone example of Fig. 6.7. In humid climates parts of a watershed can become saturated with water for a variety of reasons. For example, subsurface water flow converges towards streams in a watershed, often leading to a zone around the stream being saturated with water (Fig. 6.8). Rain falling on this saturated zone cannot infiltrate, and this zone becomes a source of what is called 'saturation overland flow', in contrast to infiltration-excess or Hortonian flow.

The extent of the saturated zone in Fig. 6.8, for which $I = 0$, can grow or shrink over time in response to climatic trends. Thus the source area for saturation overland flow is variable, though in general the change in this source area within a rainfall event will be considerably less than at longer-term or seasonal time scales. Surface water flow generated from saturated zones is commonly referred to as 'saturation-excess overland flow', or simply as 'saturation overland flow'. For the unsaturated remainder of the watershed outside the saturated zone, I can be equal to P. Denote the average infiltration rate for the watershed as a whole by $\bar{I}$. It follows that $\bar{I}$ would increase with rainfall rate, at least to some upper limit.

Another source of evidence of spatial variability in infiltration rate comes from hydrological studies at field and plot scales. Data from experiments at plot scale will be used because of their relative hydraulic simplicity compared with the hydrological behaviour of large watersheds. The data used are from cultivated situations where infiltration-excess overland flow is the origin of measured plot runoff. The concept of 'excess rainfall rate' is useful in discussing this situation when the infiltration rate is not high enough for the soil to accept all the rainfall

Figure 6.9 The response of
the infiltration rate to the
rainfall rate for a field of
spatially uniform infiltration
characteristics.

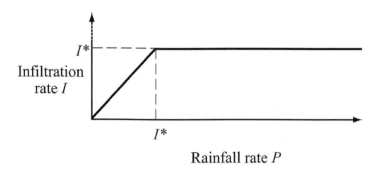

reaching the soil surface. The excess rainfall rate, R, is the excess of the
rainfall rate over the infiltration rate, so by its definition

$$R = P - I, \quad P \geq I \tag{6.3}$$

This excess rainfall rate, R, implies accumulation of water on the
soil surface, and some can be stored on the soil surface. Following the
filling of such storage, with any slope on the land water will tend to flow
downhill under the component of the gravitational force on the water
which acts in the downslope direction. Resistance is offered to any such
overland flow, and so the excess rainfall generated at any point on a plot
of ground will take some finite time to travel to the end of the plot, the
time taken depending on the velocity of the flow.

Consider a field or plot enclosed with hydrological boundaries so
that no flow onto or out of the plot is allowed except at the downslope
end of the plot. Such a plot is illustrated in Fig. 4.4. Let us assume that
the plot has a uniform infiltration rate I^* across the entire plot. Then,
if $P < I^*$, the actual infiltration rate $I = P$, so, from Eq. (6.3), $R = 0$,
indicating there will be no runoff from the plot. However, if $P > I^*$, then
the infiltration rate is given by I^*, its assumed constant maximum value,
unaffected by any variation in P (provided that $P > I^*$). This type of
infiltration response for a field of uniform infiltration rate is illustrated
in Fig. 6.9.

In Fig. 6.9, for $P < I^*$, the actual infiltration rate I is less than the
potential infiltration rate, because it is limited by the rainfall rate. Thus,
within this range of P, $I = P$. It is only when P exceeds I^* that the actual
infiltration rate achieves the potential value of the infiltration rate, so that
I is a constant and equal to I^* (Fig. 6.9).

Note that we have now made a necessary distinction between the
actual and potential infiltration rates, though they can be the same, as
when $P > I^*$ in Fig. 6.9. Since the field plot so far considered is assumed

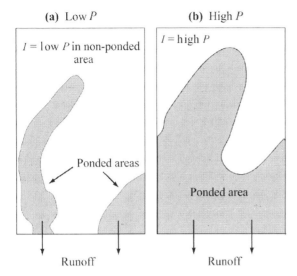

(a) Low P

I = low P in non-ponded area

Ponded areas

Runoff

(b) High P

I = high P

Ponded area

Runoff

Figure 6.10 The extent of ponded areas shown shaded for a runoff plot (a) at low rainfall rate and (b) at high rainfall rate.

to be of uniform infiltration characteristics, the infiltration rate at every point in the field is the same as the average value for the entire plot.

Now consider a field plot in which the potential infiltration rate, I, is variable across the field, so that some areas are ponded and others not (Fig. 6.10). Excess rainfall (and hence runoff) will then be generated only for those parts of the field or plot where $P > I$; it is these parts (shown shaded in Fig. 6.10), where $R > 0$ (Eq. (6.3)).

In the remainder of the plot (shown unshaded in Fig. 6.10), where the potential infiltration rate is greater than P, no ponding occurs, $R = 0$; so no runoff is generated.

As we move focus from Fig. 6.10(a), where the rainfall rate P was low, to Fig. 6.10(b), where P is higher, we see that the ponded area generating runoff has expanded in area. This expansion occurs since at a higher value of P the fraction of the field with a potential I greater than P is lower. However, during the expansion of the shaded ponded area between (a) and (b) in Fig. 6.10, the actual rate of infiltration into this area will have increased. In the non-ponded area of (b) the actual infiltration rate will also have increased, since in this area the infiltration and rainfall rates are equal. For both these reasons the actual infiltration rate averaged aver the whole field, denoted $\overline{I}$, will have increased as the rainfall rate increased. This effect of a dependence of $\overline{I}$ on P for a field of non-uniform potential infiltration rate holds over the entire range of P up to a maximum value at which the entire field is ponded and generating runoff.

Figure 6.11 Illustrating the response of the average field infiltration rate $\bar{I}$ to the rainfall rate P for a field of non-uniform potential infiltration rate.

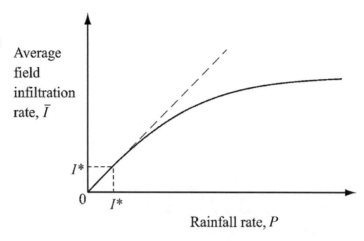

Thus it follows that, for a plot of non-uniform infiltration character-istics, the infiltration rate at any point in the field is no longer the same as the average value for the entire field, $\bar{I}$. Furthermore, $\bar{I}$ will depend on P in a positive way over the entire range of P, not just in the limited region of Fig. 6.9 where $P < I^*$.

Also in a field of non-uniform potential infiltration rate, there can be a period of general acceptance of rainfall across the field prior to the occurrence of any ponding. During this period the infiltration rate is determined by the rainfall rate, so that $I = P = \bar{I}$. However, unlike the response in Fig. 6.9 for a field of uniform potential infiltration rate, when this rate is non-uniform the field-average infiltration rate $\bar{I}$ will follow changes in P over the whole range of P. This type of response is illustrated in Fig. 6.11, where $\bar{I} = P$ up to $P = I^*$, after which $\bar{I}$ falls away from P as it rises, the reason being that increasing fractions of the field become ponded, so reaching their potential infiltration rate.

That overland flow is generated only from the fraction of the plot which experiences ponding is an example of what is commonly referred to in the hydrological literature as the 'fractional-area-runoff' concept (Kirby, 1978). This concept recognises that, even at modest plot scales, and most certainly at watershed scales, not all the land surface typically generates runoff, but only some fraction of it does.

At the field scale, this positive response of the overall infiltra-tion rate to an increase in rainfall rate foreshadowed in the conceptual Fig. 6.10 is very commonly found in practice, Fig. 6.12 providing an example of this. Note that there is a period of about 30 minutes for which $\bar{I} = P$, since no runoff occurred. Thereafter changes in $\bar{I}$ and P are closely synchronised.

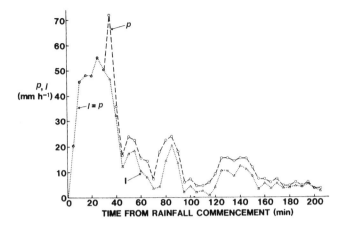

Figure 6.12 The variation with time both in rainfall rate P and in spatial-average infiltration rate $\bar{I}$ for a cultivated 2-ha field in the Darling Downs, Queensland, Australia. (After Rose (1985).)

A case study of infiltration measurement at the plot scale will be given in the following section, illustrating how the measurement data can be interpreted using the concepts of spatially variable infiltration rate outlined previously in this section. A simple mathematical model that allows prediction of the responsive link between infiltration and rainfall rates which is to be expected if the infiltration rate is spatially variable will be developed later.

6.4 A case study of infiltration rate at the plot scale with infiltration-excess overland flow

The effective infiltration rate of a field or plot cannot be measured in the direct manner with a ring infiltrometer, which is suitable only for use over a small area of order $1\,\mathrm{m}^2$ (Fig. 6.3). At the scale of a runoff plot, the infiltration rate has to be inferred from measurement of the rate of rainfall received by, and the rate of runoff from, the plot. Such measurements can be made conveniently for a hydrologically defined plot as illustrated in Fig. 4.4 and discussed in Section 4.2.

The case study now to be described is part of an analysis of field data collected under natural rainfall at Goomboorian in south-east Queensland, Australia, between 1992 and 1995. The runoff plots, located on a pineapple farm, were of length 36 m, area $108\,\mathrm{m}^2$ and slope 5%. The data presented are for a plot maintained in a bare fallow state (a reference or control treatment with which the hydrological and erosion-reduction effects of other management methods could be compared).

The soil type at the site was a loamy sand, thought to be so erodible that it should not be cultivated. However, as part of a multi-country

Figure 6.13 Indicating the lag time between peaks in the rainfall record (called a 'hyetograph') and the corresponding peaks in the runoff-rate record (called a 'hydrograph').

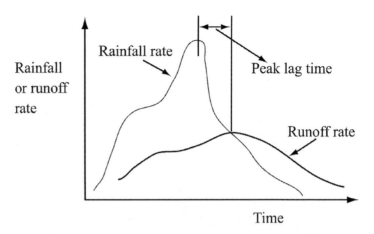

research project on sustainable agriculture in tropical and subtropical environments (Coughlan and Rose, 1997), land-management methods that led to very low rates of soil erosion, even under cultivation, were developed. The tipping-bucket technology used to measure rates of rainfall and runoff was as shown in Figs. 4.3 and 4.5, respectively. Both rates were recorded at one-minute intervals with an electronic data recorder during rainfall events. The standard error possible in rate measurement was about $1\,\mathrm{mm\,h^{-1}}$.

For a field at any site, a certain amount of rain needs to fall before any measurable runoff occurs. This issue will be discussed further in Chapter 7. In considering infiltration in the rest of this chapter, it will be assumed that this threshold amount of rain has already fallen.

To determine the infiltration rate for the runoff plot as a whole using measurements of rainfall and runoff rate, it must be recognised that it takes time for excess rainfall generated anywhere on the plot to reach the runoff measuring site located at the bottom of the plot (Fig. 4.4). For this reason, a burst of rainfall, indicated by a peak in the rainfall-rate record, is not picked up until some time later as a peak in runoff rate. The period of time between the related peaks in rainfall and runoff is commonly called the 'peak lag time' because the runoff response lags behind the rainfall whose rate is driving the rainfall–runoff system. This lag time is illustrated in Fig. 6.13.

There are other ways of expressing hydrological lag other than the time between peaks shown in Fig. 6.13; however, this method is most useful when such peaks are obvious.

For the bare fallow plot at the case-study site, the lag time between such peaks was less than 10 minutes, though the lag time between the two recorded peaks decreased somewhat as the magnitude of peak rainfall increased (Yu *et al.*, 2000). Thus, for this site, within 10 minutes the

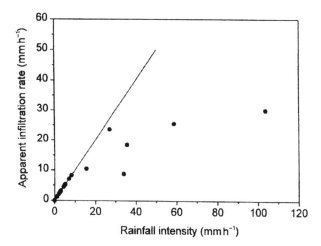

Figure 6.14 The apparent infiltration rate (I_a) as a function of the rainfall rate during a thunderstorm on 21 November 1992 at Goomboorian, Queensland, Australia. Both I_a and P are averaged over 10-minute periods, long enough for most water to traverse the 36-m-long plot at the experimental site. The straight line shows the constraint of the apparent infiltration rate equalling the rainfall rate. (After Yu *et al.* (1997).)

peak effect of a rainfall burst on the plot had reached the lower boundary of the plot. Whilst this does not mean that all rainfall delivered during the burst has reached the bottom of the plot, it does imply that, if the 1-minute data are averaged over a period of 10 minutes, then the major influence of the hydrological lag illustrated in Fig. 6.13 will be removed. The advantage of such time averaging is that the average runoff rate over this lag period will be a good approximation to the excess rainfall rate, R, defined in Eq. (6.3). Making this good approximation that, for appropriately time-averaged quantities, $R = Q$, and denoting the resulting approximate infiltration rate by I_a, it follows from Eq. (6.3) that

$$I_a = P - Q \qquad (6.4)$$

The term I_a in Eq. (6.4) is commonly called the 'apparent infiltration rate'. If all fluxes are steady, then I_a is exactly equal to the infiltration rate, since then $Q = R$ exactly. However, even with natural rainfall of varying rate, provided that P and Q are time averaged over a period greater than the hydraulic lag, I_a is a good approximation to the infiltration rate for the entire plot, $\bar{I}$. Without such time averaging, I_a is a very poor indicator of the infiltration rate I because of the substantial difference between R and Q caused by hydraulic lag.

For the case-study site, Fig. 6.14 shows that, with 10-minute averaging, the apparent infiltration rate, I_a, is strongly and positively dependent on the rainfall rate (or intensity).

Yu *et al.* (1997) reported extensive field data from a variety of countries consistent with the type of relationship for the time-averaged apparent infiltration rate shown in Fig. 6.14. This apparent infiltration rate based on time-averaged rates of P and Q is a very good approximation from spatially averaged infiltration rates for the plots as a whole, or $\overline{I}$. The type of relationship shown in Fig. 6.14 is characteristic of field data from a wide variety of sources and countries, indicating that it is a quite general form of relationship between $\overline{I}$ and P. As discussed in Section 6.3 and illustrated in Fig. 6.11, such a relationship is expected if there is spatial variation in infiltration rate across the plot.

The challenge addressed in the following Section 6.5 is that of developing a general model of infiltration that has these characteristics found in field-scale data and, where it is possible to monitor rainfall and runoff rates, always appears to be of the type illustrated in Fig. 6.14. Interpreting the hydrological behaviour of large watersheds, where the spatial variation of rainfall rate over the watershed can be important, presents other challenges not addressed in this book.

6.5 A general model of spatially variable infiltration at the field-plot scale

As argued in the previous section, when values of runoff and rainfall rates Q and P are averaged over a time period greater than the hydrological time lag for runoff from a plot, the apparent infiltration rate, I_a, provides a very good indication of the spatially averaged rainfall rate $\overline{I}$. Thus the general form of the relationship between I_a and P illustrated in Fig. 6.14 also represents the general form of the relationship between $\overline{I}$ and P. Notice that this is of the same form as that expected on conceptual grounds for a non-uniform soil and illustrated in Fig. 6.11. The challenge now is to represent this commonly found form of relationship by a convenient mathematical expression that defines the parameter involved in defining its shape. This is a simple example of a common technique used in mathematical modelling.

The general shape of the curve that would provide an approximate fit to the experimental points plotted in Fig. 6.14 is one that increases rapidly at first, but whose slope continues to decrease as the rainfall rate or intensity increases. A relationship of this general form can be developed by considering relationships that occur quite often in the environmental sciences and elsewhere. For example, a population of reproducing organisms not suffering death by predation can increase with time in the manner shown in Fig. 6.15(a). However, if the same initial population experiences severe predation, so that population numbers decline, this decline is usually of the form shown in Fig. 6.15(b). Mathematical functions that have these general shapes are shown in Figs. 6.15(c) and (d).

The time abscissa in Figs. 6.15(a) and (b) has been replaced in Figs. 6.15(c) and (d) by the general variable x. The increasing function

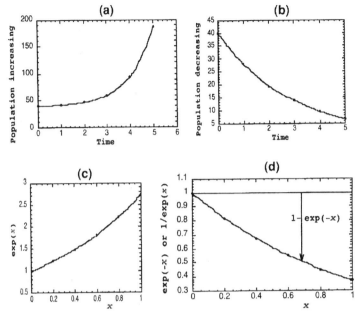

Figure 6.15 Parts (a) and (b) show possible forms of increasing and decreasing populations of organisms with time. Parts (c) and (d) are mathematical functions of similar form to (a) and (b), respectively. Note that $\exp(x)$ can alternatively be written e^x. Also $\exp(-x)$ can be written e^{-x}, and is simply $1/\exp(x)$. In (d) the function $1 - \exp(-x)$, which increases with x, is shown by the downward-pointing arrow.

in Fig. 6.15(c) is called an 'exponential function'. This function consists of a number, e, which has the value 2.718, raised to a power x, and so is written e^x, or alternatively and equivalently as $\exp(x)$. Familiarity with the form of this function can be obtained using your calculator or computer, and seeing how the function $\exp(x)$ increases as x is increased. Note that anything raised to the power zero has the value unity (see Figs. 6.15(c) and (d)). The decreasing function in Fig. 6.15(d) is $1/e^x$, which is written e^{-x}, or alternatively as $\exp(-x)$, and is often called 'the inverse exponential function'.

We haven't yet obtained a mathematical expression with a shape similar to the data points in Fig. 6.14, but this can be obtained with one further step from Fig. 6.15(d). The downward-pointing arrow in Fig. 6.15(d) gives the value of $1 - \exp(-x)$, which can be seen to increase with x, rapidly at first and then more slowly, as shown in Fig. 6.16(a). Note that this figure does have the general form shown by the data in Fig. 6.14. Furthermore, Fig. 6.16(a) can be made directly applicable to the data in Fig. 6.14 if the function $1 - \exp(-x)$ is replaced by $\overline{I}$, and the value of unity is replaced by I_m, the maximum possible value of $\overline{I}$ (Fig. 6.16(b)). This maximum value, I_m, is gradually approached as $\overline{I}$ increases. (Clearly there must be some upper limit to $\overline{I}$ for any plot.) These changes to Fig. 6.16(a) are shown in Fig. 6.16(b).

The further adaptation of Fig. 6.16(a) made in Fig. 6.16(b) is to replace the general mathematical variable x, which has no physical

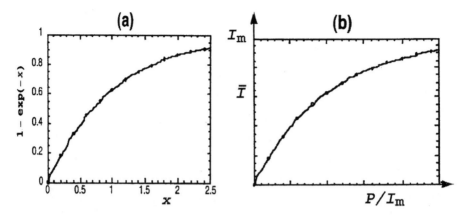

Figure 6.16 (a) The variation with x of the mathematical function $1 - \exp(-x)$. (b) Using the same mathematical form as in (a) to represent the variation with normalised rainfall rate, P/I_m, of the spatial average infiltration rate, $\bar{I}$.

dimensions (see Section 1.6), by the physically relevant rainfall rate, P, made non-dimensional by division with I_m, a constant for any particular plot or field. Thus the abscissa is now P/I_m (Fig. 6.16(b)).

It follows from Figs. 6.16(a) and (b) that the mathematical expression which well represents the experimental data on the spatially averaged infiltration rate, $\bar{I}$, is

$$\bar{I} = I_m[1 - \exp(-P/I_m)] \tag{6.5}$$

Example 6.2

To be able to use the infiltration equation (6.5) predictively, the single parameter I_m must first be determined. Using the data given in Fig. 6.14, determine an appropriate value for I_m, thus allowing prediction of the infiltration rate for any rate of rainfall at this site.

Solution

Fitting a curve of the form given by Eq. (6.5) to experimental data such as those given in Fig. 6.14 involves choosing an optimum value of I_m on some basis. Examining by eye the curve fit provided by the equation for a range of chosen values of I_m can lead to useful results. However, a more scientifically based and reproducible result can be obtained by evaluating the optimum value of I_m using a computer-implemented non-linear curve-fitting procedure that minimises the error in the fit in an agreed manner. Yu *et al.* (1997) used such a curve-fitting procedure to show that the data in Fig. 6.14 could be fitted well using $I_m = 50 \, \text{mm h}^{-1}$, with the result shown in Fig. 6.17.

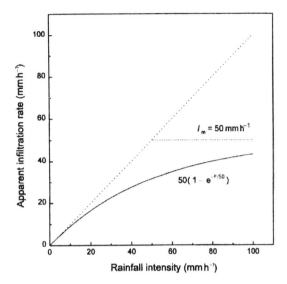

Figure 6.17 The relationship between the spatial mean infiltration rate $\bar{I}$ and the rainfall rate or intensity P fitted to the data of Fig. 6.13 using the exponential form of dependence given by Eq. (6.5). The parameter I_m is selected as $50\,\mathrm{mm\,h^{-1}}$. (After Yu *et al.* (1997).)

Denoting the spatially averaged value of the excess rainfall rate by $\bar{R}$, it follows from Eq. (6.3) that

$$\bar{R} = P - \bar{I} \tag{6.6}$$

In Eq. (6.6) it is assumed that P is a spatially averaged value for the runoff plot, which is a reasonable assumption provided that P is measured at a site close to the plot. From Eqs. (6.5) and (6.6), it follows that

$$\bar{R} = P - I_m[1 - \exp(-P/I_m)] \tag{6.7}$$

Equation (6.7) provides a mathematical model of the generation of excess rainfall rate for a plot or area of land where infiltration is spatially variable and where the form of that variability is described by the function given in Eq. (6.5).

From Eq. (6.5), since $\bar{I}$ cannot exceed I_m, the parameter I_m can be interpreted as the *maximum* possible value of the spatial *mean* infiltration rate $\bar{I}$ for the complete plot area. This maximum possible value would occur when the entire field is saturated or experiencing ponding, and runoff is being generated from the entire area of the plot, not just from a partial area of the plot, which commonly would be the case.

Although it follows from Eq. (6.5) that $\bar{I}$ will always be less than I_m, it is helpful to note that I_m is *not* the maximum value of I at any point on the plot. When $\bar{I} = I_m$, it can be shown that the probability of I for any point on the plot exceeding I_m is $e^{-1} = 0.37$, or 37%.

Commonly there would be some areas of a plot where no excess rainfall is produced, namely the local value of excess rainfall rate is $R = 0$.

Figure 6.18 The function
e^{-I/I_m} is a mathematical
description of the relative
likelihood of any particular
value of I.

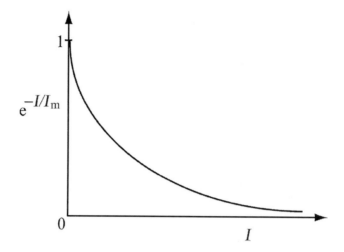

Recognising the dynamic nature of rainfall, it is possible for water to be
temporarily stored on the land surface even if $P < I$, the local value of
the infiltration rate. For such a restricted time period, $R = P - I$ will be
negative, leading to a decline in the amount of water stored temporarily
on the surface until, at its disappearance, $R = 0$. Owing to the transient
nature of such storage, it is of adequate accuracy to assume that $R = 0$
if $P < I$, and $R = P - I$ if $P > I$.

The implications of Eqs. (6.5) and (6.7) can be illustrated graphically
as in Fig. 6.18. It can be shown (Yu *et al.*, 1997) that Eqs. (6.5) and (6.7)
are consistent with the relationship between e^{-I/I_m} and I shown in
Fig. 6.18. The higher the rate of infiltration, I, the less frequent its occurrence, and the form of the function e^{-I/I_m} in Fig. 6.18 actually gives a
description of the relative frequency or likelihood of any particular value
of I.

Figure 6.19 shows a particular value of the naturally time-varying
rainfall rate P superimposed on the abscissa of Fig. 6.18. It can be shown
that, for the particular plotted value of P, the lighter-toned area under the
curve is the infiltration rate, and the upper darker-toned area is the excess
rainfall rate. Imagine the point P sliding to the right or left horizontally,
as would occur when the rainfall rate varies during an event. Then, using
Fig. 6.19, note how the two differently toned areas vary in size in response
to this movement. This size variation indicates how the infiltration rate
and excess rainfall rate vary in response to changes in P.

In Fig. 6.19 the length A_1 above the function can be shown to represent the fraction of the plot area generating excess rainfall for that
particular value of P. The complementary length A_2 represents the fractional area with no excess rainfall because P is less than the possible

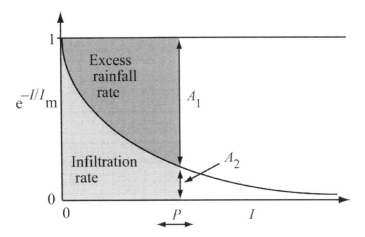

Figure 6.19 Showing the relative frequency distribution of the non-dimensional infiltration ratio I/I_m, as a function of the general infiltration-rate variable, I. At a particular rainfall rate, P, the lighter area represents the spatially averaged infiltration rate $\overline{I}$, and the darker area represents the excess rainfall rate, $\overline{R}$. The sum of both shaded areas represents the input rainfall rate, P. Thus each differently shaded area indicates the relative partition of rainfall into infiltration or excess rainfall (the latter resulting in runoff from the plot).

infiltration rate. Again, note how A_1 and A_2 vary in magnitude as P varies. At low rainfall rates it might be only a small fraction of the plot that generates excess rainfall, whereas at high rates most of the plot could be generating excess rainfall. What the simple model does not describe is where within the area of the plot excess rainfall is being generated.

It can be shown that the lengths A_1 and A_2 in Fig. 6.19 are given by

$$A_1 = 1 - e^{-P/I_m} \tag{6.8}$$

and

$$A_2 = e^{-P/I_m} \tag{6.9}$$

so that, as required, the sum of the two component length fractions, A_1 and A_2, is unity (indicating the whole plot area).

That excess rainfall is generated from some fraction of the plot area is commonly described as the 'partial-area' concept of runoff generation. In some situations those areas generating runoff may be fully saturated with water, as can commonly occur at the lower section (or 'toe') of a hill slope. In other cases, commonly when rainfall is intense and soil not well protected, it is limitations in infiltration rate that lead to ponding and runoff generation. The infiltration rate of the soil surface can be reduced by breakdown of the structure of the soil surface by rainfall impact, leading to infilling of soil pores with solids, especially if rainfall follows mechanical cultivation of a type that seriously damages soil structure. When the infiltration rate is greatly constrained by such characteristics of the surface layer, this is often referred to as a 'surface seal'.

Whilst it is vitally important to be able to describe the infiltration rate in a manner that recognises the dominant role of spatial variability in determining the potential rate, other challenges remain to be resolved in order to develop a model allowing the runoff rate to be predicted from knowledge of the rainfall rate. These challenges will be addressed in the following Chapter 7.

Main symbols for Chapter 6

I Infiltration rate
$\bar{I}$ Spatial average value of I
I_m Maximum possible value of the spatial mean infiltration rate for a field
P Rainfall rate
Q Volumetric runoff rate per unit area $(= P - I)$
R Excess rainfall rate (Eq. (6.3))

θ Volumetric water content
Σ Summation symbol

Exercises

6.1 In the ring-infiltrometer investigation described in Example 6.1, a second infiltrometer ring was located a few metres away where the approximately steady infiltration rate ultimately achieved was $3 \, mm \, h^{-1}$. Since the soil type was the same for the two infiltrometer rings, assume that the value of b determined in Example 6.1, namely $880 \, mm^2 \, h^{-1}$, also applied to the second infiltrometer site. With these values of I_c and b, use Eq. (6.1) to calculate I as a function of ΣI. At the commencement of infiltration, when $\Sigma I = 0$, I is indicated by Eq. (6.1) to be infinitely large. Thus, carry out your calculations commencing with an initial value of $\Sigma I = 10 \, mm$. Continue the calculation of I with an adequate number of assumed increasing values of ΣI to define the shape of the curve obtained when I is plotted as ordinate against ΣI as abscissa.

6.2 The figure in Example 4.2 in Section 4.2 of Chapter 4 gives the rainfall rate as a function of time in histogram form for seven consecutive 5-minute periods. Count the number of these periods when the

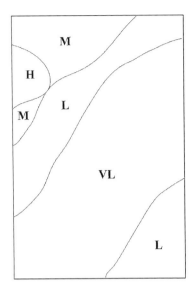

Figure 6.20 Illustrating a possible spatial variation in infiltration rate across a field plot, the data being grouped into four infiltration classes.

H - high

M - medium

L - low

VL - very low

rainfall rate was ≥ 5 mm h^{-1}, ≥ 10 mm h^{-1}, etc. up to ≥ 35 mm h^{-1}. Plot the number of such periods as ordinate against the magnitude class for rainfall rate (i.e. ≥ 5, ≥ 10, ..., ≥ 35 mm h^{-1}), which gives a frequency diagram in histogram form. Then divide each number or frequency by the highest frequency to yield a relative frequency, with 1 as the highest relative frequency. Then plot each relative frequency against the same rainfall-rate magnitude class as in the previous figure, now regarding this abscissa as the rainfall rate, P.

Finally, seek to fit a curve to this relative-frequency plot using the mathematical form e^{-kP} or exp($-kP$). This form is similar to Fig. 6.15(d). Using a trial-and-error approach, vary the parameter k to obtain what appears to be the best fit between the relationship and the data. (Alternatively a computer-based curve-fitting program can be used.)

6.3 Figure 6.20 is a plan view of a runoff plot showing how the infiltration rate (as measured by a ring infiltrometer) varies spatially over the plot. This spatial variation is represented in Fig. 6.20 by areas indicating four infiltration-rate classes from high (H) to very low (VL). Draw a regular grid over Fig. 6.20 and record the number of occurrences when grid intersections lie over each of the four classes of infiltration rate shown. Plot these data on the number of grid intersections per infiltration class as ordinate against the mean infiltration rate of the four classes, assuming that the arbitrary range of each infiltration class is as in the following table.

Class descriptor	Infiltration-rate range $(\text{mm}\,\text{h}^{-1})$	Mean infiltration rate $(\text{mm}\,\text{h}^{-1})$
H	12–24	18
M	6–12	9
L	3–6	4.5
VL	0–3	1.5

As in Exercise 6.2, divide the frequency of counted intersections in each infiltration-rate class by the highest frequency, thus producing relative-frequency data. Plot these relative-frequency data against the mean infiltration rate $\overline{I}$ for each class given in the table. Then fit these data with the mathematical form $e^{-\overline{I}/I_m}$ or $\exp(-\overline{I}/I_m)$, varying the value of the parameter I_m until a good fit to the data appears to be achieved. (Note that the parameter I_m has the meaning of the maximum value of the spatial mean infiltration rate, as discussed in Section 6.5.)

6.4 Draw a diagram of the form given in Fig. 6.19, given the following information. When $I = I_m$, the function plotted will have the value e^{-1} or 0.37; assume that this value occurs when $I = I_m = 50\,\text{mm}\,\text{h}^{-1}$. As in Fig. 6.19, place a pen or pencil vertically (or parallel to the ordinate) on this figure at a particular value of the abscissa, to indicate a particular value of the rainfall rate, P. Then move the pen left and right to simulate fluctuation in rainfall rate and note how the areas indicating the infiltration rate and excess rainfall rate in Fig. 6.19 vary as P varies. For several chosen values of P, measure the lengths denoted by A_1 and A_2 in Fig. 6.19 (recognising that $A_1 + A_2 = 1$). Check that, for each chosen value of P, the value of $A_1 = 1 - e^{-P/I_m}$, and the value of $A_2 = e^{-P/I_m}$, as given in Eqs. (6.8) and (6.9). Discuss the physical meaning of A_1 and A_2.

6.5 A method very widely used to estimate runoff from total rain falling on small-to-medium-sized watersheds is that developed by the US Soil Conservation Service (USDA Soil Conservation Service, 1986). It is widely referred to as the SCS method, and commonly thought to be without theoretical foundation, but Yu (1998) has shown the method to be compatible with the assumption of an infiltration model described by Eq. (6.5). Since symbols P and Q have been used to describe rates, the symbols ΣP and ΣQ will be used to indicate total amounts for a rainfall event. Let T denote the duration of runoff in an event, with I_m having the meaning ascribed to it in this chapter. F_o is the threshold amount of rain which needs to fall before runoff commences. Then, in terms of these symbols, the SCS method uses the following relation for ΣQ:

$$\Sigma Q = \frac{(\Sigma P - F_0)^2}{\Sigma P - F_0 + I_m T} \tag{6.10}$$

where in this equation we have taken the liberty of interpreting as $I_m T$ the term in the SCS equation which is described as the 'maximum retention'.

Assuming that $\Sigma P = 65\,\mathrm{mm}$, $F_0 = 9\,\mathrm{mm}$, $I_m = 29\,\mathrm{mm\,h^{-1}}$ and $T = 0.6\,\mathrm{h}$, calculate ΣQ.

References and bibliography

Clothier, B. E. (1988). Measurement of soil physical properties in the field: a commentary. In *Flow and Transport in the Natural Environment: Advances and Applications*, eds. W. L. Steffan and O. T. Denmead. Heidelberg: Springer-Verlag.

Clothier, B. E., and Heiler, T. D. (1983). Infiltration during sprinkler irrigation: theory and field results. In *Advances in Infiltration*. Chicago, Illinois: American Society of Agricultural Engineers, pp. 275–284.

Coughlan, K. J., and Rose, C. W. (eds.) (1997). A new soil conservation methodology and applications to cropping systems in tropical steeplands. Technical Report 40. Canberra, ACT: Australian Centre for International Agricultural Research.

Hillel, D. (1998). *Environmental Soil Physics*. New York: Academic Press.

Kirby, M. J. (ed.) (1978). *Hillslope Hydrology*. New York: Wiley.

Marshall, T. J., Holmes, J. W., and Rose, C. W. (1996). *Soil Physics*, 3rd edn. Cambridge: Cambridge University Press.

Neilsen, D. R., Biggar, J. W., and Ehr, K. T. (1973). Spatial variability of field-measured soil-water properties. *Hilgardia* **42**, 215–259.

Rose, C. W. (1985). Developments in soil erosion and deposition models. *Adv. Soil Sci.* **2**, 1–63.

USDA Soil Conservation Service (1986). Urban hydrology for small watersheds. Technical release No. 55. Washington: USDA.

Yu, B., Rose, C. W., Coughlan, K. J., and Fentie, B. (1997). Plot-scale rainfall–runoff characteristics and modelling at six sites in Australia and Southeast Asia. *Trans. Am. Soc. Agric. Eng.* **40**, 1295–1303.

Yu, B., Rose, C. W., Ciesiolka, C. C. A., and Cakurs, U. (2000). The relationship between runoff rate and lag time and the effects of surface treatments at the plot scale. *Hydrol. Sci.* **45**, 709–726.

Yu, B. (1998). Theoretical justification of the SCS method for runoff estimation. *J. Irrigation Drainage Eng.* **124**, 306–310.

7
Overland flow on watersheds

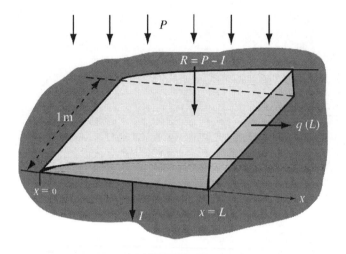

7.1 Introduction

There are important interactions between surface and subsurface hydrology, as indicated by Fig. 4.7 and considered further in Chapters 9 and 10. Also, as illustrated in Fig. 4.7, there are multiple pathways by which water received by a watershed can be dispersed within it or be lost from it. Some of these pathways are discussed in Chapter 9. Thus it may be regarded as somewhat artificial, even though it is useful, to consider overland flow on watersheds separately, as is done in this chapter. However, the reason for first considering the characteristics and dynamics of overland flow on watersheds in this text is that it is this hydrological component which is of particular concern to environmental land managers. Overland flow commonly dominates the dislodgement and delivery to rivers of sediment and chemicals, which may either be associated with the eroding sediment, or alternatively come from land disposal of contaminants. Overland flow is thus a major contributor to land degradation through erosion by water and also to the degradation of stream-water quality through the delivery of sediment and chemicals to streams.

As discussed in Section 6.3, overland flow can arise for a number of different reasons. In humid regions of the world a common source of overland flow occurs where precipitation falls on an already saturated part of a watershed, as illustrated in Fig. 6.8. In a humid climate rainfall commonly exceeds evapotranspiration for a significant part of the year. Hence the groundwater, shown in Fig. 4.1, is often close to the soil surface, and can also intersect it to form saturated zones (Fig. 6.8). No infiltration can take place in such saturated zones, and, following rainfall, saturation overland flow occurs as described in Section 6.3. It is also possible in land with a changing surface slope for water to emerge from a saturated zone, a process sometimes called 'exfiltration' (in contrast to infiltration).

If there is no infiltration into or exfiltration from a saturated zone, then, from Eq. (6.3), the excess rainfall rate (R) will be equal to the rainfall rate (P), or $R = P$. However, at a watershed scale, as shown in Fig. 6.8, there will generally be parts of a watershed that are saturated and other parts that are not saturated, where infiltration can occur.

In more arid, semi-arid, or sub-humid climates the intersection of groundwater with the soil surface is much less common than it is in humid climatic zones, though it can occur. For example, even in the Sahara Desert there are oases where the water table is at or near the soil surface. In drier regions of the world the most common cause of overland flow is that the infiltration rate of the surface, though not zero, is exceeded by the rainfall rate, in which case R is less than P (Eq. (6.3)). In Section 6.3, such overland flow generated when $I < P$ was called infiltration-excess (or Hortonian) overland flow. Especially in strongly arid regions of the world, sections of the landscape where soils have extremely low infiltration rates can develop, this being illustrated in Fig. 6.7. Thus, both in humid regions and in arid regions, parts of a watershed can exhibit zero or extremely small infiltration rates, though for very different reasons. In either case, other parts of the watershed can have significant infiltration rates.

Many environmental issues involve hydrology at the watershed or catchment scale. Local, regional and over-bank river flooding can all cause great damage and human suffering through loss of property, infrastructure and crops and the interruption to support services. The transport of waterborne contaminants and the erosion and transport of sediment and associated chemicals involve surface hydrology, but also involve processes that will receive more attention in later chapters. Downslope net deposition of sediment exaggerates the damage caused by water alone. Such hydrologically driven sediment flows are a common feature of natural environmental processes, and the formation of extensive and productive alluvial flat lands is one important and beneficial outcome of such processes. Occasional flooding of rivers has also been shown to be beneficial, and is sometimes crucial to the well-being or even survival of natural populations of flora and fauna, especially fish. However, human-implemented changes in land use and land characteristics, and the construction of dams and weirs, have all substantially altered the nature and frequency of flooding and its consequences.

Despite significant improvements having been made in some countries in recent times, current methods of industrial production, mining and other human activities usually result in the release of pollutants with the potential to contaminate land, lakes, rivers and the sea well away from the site of contaminant generation. The extensive distribution of such contaminants commonly is due to hydrological processes; hence the discussion of such processes in this and later chapters. Such environmental issues are now much better understood and recognised, and serious efforts are being made by some industries to mitigate such undesirable problems associated with their activities. Nevertheless, news of serious contamination is almost a daily feature in many, if not most, countries, reports ranging from contaminant release by highly technical industry to the uncontrolled spread of raw sewage.

Some of these consequences of hydrological processes in watersheds will be considered in the next three chapters, Chapters 8–10. Meanwhile, in this chapter, some quantitative understanding of surface hydrological processes at the field and watershed scales will be sought.

7.2 Runoff-model components

As mentioned in Section 6.4, for any given area of land under any particular condition of land cover and antecedent moisture content, a certain amount of rainfall must reach the land surface before runoff from the area commences. This amount or depth of rainfall is denoted F_o.

In Section 6.5 it was shown that, at the field scale, a mathematical model of the infiltration process was able to predict the infiltration rate adequately under any rainfall rate, provided that the one parameter in this model could be determined. The aim of this chapter is to build on Chapter 6 and develop the capacity to predict the runoff rate from the rainfall rate, which can be readily measured. It is the runoff rate which is the driving variable, not only for hydrological processes, but also for other related processes of environmental interest, such as the overland transport of sediment and chemicals considered in later chapters. Whilst meteorological science has developed some capacity to predict the rainfall rate, in this book the rainfall rate will be assumed to be a measured quantity.

In general, three components are needed in a model relating the rate of runoff from a watershed to the rainfall it receives. These three components are the following.

1. A value for the threshold amount of rainfall required to generate runoff, F_o.

2. A sub-model describing the infiltration rate at the field scale. We will use Eq. (6.5) since it appears to be more realistic at that scale than are infiltration equations derived using one-dimensional theory, such as Eq. (6.1).
3. A way of describing the time lag due to the fact that overland flow takes time to move over the land surface. This time lag between the generation of excess rainfall on a field or a watershed and its arrival as runoff at some down-slope or downstream position is illustrated in Fig. 6.13 in Section 6.4. Representation of this hydrological lag provides the third part of a complete rainfall–runoff model, and is commonly called 'runoff routing'. The development of a runoff-routing procedure will be given in Section 7.6.

The magnitude of the hydrological lag, and hence the importance of the role of runoff routing, increases with the scale of the overland flow. The lag increases from a number of minutes for a small plot, which may be a fraction of a hectare in area, to hours for a river reach and to days or weeks for a large watershed. It is this lag time that fortunately often allows an early warning of the timing and severity of downstream flooding to be given. For flow distances of the order of thousands of kilometres over very flat land, lag times can extend to months. In addition to spatial scale, the resistance offered to overland flow also has a significant effect on the hydrological lag time.

At the scale of an agricultural plot or small watershed, the effect of hydrological lag is quite discernible, but is not dominant over the effects of initial rainfall retention (F_o) and the subsequent effect of the plot's infiltration characteristics. Data collected at such a plot scale will be analysed in Section 7.3. Initially the effect of hydrological lag will be neglected in the analysis; later it will be included, allowing the effect of lag to be seen. Following a consideration of factors affecting overland flow in Section 7.4, the dynamics of overland flow will be considered in Section 7.5. This provides the basis for a model of hydrological lag or runoff routing that will be developed in Section 7.6. Then this model will be used to reanalyse the data set considered in Section 7.3, but now the effect of hydrological lag will be included and the improvement in prediction evaluated.

7.3 A small-scale runoff model neglecting hydrological lag

Neglecting hydrological lag, which is listed in Section 7.2 as the third part of the complete rainfall–runoff model, implies that hydrological lag is assumed to be negligible. It also implies, using terms defined in Section 6.4, that the runoff rate Q is equal to the excess rainfall rate R, where, by definition, $R = P - I$. In Example 7.1, which follows, the plot scale is quite small, so that the error introduced by assuming that $Q = R$ should not be great.

Analysing the same data set, first ignoring lag in Example 7.1, and then later in Example 7.6 including it, will allow evaluation of the effectiveness of using the initial absorption and infiltration model components alone in this small-plot context. This two-step procedure will also reveal the improvement in model performance obtained by including the effect of lag, even with small-scale data.

Example 7.1

The data used in this example are from a commercial pineapple farm different from that which sourced the case study of Section 6.4. The data are from a pineapple farm at Imbil, south-east Queensland, Australia. Rainfall and runoff rates were measured at 1-minute time intervals using the tipping-bucket technology shown in Figs. 4.3 and 4.5, with data-logger recording. Experimental details are given in Ciesiolka $et\ al.$ (1995a, b). These experiments were part of a multi-country study involving a number of south-east Asian countries and Australia described in Rose (1995). Runoff was from a bare soil plot of length 12.2 m and slope approximately 33%. Values of the two model parameters (F_o and the infiltration model parameter I_m in Eq. (6.5)) were obtained by optimising the fit of the model to the data set. This process yielded $F_o = 8.2$ mm and $I_m = 29.2$ mm h^{-1}.

The first column (denoted i) in Table 7.1 indicates that there was 13 minutes of rainfall prior to the commencement of runoff at $i = 14$ min. The second column of Table 7.1 gives the measured rainfall rate P, and the last column shows the measured runoff rate, Q_{meas}, both being 1-minute data. The third column gives $\overline{I}$ calculated using Eq. (6.5). Q is then calculated, and denoted Q_{calc} in Table 7.1, by assuming that $Q \equiv Q_{calc} = P - \overline{I}$, an equation that would be accurate only if $R = Q$, implying that there is no hydraulic lag. Because of the small plot length (12.2 m) and substantial slope (33%), the neglect of hydraulic lag has restricted the error between Q_{calc} and Q_{meas}, though this error is still quite noticeable, as is apparent from Table 7.1 and its graphical presentation in Fig. 7.1.

It can be seen from Fig. 7.1 that, despite the neglect of lag effects, there is a degree of similarity between the measured and calculated values of Q. That this agreement is reasonably good is entirely due to the limited scale of the plot from which these data were derived. With a substantially longer slope length than 12 m, the discrepancy between Q_{meas} and Q_{calc} obtained by assuming Q_{calc} to be equal to R (or $P - \overline{I}$) could become enormous.

Table 7.1 *A comparison of runoff rate Q_{calc}, calculated by neglecting hydrological lag, and with Q measured (Q_{meas}) (data are from Walker's pineapple farm, Imbil, Queensland, Australia; column i denotes minutes from commencement of rainfall)*

i (min)	P (mm h^{-1})	$\bar{I}$ (mm h^{-1})	Q_{calc} (mm h^{-1})	Q_{meas} (mm h^{-1})
14	47.10	23.38	23.72	4.430
15	123.0	28.77	94.23	33.23
16	135.5	28.92	107.5	58.91
17	147.9	29.02	118.9	101.3
18	110.6	28.54	82.02	87.11
19	135.5	28.92	107.5	87.62
20	123.0	28.77	94.23	87.61
21	135.5	28.92	107.5	101.8
22	110.6	28.54	82.02	87.61
23	98.15	28.19	69.96	73.33
24	85.76	27.65	58.11	59.92
25	85.76	27.65	58.11	60.42
26	48.76	23.70	25.06	60.42
27	24.27	17.48	7.788	35.22
28	48.76	23.70	25.06	37.21
29	48.76	23.70	25.06	35.22
30	12.10	9.906	2.194	22.45
31	12.10	9.906	2.194	37.70
32	12.10	9.906	2.194	11.53
33	6.050	5.464	0.5856	23.93
34	6.050	5.464	0.5856	11.78
35	24.27	17.48	7.788	11.78
36	6.050	5.464	0.5856	11.04
37	6.050	5.464	0.5856	11.78
38	3.020	2.869	0.1509	11.78
39	3.020	2.869	0.1509	11.90
40	3.020	2.869	0.1509	5.700
41	3.020	2.869	0.1509	5.700
42	0.1700	0.1695	0.0005	11.90
43	0.1700	0.1695	0.0005	5.820
44	0.1700	0.1695	0.0005	5.820
45	0.1700	0.1695	0.0005	5.820
46	0.1700	0.1695	0.0005	5.820
47	0.1700	0.1695	0.0005	3.840
48	0.1700	0.1695	0.0005	3.840
49	0.1700	0.1695	0.0005	3.840

The number of figures given in the spreadsheet output is clearly physically ridiculous, and it is good practice to reduce this number to give meaningful values. The accuracy of measurement is probably no better than 0.1 or even 1 mm h^{-1}.

$\bar{I}$ is calculated using Eq. (6.5) with $I_m = 29.2$ mm h^{-1}. A computer spreadsheet program is most convenient for such calculations.

Because of the neglect of lag effects, it is assumed that $Q = R = P - \bar{I}$.

Figure 7.1 A comparison of Q_{calc} and Q_{meas} from Table 7.1 for the duration of the runoff event on a pineapple farm at Imbil, Queensland, Australia. Measured values of Q are joined by the continuous line; Q_{calc} is shown by the dashed line.

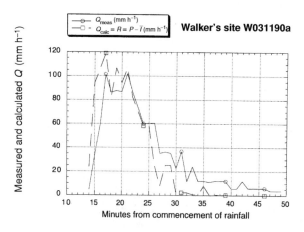

Figure 7.2 The time variation in rainfall rate P (upper solid line) and calculated mean infiltration rate $\bar{I}$ (lower dashed line).

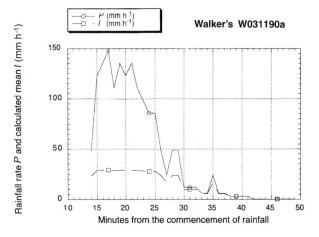

The effect on Fig. 7.1 of the inclusion of hydrological lag in the rainfall–runoff model will be investigated in Section 7.6.

Figure 7.2 shows that the relationship between the spatial mean infiltration rate, $\bar{I}$, calculated using Eq. (6.5), and the rainfall rate. The responsiveness of $\bar{I}$ to P, which is a consequence of Eq. (6.5), can be seen in Fig. 7.2.

It can be seen from Fig. 7.2 that, in addition to $\bar{I}$ responding to the value of P, $\bar{I}$ also tends to decrease with time, and hence also with the cumulated infiltration, denoted ΣI in Chapter 6. Hence the manner of variation of $\bar{I}$ is not obviously in disagreement with that predicted by the Green–Ampt type of relationship given in Eq. (6.1). Yu *et al.* (1999) compared the ability of Eqs. (6.1) and (6.5) to interpret data from six experimental sites in tropical and subtropical regions of Australia and south-east Asia. They found that Eq. (6.5), which is based on the spatial variability of I, provided a more adequate interpretation of the data than did Eq. (6.1).

These comments illustrate the general point that it is often the case that one can say only that one model of a complex system is more efficient than another. Even less efficient models can still reflect some of the observed behaviour of interest, which is the case with the Green–Ampt equation.

Especially in a water-supply context, there is considerable interest in being able to predict the total runoff, ΣQ, which can be expected from a watershed receiving a total amount of rainfall, ΣP. For small-to-medium-sized watersheds, a method developed by the US Soil Conservation Service (SCS) (USDA Soil Conservation Service, 1986) is widely used for this purpose. Yu (1998) has shown that this SCS method of total-runoff prediction, as it is commonly referred to, is compatible with calculation of the spatial average infiltration rate, $\bar{I}$, using Eq. (6.5). If the further assumption that rainfall rates are exponentially distributed is made, then the SCS equation is found to follow analytically from Eq. (6.5). There is general experimental support for this assumption that rainfall rates are distributed in a negative exponential manner, in a form similar to that shown in Fig. 6.15(d). (This support is consistent with common experience that high rainfall rates are less common than lower rates.)

The SCS method for predicting the total runoff or 'catchment yield' of water is widely used. Boughton (1989) and others have described limitations of the SCS method. Though choice based on experience, or preferably on experimental data from the watershed of interest, is required in selecting a value for the parameter involved (called a curve number in the method's documentation), the method is commonly successful at a range of watershed scales. The method has been used with data on watersheds from 0.25 ha to 1000 km^2 in area, use being most common on small-to-medium-sized rural catchments (Boughton, 1989). This wide experience of use of the SCS method in a number of countries thus provides further support for the utility of the infiltration-rate equation (6.5), at least at such scales. Since only the total catchment yield, ΣQ, is predicted using the SCS method, rather than the runoff rate Q as a function of time, the issue of hydrological lag is not particularly relevant in its use.

In summary, as discussed in Chapter 6 and earlier in relation to the SCS catchment-runoff model, there is a wide range of experimental support from many countries for the utility of Eq. (6.5) for describing

Figure 7.3 Flow of water
over an impermeable surface
at average velocity $V\,(\mathrm{ms}^{-1})$.

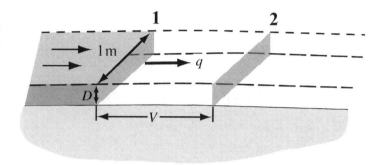

Figure 7.3 Flow of water over an impermeable surface at average velocity $V\,(\mathrm{ms}^{-1})$.

infiltration-rate characteristics at scales from plots to watersheds of modest size. Future research no doubt will be able to define better the limits of application of Eq. (6.5). There is certainly the possibility of improvement, though such improvement is likely to require a more parameter-intensive model. Thus, in addition to its demonstrated utility at field scales, other reasons why Eq. (6.5) is used in this book to describe infiltration-rate behaviour are its mathematical simplicity and parameter frugality.

As shown in Chapter 9, many processes of environmental significance at the watershed scale depend on, or are driven by, the rate of runoff, Q, rather than just depending on its accumulated value, ΣQ. In order to interpret dynamic data and predict Q, hydraulic lag becomes increasingly important with enlargement in the scale of the watershed considered. To develop theory representing the effect of hydraulic lag, we must first consider the hydraulics of overland flow of water and the dynamics involved in the generation of rainfall excess and its downslope transport. These topics will be developed in the Sections 7.4 and 7.5 which follow.

7.4 Overland flow of water

Volumetric flux of water

Suppose that water is flowing over an impermeable road surface after rainfall has ceased. The flow could be quite extensive, but consider a section of the flow which is 1 m wide measured normal to the flow (Fig. 7.3). Since V is the velocity of flow, the cross-section of water shown initially at position 1 in Fig. 7.3 will be at position 2 just one second later, having travelled a distance V metres during that second. If the flow is uniform and of depth D, then the volumetric flux of water across the unit-width cross-section, q, also called the 'unit flux', is given by the volume of water between the two sections 1 and 2 of Fig. 7.3. Thus it follows that

$$q = DV \quad (\text{m}^3 \, \text{m}^{-1} \, \text{s}^{-1} \text{ or } \text{m}^2 \, \text{s}^{-1}) \tag{7.1}$$

While it was convenient in deriving Eq. (7.1) for water to be assumed to be flowing over an impermeable surface (Fig. 7.3), the result given by Eq. (7.1) in no way depends on this assumption, and this equation holds quite generally.

Resistance to overland flow

The reason why water flows downhill over a sloping land surface follows from Bernoulli's equation, Eq. (3.16). For overland flow the static pressure, p, at the air/water surface is the pressure exerted by the atmosphere. The static pressure increases linearly with depth beneath the surface as described by Eq. (3.2). Thus, for a uniform flow of water over a plane surface, there is no gradient in static pressure in the direction of flow, so it is not this term in Bernoulli's equation that causes downhill flow to occur. Rather, the reason for downslope flow is the decrease in the potential-energy term $\rho g z$, where the height z above some datum is referred to as the 'gravitational head'. The potential energy released to the flow from a downslope decrease in z can either be converted into an increase in the kinetic-energy term $\frac{1}{2} \rho V^2$, or be used in overcoming the resistance to flow resulting in energy loss described by the term E_{loss} in Eq. (3.17).

Thus the velocity of flow achieved depends on two factors:

- the downslope rate of decrease in potential energy of the flowing water, and thus on factors such as land slope and depth of water; and
- the resistance experienced by the flow.

The roughnesses of most surfaces, and the Reynolds numbers involved, are such that environmental flows are often turbulent or near turbulent in character. If so, the frictional resistance offered by the surface is transmitted up from the surface into the flow by the turbulent eddies described in Chapter 3. Let us now consider the forces acting on an elementary volume of flowing water of unit width (Fig. 7.4).

The downslope force, F_d, driving the motion of water is the component of the weight of the volume element which acts in the downslope direction (Fig. 7.4). The magnitude of $F_d = W \sin \alpha$, where W is the weight of water in the elementary volume of length Δx (Fig. 7.4) and α is the land slope. Thus

$$F_d = \rho g D \sin \alpha \cdot \Delta x \tag{7.2}$$

Figure 7.4 The downslope force, F_d, acting on the elementary volume of flow of unit width and length Δx.

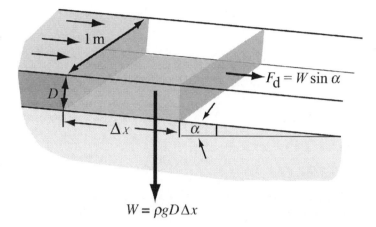

$$W = \rho g D \Delta x$$

Figure 7.5 The upslope force F_u acting on the elementary flow volume due to the shear stress, τ, between the flow and the surface.

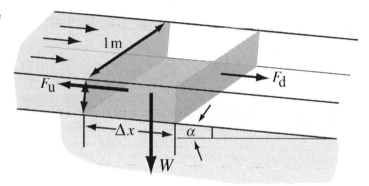

The resistance force which acts upslope, F_u, is due to the shear stress, τ, acting between the flow in the volume element and the land surface over which it is flowing, as shown in Fig. 7.5. From Fig. 7.5,

$$F_u = \tau \, \Delta x \tag{7.3}$$

If there is a difference in magnitude between F_u and F_d then the volume of water will accelerate (if $F_d > F_u$), or decelerate (if $F_d < F_u$). There is a wide range of situations in environmental flows in which the net force is quite small, so that it is a good approximation to adopt the assumption made in what is referred to as 'normal flow', which is that $F_d = F_u$. Thus, for normal flow,

$$\tau = \rho g D \sin \alpha$$

or

$$\tau = \rho g D S \tag{7.4}$$

where the slope $\sin \alpha$ is denoted by S.

Since shear stress τ is of the nature of a drag force as described in Section 3.6, a French engineer, Chezy (1718–1798), assumed that

$$F_u = C_d \cdot \Delta x \cdot \tfrac{1}{2}\rho V^2 \tag{7.5}$$

using the wetted area, Δx, as the effective area involved.

Example 7.2

When the shear stress between flowing water and loose sand grains exceeds a critical value, τ_{cr}, which depends on the diameter d of the grains, then the sand grains begin to be rolled along the bed of sand. For example, if $d = 1$ mm, $\tau_{cr} = 0.5$ Pa approximately. If sea water of density $1010\,\mathrm{kg\,m^{-3}}$ is flowing 'normally' over a bed of sand of grain size 1 mm and slope $S = 0.005$, calculate the depth of flow at which the sand grains will commence to move.

Solution
From Eq. (7.4),

$$D = \tau/(\rho g S)$$

Since

$$\tau = \tau_{cr} = 0.5\,\mathrm{Pa} \quad (\text{or } \mathrm{N\,m^{-2}})$$

then

$$D = 0.5/(1010 \times 9.8 \times 0.005)$$
$$= 1.01 \times 10^{-2}\,\mathrm{m}$$

or about 1 cm.

Equating the expressions for F_u (Eq. (7.5)) and F_d (Eq. (7.2)) and solving for the flow velocity V gives

$$V = (2g/C_d)^{1/2} S^{1/2} D^{1/2} \tag{7.6}$$

With the term $(2g/C_d)^{1/2}$ written as C, Eq. (7.6) is called 'Chezy's equation'. Thus 'Chezy's C' is inversely proportional to the drag coefficient involved, C_d.

An Irish engineer, Manning (1816–1897), carried out an extensive series of hydrological experiments and found that the best fit to his data was given by a relationship very similar to, but not exactly the same as, that of Chezy. Manning found that a power on the depth D of $\tfrac{2}{3}$ gave a better fit to his data than the value of $\tfrac{1}{2}$ used by Chezy. He also introduced a resistance term, called 'Manning's n', which is directly related to the drag coefficient and so inversely related to Chezy's conductivity term C. This gives what is now called 'Manning's equation' for the flow velocity:

$$V = (S^{1/2}/n)D^{2/3} \tag{7.7}$$

This equation will be used where relevant in the remainder of this book. The magnitude of Manning's resistance term, n, is not constant, but typical values for various overland, stream and river flows are given in texts such as Chow (1959) and Dingman (1984). From Eq. (7.7), n can be given units of $m^{-1/3}$ s. For flow over well-inundated unvegetated surfaces, values of n from 0.02 to 0.06 $m^{-1/3}$ s are commonly given (V and D also being in SI units). Whilst units are not always given for values of Manning's n, a numerical factor is required in Eq. (7.7) if SI units are not employed; this is done so that the values of n are the same regardless of the system of units employed.

From Manning's equation, Eq. (7.7), it follows that the velocity of overland flow depends on three different kinds of factors.

- The factor $S^{1/2}$ in Eq. (7.7) reflects the topography of the landscape. In general S will vary with downslope distance, and multi-dimensionally, but the contribution from slope will still be proportional to $S^{1/2}$.
- Manning's roughness coefficient, n, is in the denominator of Eq. (7.7). This indicates that the rougher the land surface over which the water is flowing (indicated by higher values of n), the slower the flow will be.
- The depth of overland flow will increase with downslope distance in general, as the volume of rainfall causing runoff accumulates. Thus the factor $D^{2/3}$ in Eq. (7.7) is effectively a landscape scale factor, indicating that the flow velocity V will increase in proportion to $D^{2/3}$.

Examples of flow velocity achieved are given in Example 7.3.

Example 7.3

Assume that overland flow has commenced at a position on a hill of uniform slope where the downslope vector x will be taken to have its origin (i.e. $x = 0$, see the figure). The slope of the land is $S = \sin\alpha = 0.05$ (or 5%). Assume further that a steady state of overland flow has been achieved, so that there is no change with time in the volume of water stored on the plane between $x = 0$ and $x = L$ (see the figure). Hence mass conservation of water requires that the unit flux $q = RL$, where R is the steady rate of rainfall excess (30 mm h^{-1}) and L is the downslope length at which q is measured.

Steady excess rainfall
$$R = 30 \text{ mm h}^{-1}$$

Calculate the velocity of overland flow, V, for $L = 10$ m and 100 m, assuming a constant Manning's n of 0.03 m$^{-1/3}$ s.

Solution

From Eq. (7.1), unit flux $q = RL = DV$. Thus $D = RL/V$, so, from Manning's equation, Eq. (7.7),

$$V = (S^{1/2}/n)(RL/V)^{2/3}$$

so

$$V = (S^{1/2}/n)^{3/5}(RL^{2/3})^{3/5}$$

or

$$V = (S^{3/10}/n^{3/5})(RL)^{2/5}$$

On substituting the given numerical values into this expression for V,

$$V = 0.078 \text{ m s}^{-1} \quad \text{at } L = 10 \text{ m}$$

and

$$V = 0.196 \text{ m s}^{-1} \quad \text{at } L = 100 \text{ m}$$

Though these calculated velocities of flow are substantial, clearly it will take quite some time for excess rainfall to reach the bottom of the hillslope measuring site at $x = L$. The mean time between excess rainfall being generated and its arrival at the measuring site lower down the hillslope is the hydraulic lag discussed in Section 6.4 and illustrated in Fig. 6.13 using the rainfall rate P rather than the excess-rainfall rate $R = P - I$.

For relatively shallow flows over hydraulically rough surfaces (such as soil or gravel), the value of Manning's n is found to decrease as discharge and flow depth increase. This is expected since, as the flow depth increases, surface roughness would become less effective at retarding flow. However, as discharge increases, eroding surfaces can develop irregular bed forms, which increase the value of Manning's n. Whilst Manning's equation had its origins in dealing with turbulent flows, it is commonly used even in low-flow situations. Especially if flow is so shallow that it does not overtop vegetation or soil-roughness elements, or if the surface over which flow occurs has irregularities of significant size compared with the flow depth, then estimation of the appropriate value of Manning's n on the basis of handbook values could be considerably in error, so it should be determined experimentally. When flows are shallow, which is not uncommon in situations of environmental relevance, it is difficult to measure the flow depth D accurately. In this context it is more accurate to determine Manning's n by measuring V (e.g. by dye tracing) and q. Then, from Eqs. (7.7) and (7.1), Manning's n can be calculated from

$$n = (S^{1/2}q^{2/3})/V^{5/3} \qquad (7.8)$$

Manning's equation is widely used, for example in the design of channels to carry water in urban water supplies, in irrigation, etc. Example 7.4, given later, will illustrate how it is used in the design of structures to carry excess runoff water away safely in order to prevent soil erosion from occurring during rainfall events when significant runoff is generated.

The ability of overland flow to cause soil erosion is commonly related to the rate of working of the shear stress which acts between the flowing water and the surface over which it is flowing. From Eq. (3.8), the work done, W, is given by the product of the force F and distance moved, s (i.e. Fs). Now consider F to be the force exerted by overland flow acting over an area of surface A; then the work done per unit area by this force is

$$W/A = Fs/A$$

where $F/A = \tau$, the shear stress between the land surface and its overland flow. Thus $W/A = \tau s$. Then dividing this equation by the time t during which work W is done yields

$$W/(At) = \tau(s/t)$$

Since s/t is the velocity of flow, V, this equation can be written as

$$\Omega = \tau V \quad (\text{W m}^{-2}) \qquad (7.9)$$

where Ω is the rate of work done per unit bed area, or the rate of working of the shear stress, and is referred to as the 'stream power', or, more fully, the stream power per unit bed area. In physics, the rate at which work is done is defined as 'power', measured in watts (W). Thus the SI unit of Ω is W m^{-2}, as shown in Eq. (7.9).

Application of the concept of stream power and Manning's equation is illustrated in the following example.

Example 7.4

A 'grassed waterway' is a soil-conservation structure designed to conduct excess rainfall safely downslope to a local stream. In designing such a waterway on a particular land-holder's property, it is estimated that the maximum volumetric flow rate which the waterway will be required to conduct is 3 cubic metres per second (i.e. 3 m^3 s^{-1} or 3 cumecs). The slope of the waterway is 7% (i.e. $\sin\theta = 0.07$, where θ is the angle of slope of the land to the horizontal).

To minimise the danger of erosion of soil from the waterway, it is required that the stream power (given by the product τV as in Eq. (7.9)) should not exceed 500 W m^{-2}.

Calculate the maximum depth of water flow which satisfies the erosion criterion limiting the maximum allowable stream power to 500 W m^{-2}.

Using this maximum allowable design flow depth and the corresponding flow velocity, also calculate the width, W, of grassed waterway required to deliver safely the design maximum expected volumetric flow of water of 3 m^3 s^{-1}.

Assume that the density of flowing water is 1000 kg m^{-3}, and take $g = 9.8$ m s^{-2}. Manning's equation can be assumed, taking the value of Manning's roughness coefficient, n, to be 0.05 m$^{-1/3}$ s for the waterway.

Solution

$$\Omega = \tau V$$

which, from Eqs. (7.4) and (7.7), can be written

$$\Omega = (\rho/n)gS^{1.5}D^{1.667}$$

For $\Omega = 500$ W m^{-2},

$$500 = (1000/0.05) \times 9.8 \times 0.07^{1.5} \times D^{1.667}$$

So

$$D = 0.304\,\text{m}$$

which is the maximum allowable depth of water if the limit on stream power set by the danger of erosion is not to be exceeded.

The total volumetric flow rate down a waterway of width W is given by DVW, and this must allow a flow rate of $3\,\mathrm{m^3\,s^{-1}}$, with V given by Eq. (7.7).

Thus the necessary width is

$$W = 3/(DV) = 3n/(DS^{1/2}D^{2/3})$$

giving

$$W = 4.1\,\mathrm{m}$$

A grassed waterway of this width should carry the maximum design flow rate without eroding.

In order to develop a model for predicting the hydraulic lag between the time-varying runoff rate and rates of rainfall (and excess rainfall) responsible for it, we need to consider the dynamics of overland flow further. For simplicity we focus discussion on a plane hillslope of constant roughness and slope. An understanding for this simple system allows the implications for more complex landscapes to be appreciated, at least in qualitative terms.

7.5 The dynamics of overland flow

The commencement of overland flow on land of uniform slope and hydraulic roughness will be considered to occur at $x = 0$, the vector x being in the downslope direction. The volumetric flux per unit slope width, or unit flux, q, is measured at the distance downslope given by $x = L$, and is denoted $q(L)$ in Fig. 7.6. The rate of runoff per unit area of flow, Q, is defined by

$$Q = q(L)/L \tag{7.10}$$

The principle of mass conservation applied to any system enclosed by boundaries implies that any mass entering the system across its boundaries must either leave the system across other boundaries, or else lead to an increase in mass within the system. Let us apply this general principle to the system in Fig. 7.6 consisting of a 1-m wide wedge of water, commencing at $x = 0$ and with a measurement boundary at $x = L$. The fixed lower boundary to this system is the soil surface, and the upper boundary is the upper surface of the overland flow (Fig. 7.6). This upper boundary can rise if the rate of net input of water, given by RL, exceeds the rate of overland flow at the exit, given by $q(L)$ or QL (Eq. (7.10)). Alternatively,

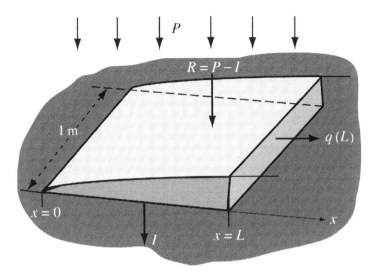

Figure 7.6 A 1-m wide downslope segment of overland flow on a plane hillside, with commencement of flow denoted by $x = 0$. With effective or spatially averaged rates of rainfall, P, and infiltration, I, the excess-rainfall rate $R = P - I$ leads to a build up of flowing water on the plane land surface whose unit flux, $q(L)$, is measured at $x = L$.

the upper boundary surface to the overland flow will decline with time if $RL < QL$, or $R < Q$.

Denote the volume of water stored on the plane per unit area between $x = 0$ and $x = L$ by S_t (Fig. 7.6). The value of S_t will depend on the average depth of water on the plane. The average rate of increase of stored water volume resulting from a rate of increase in depth of water is given by $\Delta S_t / \Delta t$, where ΔS_t is the increase in stored water volume per unit area during time interval Δt. Applying the mass-conservation principle to the overland-flow system depicted in Fig. 7.6 requires that

$$RL = QL + L\Delta S_t / \Delta t$$

or, dividing by L,

$$R = Q + \Delta S_t / \Delta t \qquad (7.11)$$

Consider what is likely to happen during the earlier periods of a runoff event. From Example 7.3, the velocity of flow, V, at $L = 100\,\mathrm{m}$ for the conditions described in this example was $0.196\,\mathrm{m\,s^{-1}}$. Using the expression for V given in the solution for this example, the velocity halfway down the plane, at $L = 50\,\mathrm{m}$, is $0.148\,\mathrm{m\,s^{-1}}$. Thus the lag time, described as the average time for water to travel over the plot to the exit, would be of order $100/0.148 = 674\,\mathrm{s}$ or about 11 min. In consequence of this time lag, during the early stage of build up of water depth on a plane land surface it would be expected that the rate of increase of storage ($\Delta S_t / \Delta t$ in Eq. (7.11)) would be considerably greater than Q. It follows from Eq. (7.11) that R would be comparably greater than Q.

Figure 7.7 If the rainfall rate,
P, rises as shown, then for
early times the variation in P, R
and Q can be as indicated.

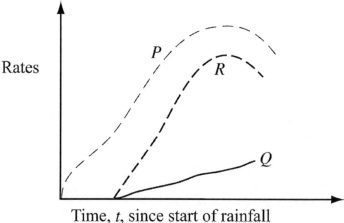

Figure 7.7 If the rainfall rate, P, rises as shown, then for early times the variation in P, R and Q can be as indicated.

Figure 7.6 depicts the spatial variation in a segment of overland flow at a particular time. For early times, with both P and R increasing, Fig. 7.7 illustrates the temporal variation in P, R and Q. Runoff does not commence until some time after rainfall commences, and the amount of rainfall received during this delay period was denoted F_0 in Section 7.1. As shown in Fig. 7.7, both R and Q commence at essentially the same time, but, because the rate of increase in water storage on the surface is considerably greater than Q, R is correspondingly greater than Q (Eq. (7.11)).

The water storage per unit area, S_t, is given by the depth of water in the 1-m wide flow of Fig. 7.6. Clearly S_t varies with x, its maximum value corresponding to the depth of water at $x = L$, namely $D(L)$. From Fig. 7.6, the average value of S_t for the overland-flow volume between $x = 0$ and $x = L$ will be some fraction of $D(L)$. However, as will be shown later, the general form of dependence of S_t on relevant factors can be developed very directly by making the simplifying approximation that $S_t = D(L)$. For simplicity we will now write $D(L)$ as D, and, remembering that

$$q = DV \quad \text{(Eq. (7.1))}$$
$$= QL \quad \text{(Eq. (7.10))}$$

then, from the approximation noted,

$$S_t = LQ/V$$

From Manning's equation, Eq. (7.7), $V = KD^{2/3}$, where $K = S^{1/2}/n$. Hence

$$S_t = LQ/(K D^{2/3}) \tag{7.12}$$

From Eqs. (7.1) and (7.10), and also noting that $q = KD^{5/3}$, it follows that

$$D^{2/3} = (QL/K)^{2/5}$$

and so from Eq. (7.12),

$$S_t = (LQ/K)^{3/5} \tag{7.13}$$

Equation (7.13) has been derived assuming that $S_t = D(L)$, a constant independent of x (though varying with time). Rose et al. (1983) developed a general approximate analytical solution to the basic equation describing overland flow on a plane land surface, recognising that D varies with x as well as with t. This solution, also using Manning's equation, showed that

$$S_t = \tfrac{5}{8}(LQ/K)^{3/5} \tag{7.14}$$

Notice that the only difference between Eqs. (7.13) and (7.14) is the numerical factor $\tfrac{5}{8}$, a factor expected to be <1 since obviously the average depth of water on the surface is $<D(L)$. Thus, using the approximate solution of Rose et al. (1983),

$$S_t = K_1 Q^{3/5} \tag{7.15}$$

where

$$K_1 = \tfrac{5}{8}(L/K)^{3/5} \tag{7.16}$$

Although Eq. (7.11) was derived assuming that overland flow took place on a plane uniform land surface (as in Fig. 7.6), it has been used widely in engineering hydrology to analyse and predict the dynamics of overland flow across natural watersheds with naturally complex topography and surface coverage. In this wider context of application, some of the assumptions or relationships used in deriving Eq. (7.14) would not be entirely appropriate, and modification of some of the parameters or powers involved may allow a better description of the hydrological behaviour of natural watersheds. In analysing watershed hydrographs, a generalised form of Eq. (7.15) has received considerable investigation, this form being

$$S_t = K_1 Q^m \tag{7.17}$$

where the power $\tfrac{3}{5}$ has been generalised by the parameter m. In Eqs. (7.16) and (7.17) K_1 is conceived to be a function of catchment properties or characteristics. Whilst Eq. (7.16) for K_1 might not be appropriate in detail, it shows that scale factors, such as L, and also slope and hydraulic roughness characteristics will affect the value of K_1. Consensus is growing that, in order to describe flood routing in natural watersheds, the appropriate value of m, the power on Q in Eq. (7.17), may be close to 0.8 (Pilgrim, 1987).

This adaptation of the simple physically based model expressed in Eqs. (7.15) and (7.16) for use in flood prediction in natural watersheds illustrates the general experience that, provided that a model captures the interaction of the salient or dominant processes, even if not exactly, the model can often be most usefully adapted to contexts more complex than can be represented in any detail in a simple physically based model.

Since in hydrological engineering practice the value of the parameter m in Eq. (7.17) is often taken to be 0.8, it is not an unreasonable approximation to take $m = 1$. This approximation provides simplification of the theory, and so will be used in the remainder of this chapter. In engineering practice K_1 is also regarded as a parameter to be determined from field experimental data, and the value so derived will depend somewhat on the value chosen for m. (Such interaction between parameter values is a common issue in mathematical modelling.) Adopting $m - 1$, Eq. (7.17) becomes

$$S_t = K_1 Q \tag{7.18}$$

Substituting for S_t from Eq. (7.18) into the mass-conservation equation (7.11) yields

$$R = Q + K_1 \Delta Q / \Delta t \tag{7.19}$$

where ΔQ is the increase in Q which occurs over the time interval Δt.

At early times in a runoff event Q will increase with time somewhat as illustrated in Fig. 7.7. For all this early part of the time history of Q in Fig. 7.7 a step increase Δt in time will lead to a step increase ΔQ in Q. Thus the ratio $\Delta Q / \Delta t$ is positive, and this ratio is called the 'slope' of the relationship between the two variables. It follows that, in this context, the term $K_1 \Delta Q / \Delta t$ in Eq. (7.19) will also be positive, and so from this equation $R > Q$ as shown in Fig. 7.7.

As time proceeds during the runoff event under consideration, the time will come when P (and hence R) will decline. A simplified sketch of a possible time history of R and Q for a complete rainfall–runoff event is shown in Fig. 7.8, which is an extension in time of Fig. 7.7.

From Figs. 7.7 and 7.8 it can be seen that there is an earlier period during which R quite substantially exceeds Q, the maximum value of R being at point B in Fig. 7.8. Following Eq. (7.19), the two component terms which added together form R are both shown in Fig. 7.8. These two components are Q (measured directly as a function of time) and $K_1 \Delta Q / \Delta t$. Thus, if K_1 is known, then R can be calculated from Q and the slope, $\Delta Q / \Delta t$, of the relation between Q and t (Fig. 7.8).

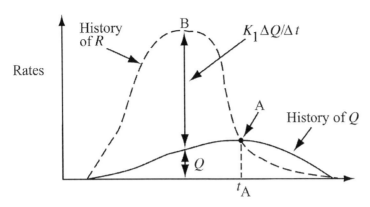

Figure 7.8 The time history of R is calculated from the history of Q using Eq. (7.19). The schematic time variation in Q reaches a maximum (at A) at time t_A.

Time, t, since start of rainfall

Figure 7.8 shows that, in response to the hydraulic lag discussed earlier, the maximum in Q (at A in Fig. 7.8) occurs well after the maximum in R (and that in P). Indeed, from Eq. (7.19), R will equal Q when $\Delta Q/\Delta t = 0$, which defines the maximum in Q versus time, at A in Fig. 7.8. Thus the variations with time of R and Q are predicted to intersect at the maximum of Q (i.e. at A in Fig. 7.8, at time t_A.)

For times beyond t_A, Q decreases with time, t, so the slope $\Delta Q/\Delta t$ is now negative. Hence, from Eq. (7.19), R is now less than Q by the magnitude of $K_1 \Delta Q/\Delta t$ (Fig. 7.8).

From Eq. (7.15) it follows that the magnitude of K_1 determines how strongly the storage S_t depends on Q. It can also be shown that it is the magnitude of K_1 which determines the duration of the hydraulic time lag. Noting that $K = S^{1/2}/n$, it follows from Eq. (7.16) that K_1 is proportional to $(Ln)^{3/5}/S^{3/10}$. Thus lag will increase with the scale of flow and the hydraulic roughness experienced by the flow. Also, the smaller the slope of the land, the longer the lag. These are the reasons why, for low-slope rivers in large watersheds, and especially with overtopped stream banks where water has to flow through vegetation, the flood peak downstream in the watershed may arrive many hours or days after the main rainfall event has ceased or eased. From Eq. (7.16) this time lag clearly increases with the spatial scale of the watershed.

In a large natural watershed, factors such as slope length, slope and roughness will clearly vary spatially in complex ways. Thus, although Eq. (7.16) for K_1, which is based on flow over a plane land surface, would be unable to predict the hydraulic lag accurately in a complex watershed, the form of the equation provides a most useful conceptual guide to the factors involved and the manner of their effects on lag. In hydrological practice, where a watershed hydrograph is available, the magnitude of

K_1 is determined by a process of fitting Eq. (7.18) (or Eq. (7.17)) to the hydrograph data in association with Eq. (7.19) (or the appropriate modification of Eq. (7.19) if Eq. (7.17) is used for S_t).

There are several of widely used computer programs based on the theory discussed earlier, which greatly assist in determining the best values of the parameters involved, by using on some optimising principle to assess the best fit (see e.g. Laurenson and Mein (1988) and Boyd and Bufill (1989)). Such programs are commonly used for flood prediction in hydrological engineering practice.

Alternative methods for determination of field infiltration

As illustrated in Fig. 7.8, and discussed in association with Eq. (7.19) on which this figure is based, R can be calculated from measured values of Q, both being functions of time. This ability provides the basis of an alternative methodology for determining the time variation in effective infiltration rate at the field scale since, from Eq. (6.3),

$$I = P - R \tag{7.20}$$

Equation (7.20) shows that, if R is calculated from Q as described, and the rainfall rate P is measured, then I can immediately be calculated. This procedure was used to calculate I for a 2-hectare field with the results shown in Fig. 6.12.

There is a basic assumption in such analysis. This assumption is that all flow measured at the exit from the field arrived there by overland flow rather than by flow beneath the soil surface, which has emerged as exfiltration from the soil surface prior to measurement. It is possible in some circumstances for a flow-measurement structure to encourage such exfiltration or emergence of previously subsurface flow to the surface. Exfiltration is also encouraged in situations in which there is a substantial decrease in land slope, or where there is a sharp decrease in depth to subsoil. Figure 6.8 illustrates a saturation region low in the watershed, another situation in which exfiltration is possible.

Exfiltration was most unlikely in the field from which the results given in Fig. 6.12 were obtained. This figure illustrates a common feature of infiltration rate measured at the field scale, namely that time variations in I closely follow those in P. It is this strong dependence of I on P which is represented in Eq. (6.5), where I was written as $\overline{I}$ to emphasise the spatially averaged nature of I.

A quite different methodology from that described here was used in Section 6.4 to determine I (or $\overline{I}$). This alternative methodology involved time averaging $P - Q$ over a period greater than the relevant hydraulic lag time. These two quite different types of approach to determining the spatially averaged infiltration rate at the field scale both indicate the strong dependence of infiltration rate on rainfall rate. The best-quality data which have provided extensive evidence of this strong interdependence of I and P have been for field scales of up to a few hectares. With large natural watersheds the complexity in topography and other factors makes for less-certain analysis.

Such a strong dependence of infiltration rate on rainfall rate is not explainable with small-area infiltration models, such as the Green–Ampt equation introduced in Section 6.2. However, such small-area models do provide an explanation of the tendency for the infiltration rate to decline with time during a rainfall event, though sometimes this decline is difficult to discern if there are strong fluctuations in P and thus in I.

General comments on mathematical modelling

Spatial variation in infiltration rate implies temporal variation in the proportion of land area contributing to runoff (Fig. 6.10). However likely this may be, it is worth adding a general warning based on extensive experience with the use of mathematical models. This warning is that, however successful a mathematical model of processes may be, the task of identifing model parameters evaluated by some form of model fitting to experimental data with relevant independently measurable characteristics is commonly daunting. Until this is done, it is desirable to hold a degree of scepticism rather than believing that the model is giving a direct and completely adequate description of the processes involved.

Also there remains the important issue of the scale of application of a particular type of model. Despite the encouragement referred to in Exercises 6.4 and 7.2, it remains to be more adequately investigated how well the model of field-scale average infiltration given in Eq. (6.5) operates at much greater scales than that of a few hectares or fractions of hectares which has so far been investigated in detail. That the approach is equally successful for data from a range of countries gives support to our belief in its generality of application at this modest scale range.

7.6 A dynamic runoff model recognising hydraulic lag

As discussed and illustrated in Section 7.2, the surface hydrological behaviour of relatively small-scale plots can be interpreted in approximate terms neglecting hydraulic lag. This is because at this small scale such lag is of limited duration, often being less than or comparable to the fluctuation periods of the rainfall rate. However, as scale increases, slope decreases and surface roughness increases, so the effect of hydraulic lag becomes increasingly dominant. As noted earlier, for large watersheds and flow over extensive flat lands, the lag between rainfall and runoff further downslope is an important major feature of the runoff event.

Thus we now turn more explicitly to the third component of a rainfall–runoff model listed in Section 7.2, the first two components being provided by the threshold rainfall required to initiate runoff, F_o, and the spatially variable infiltration model given in Eq. (6.5). The third, and last, component of a rainfall–runoff model is the translation of the excess rainfall rate, R, into the runoff rate Q. It is the value of Q which determines the flow rate at any position on a plane hillslope, since it can be shown (Rose et al., 1983) to be a good approximation that, at any distance x downslope from the commencement of runoff, the unit flux, q, is given by

$$q = Qx \qquad (7.21)$$

In the previous Section 7.5 we saw that it was possible to compute R from Q recorded through time for the runoff event, but that, to do this, the lag parameter K_1 has to be known. Equation (7.16) gave an analytical expression for K_1 for a plane land surface, making some approximations. However, as discussed in Section 7.5, in all but small watersheds of simple topography and uniform cover, the value of K_1 has to be determined by a process of fitting theoretically based forms to hydrographic data. This involves adjusting the values of parameters defined in these forms, including the lag parameter K_1, to obtain the best fit between the theoretical form and the experimentally obtained data.

What is now described in this section is one efficient and effective way in which the consequence of hydraulic lag can be evaluated quantitatively.

Because of the interaction between the two parameters K_1 and m in Eq. (7.17), and despite the simple Eq. (7.16) for K_1, some dependence of K_1 on Q can be found, indicating the possibility that the lag factor K_1 can vary with time. Nevertheless, at least for runoff from plots of areas up to $300 \, \text{m}^2$, Yu et al. (1997) have shown that K_1 can be taken as a constant during runoff events with little error.

For reasons of simplicity, the approximation that $m = 1$ in Eq. (7.17) will continue to be made. Thus the storage term S_t is linearly related to Q as in Eq. (7.18), namely $S_t = K_1 Q$.

We will now put Eq. (7.19) into a form more convenient for spreadsheet calculation, or any other form of computer modelling, as follows.

Let Q_i and R_i denote the average runoff rate and excess rainfall rate for some general time interval i. For example, if Q were measured every minute, then the subscript i in Q_i would increase at minute intervals. Then in the term $\Delta Q/\Delta t$ of Eq. (7.19) the finite difference ΔQ can be written as the difference $Q_i - Q_{i-1}$, where Q_{i-1} is the value of Q for the time interval prior to Q_i.

Using these forms of expression, Eq. (7.19) can be written

$$K_1(Q_i - Q_{i-1}) = (R_i - Q_i)\Delta t$$

From this it follows that

$$Q_i = \frac{K_1 Q_{i-1}}{K_1 + \Delta t} + \frac{\Delta t \cdot R_i}{K_1 + \Delta t}$$
$$= \alpha Q_{i-1} + (1 - \alpha)R_i \qquad (7.22)$$

where

$$\alpha = K_1/(K_1 + \Delta t) \qquad (7.23)$$

Note that α is related both to the lag K_1 and to Δt, and can be regarded as a dimensionless routing parameter whose value must lie between 0 and 1.

If data on the time variations in P and Q are collected, then the two parameters I_m (Eq. 6.5) and α (Eqs. (7.22) and (7.23)) can be determined by finding which combination of parameter values gives the best fit between predictions of Q based on the model given by these equations and measured values of Q. In addition to these two parameters I_m and α, the third parameter, F_o, is required in order to complete the model, F_o being the initial amount of totally infiltrating rainfall which occurs before runoff commences.

The criterion commonly used to judge which combination of model parameter values gives the best fit to the measured data is minimisation of the sum of the errors squared between the observed and modelled values of Q. Computer programs are available to automate this process, which is called 'parameter optimisation', and Yu et al. (1997) give details and references on how this can be done.

In Example 7.1 a set of hydrological data was used to investigate the consequences of using the very simple dynamic runoff model in which lag effects were ignored, it being assumed in effect that $K_1 = 0$, and hence that $Q = R$ (Eq. (7.19)). The results were given in Figs. 7.1 and 7.2

In Example 7.6, which follows, the same set of data will be used to examine what improvement in model performance results when the existence of a hydrological lag in routing excess rainfall to its point of runoff measurement is recognised by using Eq. (7.22) as the model structure, using the routing parameter α defined in Eq. (7.23).

Figure 7.9 A comparison of Q_{calc} (w. lag) (dashed line) and measured Q or Q_{meas} (solid line), for the duration of a runoff event on a pineapple farm at Imbil, Queensland, Australia. Data are from Table 7.2.

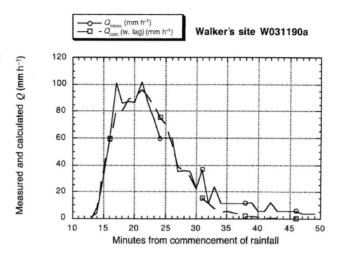

Example 7.6

Using the same set of data on P and Q (denoted Q_{meas}) as given in Table 7.1 of Example 7.1, use the hydrological model described earlier, involving Eqs. (6.5) and (7.22) and the three model parameters F_0, I_m and α, to made a revised calculation of Q (denoted Q_{calc}(w. lag)).

Optimisation of the three parameter values gave $F_0 = 8.2$ mm, and $I_m = 29.2$ mm h^{-1} as given in Example 7.1, but now lag effects will be included using $\alpha = 0.667$ (Eq. (7.23)).

Calculations using a spreadsheet program are given in Table 7.2. The results of Q_{calc}(w. lag) and Q_{meas} from Table 7.2 are compared graphically in Fig. 7.9.

On visually comparing the plot of calculated and measured Q in Fig. 7.1 (where lag effects were neglected) with that in Fig. 7.9 (with the effects of lag included), it can be seen that there is some improvement in agreement of the calculated with the measured value. The statistical measure of 'model efficiency' significantly improved from 0.74 for Fig. 7.1 to 0.95 for Fig. 7.9, the value of 1.0 for this efficiency measure indicating perfect agreement between the calculated and measured values. The effect of including lag effects would become much more pronounced for a longer plot than it is for the rather short plot length for which these data were collected (12.2 m). The plot length was short because of the danger of soil erosion on this steep (33%) slope.

Where the assumptions are appropriate, the rainfall–runoff model used in Example 7.6 can be used quite generally to predict the time history of Q given information on P, assuming that the model parameters F_0, I_m and α are known or can be assessed. It is the normal situation

Table 7.2 *A comparison of runoff rate, now calculated recognising hydrological lag,*
Q_{calc}(w. lag), *with measured Q(Q_{meas}) (data are from Walker's pineapple farm, Imbil,*
Queensland, Australia; column i denotes minutes from commencement of rainfall) for the
same data as in Table 7.1, where the calculation neglected lag effects

i (min)	P_i (mm h^{-1})	Mean I (mm h^{-1})	R_i (mm h^{-1})	Q_{calc}(w. lag) (mm h^{-1})	Q_{meas} (mm h^{-1})
13	135.46			0.00	0.40
14	47.10	23.38	23.72	7.90	4.43
15	123.00	28.77	94.23	36.65	33.23
16	135.46	28.92	106.54	59.92	58.91
17	147.95	29.02	118.93	79.57	101.28
18	110.56	28.54	82.02	80.39	86.11
19	135.46	28.92	106.54	89.10	87.62
20	123.00	28.77	94.23	90.81	86.61
21	135.46	28.92	106.54	96.05	101.78
22	110.56	28.54	82.02	91.38	86.61
23	98.15	28.19	69.96	84.25	73.33
24	85.76	27.65	58.11	75.54	59.92
25	85.76	27.65	58.11	69.74	60.42
26	48.76	23.70	25.06	54.86	60.42
27	24.27	16.48	7.79	39.18	35.22
28	48.76	23.70	25.06	34.48	36.21
29	48.76	23.70	25.06	31.34	35.22
30	12.10	9.91	2.19	21.64	22.45
31	12.10	9.91	2.19	15.16	36.70
32	12.10	9.91	2.19	10.84	11.53
33	6.05	5.46	0.59	7.43	23.93
34	6.05	5.46	0.59	5.15	11.78
35	24.27	16.48	7.79	6.03	11.78
36	6.05	5.46	0.59	4.22	11.04
37	6.05	5.46	0.59	3.01	11.78
38	3.02	2.87	0.15	2.06	11.78
39	3.02	2.87	0.15	1.42	11.90
40	3.02	2.87	0.15	1.00	5.70
41	3.02	2.87	0.15	0.72	5.70
42	0.17	0.17	0.00	0.48	11.90
43	0.17	0.17	0.00	0.32	5.82
44	0.17	0.17	0.00	0.21	5.82
45	0.17	0.17	0.00	0.14	5.82
46	0.17	0.17	0.00	0.09	5.82
47	0.17	0.17	0.00	0.06	3.84
48	0.17	0.17	0.00	0.04	3.84
49	0.17	0.17	0.00	0.03	3.84

$\overline{I}$ is calculated using Eq. (6.5) with $I_m = 29.2$ mm h^{-1}.

$R = P - \overline{I}$.

Since the lag or routing parameter α is 0.667 for this data set, Q is calculated, using Eq. (7.22), from

$$Q_i = 0.667 Q_{i-1} + 0.333 R_i$$

Because of the nature of these calculations (with the current calculated value of Q depending on the value of Q for the previous period), a computer program such as Excel^{TM} is convenient (though not essential) for calculations such as those in this table.

in environmental assessment that there are no suitable data at the particular site of interest to allow direct estimation of parameter values. If so, parameter values must be estimated on the basis of other surrogate measured or described characteristics deemed to bear some relation to the model parameters.

Even if appropriate data are available at a particular scale of measurement for estimation of parameters to use in a model at a particular site, it is uncertain how reliable such parameters may be for use at a significantly different scale, even at the same general site.

Also, as the scale of application increases, it is likely that the influence of surface three-dimensionality on overland flow will become more important, although, as we have noted previously, flood-prediction models in common engineering use have a structure consistent with flow over a plane surface. However, we know that water is generally shed by convexities and collected by concavities in landscapes, commonly resulting in the formation of streams or concentrated flows, however transient.

At larger scales, and especially in temperate climates where rates of rainfall and evaporation are typically lower than for more tropical climates, the infiltration rate can commonly exceed the rainfall rate except in those areas of the landscape where the soil profile becomes saturated with water up to the soil surface due to a confluence of subsurface flow. In such areas the 'water table' (defined by the pressure in the water being at atmospheric pressure) is located essentially at the soil surface. In these situations, surface runoff occurs only from such surface-saturated areas, and runoff generated there is commonly referred to as saturation overland flow. Thus modelling runoff requires analysis to locate those parts of the landscape which are (or will become) saturated during any given rainfall. The location and extent of such surface-saturation zones depends on topographical, soil-profile and climatic characteristics as explained by O'Loughlin (1981, 1986) and Bonell and Balek (1993).

Mathematical models of hydrological behaviour at larger basin scales than those considered explicitly in this chapter usually deal separately with overland flow over the catchment surface and with flow in the complex network of stream systems in the basin (Jakeman et al., 1998). Flow in streams and rivers will briefly be considered in Chapter 9.

Main symbols for Chapter 7

C_d Drag coefficient
D Depth of water flow
F Force

F_o	Initial ponded depth of rainfall retained prior to runoff generation
g	Acceleration due to gravity
I	Infiltration rate
$\bar{I}$	Spatial average value of I
I_m	Maximum possible value of the spatial mean infiltration rate for a field
K	$S^{1/2}/n$
K_1	A term defined in Eq. (7.16)
L	Downslope length of a field or plot
m	A power to which Q is raised in Eq. (7.17)
n	Manning's roughness coefficient
P	Rainfall rate
q	Volumetric flux per unit width, or unit flux
$q(L)$	q at $x = L$
Q	Volumetric runoff rate per unit area $(= P - I)$
R	Excess rainfall rate
S	Slope of land surface $(= \sin \alpha)$
S_t	Volume of water stored per unit area between $x = 0$ and $x = L$
t	Time
V	Velocity of water flow
W	Weight of a volume of water, or width of flow
x	Downslope distance
α	Angle of slope of land surface, or a parameter given by Eq. (7.23)
Δ	Finite difference in any quantity
ρ	Density of water
τ	Shear stress in flowing water
Ω	Stream power (Eq. (7.9))

Exercises

7.1 On the Experimental Farm of the University of the Philippines at Los Banos, Laguna, Paningbatan *et al.* (1995) measured rainfall and runoff rates on soil-erosion plots similar to that shown in Fig. 4.3. For a plot of length 12 m and average slope 18%, kept bare by hoeing, data are given in the following table for rainfall and runoff rates measured at 1-minute intervals since the commencement of rainfall. Values of the three parameters F_o, I_m and α in the hydrological model described in this chapter were obtained by fitting these data, and these are 17.8 mm, 177.8 mm h^{-1} and 0.780, respectively. Since the runoff event was of long duration, data are given only for a time segment of the total event, well after the initial abstraction F_o was satisfied. Both rainfall rates and infiltration rates are high for this tropical site.

Using the methods illustrated in association with the data given in Tables 7.1 and 7.2, calculate the values of runoff rate per unit area, Q, both neglecting and including runoff routing, as was done in Tables 7.1 and 7.2, respectively. Thus both Q_{calc} (Table 7.1) and Q_{calc}(w. lag) (Table 7.2) are to be calculated.

Data on i (minutes from the commencement of rainfall), P and measured Q (Q_{meas}) follow.

i (minutes)	70	71	72	73	74	75
P_i (mm h^{-1})	57.9	93.0	104.7	117.5	104.7	93.0
Q_{meas} (mm h^{-1})	3.2	4.0	5.8	13.6	27.5	27.3
i (minutes)	76	77	78	79	80	81
P_i (mm h^{-1})	81.2	104.7	81.2	104.7	117.5	69.5
Q_{meas} (mm h^{-1})	32.9	33.7	18.0	19.6	18.0	17.2

7.2　A method very widely used to estimate runoff from total rain falling on small-to-medium-sized watersheds is that developed by the US Soil Conservation Service (USDA Soil Conservation Service, 1986). It is widely referred to as the SCS method, and commonly thought to be without theoretical foundation, but Yu (1998) has shown the method to be compatible with the assumption of an infiltration model described by Eq. (6.5). Since symbols P and Q have been used to describe rates, the symbols ΣP and ΣQ will be used to indicate total amounts for a rainfall event. Let T denote the duration of runoff in an event, with I_m and F_o having the meanings ascribed to them in this chapter. Then, in terms of these symbols, the SCS method uses the following relation for ΣQ:

$$\Sigma Q = \frac{(\Sigma P - F_o)^2}{\Sigma P - F_o + I_m T} \tag{7.24}$$

where in this equation the liberty of interpreting as $I_m T$ the term in the SCS equation which is described as the 'maximum retention' has been taken. Whilst this substitution appears plausible, it has not been tested, this term being inferred from a 'curve number' in the SCS methodology.

Do some cutting-edge research by using the data in Table 7.1 (or Table 7.2) to test the accuracy of Eq. (7.24). In these tables the term $\Sigma P - F_o$ can be calculated by summing P_i (mm h^{-1}) (since F_o has already been subtracted), where i goes from 13 to 49 minutes, so $T = 0.6\,$h. Thus $\Sigma P = \Sigma P_i \times 0.6\,$mm. From Example 7.1, $I_{\mathrm{m}} = 29.2\,$mm$\,$h^{-1}.

You are thus in a position to be able to calculate ΣQ using Eq. (7.24). The point of this research is then to see how well or how poorly this calculated value of ΣQ compares with the experimentally measured value of ΣQ which can also be obtained by summing Q_{meas} in Table 7.1 or 7.2 and multiplying by $T = 0.6\,$h.

I think you will find that the agreement is fair (a disagreement of some 25%). The value of this research lies in the fact that there is a much greater data base of SCS curve numbers (at least in the USA) than there is of estimates of I_{m}. Note that the approach described in this chapter is compatible with the SCS curve-number approach, which had previously been thought to be purely empirically based, even though it was effective in catchments not so large that spatial variability in rainfall input becomes dominant.

7.3 (For engineering students in particular.) One of the environmentally significant responsibilities usually carried out by hydrological, civil, or environmental engineers is to provide a flood-forecasting or -prediction service. Being able to warn in advance the citizens and communities living downstream of a flooding river of the expected spatial extent, height and time of arrival of the maximum flood peak is a service that reduces human suffering and loss.

Each nation has guidelines for use in this forecasting activity. For example, in Australia the RORB catchment model (Laurenson and Mein, 1988) is given professional standing through incorporation into a guideline published by the Institution of Engineers Australia (Pilgrim, 1987). Engineering students can find equivalents to this standards publication in their own country of study.

Using the engineering flood-forecasting guideline appropriate to your country, undertake research by seeking to compare its basic structure with the theory given in Sections 7.4 and 7.5 of this chapter, to which Eq. (7.19) is important.

References and bibliography

Bonell, M., and Balek, J. (1993). Recent scientific developments and research needs in hydrological processes of the humid tropics. In *Hydrology and Water Management in the Humid Tropics – Hydrological Research Issues and Strategies for Water Management*, eds. M. Bonell, M. M. Hufshmidt and J. S. Gladwell. Cambridge: Cambridge University Press, pp. 167–260.

Boughton, W. C. (1980). A review of the USDA SCS curve number method. *Aust. J. Soil Res.* **27**, 511–523.

Boyd, M. J., and Bufill, M. C. (1989). Determining runoff routing model parameters without rainfall data. *J. Hydrol.* **108**, 281–294.

Chow, V. T. (1959). *Open Channel Hydraulics*. New York: McGraw-Hill.

Ciesiolka, C. A. A., Coughlan, K. J., Rose, C. W., Escalante, M. C., Hashim, G. M., Paningbatan, E. P. Jr, and Sombatpanit, S. (1995a). Methodology for a multi-country study of soil erosion management. *Soil Technol*. **8**, 179–192.

Ciesiolka, C. A. A., Coughlan, K. J., Rose, C. W., and Smith, G. D. (1995b). Erosion and hydrology of steeplands under commercial pineapple production. *Soil Technol*. **8**, 243–258.

Dingman, S. L. (1984). *Fluvial Hydrology*. New York: W. H. Freeman and Company.

Jakeman, A. K., Green, T. R., Zhang, L., Beavis, S. G., Evans, J. P., Deitrich, C. R., and Barnes, B. (1998). Modelling catchment erosion, sediment and nutrient transport in large basins. In *Soil Erosion at Multiple Scales – Principles and Methods for Assessing Causes and Impacts*, eds. F. W. T. Penning de Vries, F. Argus and J. Kerr. Wallingford: CABI publishing, in association with the International Board for Soil Research and Management (IBSRAM), pp. 343–355.

Laurenson, E. M , and Mein, R. G. (1988). RORB – Version 4, Runoff routing program user manual. Clayton, Victoria: Department of Civil Engineering, Monash University.

O'Loughlin, E. M. (1981). Saturation regions in catchments and their relations to soil and topographic properties. *J. Hydrol*. **53**, 229–246.

O'Loughlin, E. M. (1986). Prediction of surface saturation zones in natural catchments by topographic analysis. *Water Resources Res*. **22**, 794–804.

Pilgrim, D. H. (ed.) (1987). *Australian Rainfall and Runoff: A Guide to Flood Estimation*. Canberra, ACT: Institution of Engineers.

Rose, C. W. (ed.) (1995). Soil erosion and conservation. *Soil Technol*. Special Issue **8**, 177–258.

Rose, C. W., Parlange, J.-Y., Sander, G. C., Campbell, S. Y., and Barry, D. A. (1983). Kinematic flow approximation to runoff on a plane: an approximate analytical solution. *J. Hydrol*. **62**, 363–369.

USDA Soil Conservation Service (1986). Urban hydrology for small watersheds. Technical release No. 55. Washington: USDA.

Yu, B. (1998). Theoretical justification of the SCS method for runoff estimation. *J. Irrigation Drainage Eng*. **124**, 306–310.

Yu, B., Rose, C. W., Coughlan, K. J., and Fentie, B. (1997). Plot-scale rainfall–runoff characteristics and modelling at six sites in Australia and Southeast Asia. *Trans. Am. Soc. Agric. Eng*. **40**, 1295–1303.

8
Erosion and deposition by water

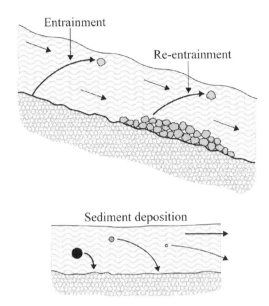

8.1 Introduction

The landforms on the earth's surface are dominated by the types of rock-cycle processes summarised in Fig. 1.1. Some possible forms of land-surface sculpturing resulting from natural erosion, transport and deposition of sediment are pictured in Fig. 1.2. Erosion of the land surface can take many forms. Especially in humid climates, mass movement of soil, as in landslides, can play a major role, whereas in arid regions erosion caused by strong winds can be the dominant erosion process. However, it is erosion due to water that will be considered in this chapter.

Seeking to document and understand the long-term and large-scale evolution of landscapes, and the role played by erosion by water in such evolution, is an area of research by geomorphologists. Some sediment lost to natural erosion processes (as illustrated in Fig. 1.2) is transported downslope, and some can be deposited as 'alluvial beds', perhaps augmented by 'colluvium' (or sediment transported by gravity). The scale of such transport depends greatly on the interactions among landscape, hydrological characteristics and characteristics of underlying rock and sediment. Whilst the attempt to understand long-term landscape evolution and the role played by erosion, transport and deposition processes is an important area of geomorphic research, our objective in this chapter is much more modest, both in space and in time scales. Though mass movement of sediment (discussed in Chapter 2) can be associated with erosion by water, it is a process not discussed in this chapter.

As discussed in Sections 1.3 and 1.4, the formation of soil which sustains all land plants (and therefore all forms of life on land) is a slow process. Hence, if erosive processes remove soil at a greater rate than that at which it can form, then, at the very least, the more fertile and productive characteristics of soil will suffer.

Acceleration of natural erosion processes commonly takes place when there are substantial reductions in vegetative or litter cover on the soil surface, and when soil is disturbed or loosened. These changes are a common outcome of most agricultural practices. Animal grazing, certain stages of forestry operations and urban, industrial and road construction can also lead to substantial soil erosion. Thus the challenge is to understand the processes of soil erosion and deposition so that ways may be found to engage in these necessary human-related activities without leading to long-term degradation of the land and water resources on which humankind depends. This challenge is a basic environmental and land-management objective. Failure to achieve this objective leads ultimately to unsustainable use of these sources of food and fibre production, to severe degradation of the wide range of environmental services provided by fully functioning ecosystems and to a diminution of biodiversity.

The objective of this chapter is to seek a physical understanding of the coexistent processes of soil erosion and sediment deposition. In the next chapter, Chapter 9, this knowledge will be applied and expanded to consider ways in which the living systems of watersheds and their rivers can be utilized and enjoyed in a more sustainable manner than is common currently.

Degradation due to accelerated rates of water-induced erosion can be recognised both at the source of erosion and also as a consequence of off-site transport of sediment. Hence these two types of degradation are described as 'on-site' and 'off-site' effects. Examples of on-site effects are the following:

(a) loss of topsoil, which can degrade ecosystem functioning and biodiversity, as well as reduce the yield or quality of agricultural products;
(b) loss of organic matter, leading to deterioration in soil structure and a decrease in biotic activity in soils; and

(c) an associated reduction in infiltration rate, which implies increases in runoff and in danger of soil erosion, as well as a decreased storage of water in the soil to support vegetative growth.

Examples of off-site effects of soil erosion are the siltation of streams, dams and roads and the deleterious growth of algae in rivers, lakes and near-shore oceans and reefs due to nutrient enrichment by eroded sediment reaching these waters. Off-site effects of erosion by water are typically at watershed scale, and will be further considered in Chapter 9. More carbon is thought to be stored in the world's soils than in its atmosphere, and a global off-site consequence of extensive soil erosion is a corresponding decrease in the ability of soils to withdraw or sequester carbon as carbon dioxide from the earth's atmosphere.

This chapter will seek an understanding of the physical processes involved in soil erosion and deposition, an understanding that can be gained by experiment and observation at the field-plot scale.

8.2 An overview of soil-erosion and deposition processes

This section briefly considers how erosion and deposition processes interact to control the concentration of sediment as it is transported downslope by overland flowing water.

When precipitation arrives as snow, the danger of accelerated soil erosion emerges during the subsequent snow-melting period. The resulting overland flow, arising either from snow melting or from excess rainfall, exerts shear stresses on the surface over which it flows as discussed in Section 7.4. Figure 7.5 shows that overland flowing water experiences an upslope force retarding its flow. Correspondingly the soil surface experiences an equal but oppositely directed force in the downslope direction. This force per unit area is a shear stress, τ. As noted in Eq. (7.9), the ability of overland flow to erode soil can be related to the rate of working of this shear stress, called the stream power, Ω, given by

$$\Omega = \tau V \qquad (8.1)$$

where V is the velocity of the overland flow. At field-plot and watershed scales, substantial soil erosion by water is dominantly due to the stream power of overland flow, though the role of the impact of raindrops can be very significant in shallow flows.

When precipitation occurs as rainfall, the impact of raindrops on a bare unprotected soil surface can play two roles contributing to soil erosion. Firstly, the impact and splash of raindrops can add directly to the concentration of sediment in the surface water (Fig. 8.1(a)). Secondly,

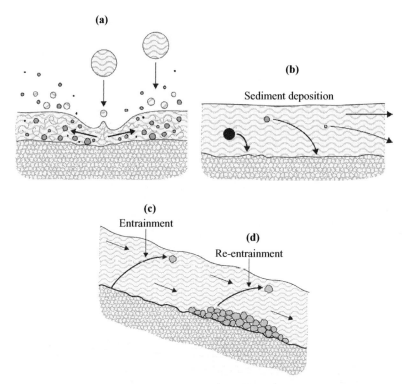

Figure 8.1 Processes involved in soil erosion and deposition: (a) rainfall detachment, (b) sediment deposition, (c) runoff entrainment of original soil-matrix material and (d) re-entrainment of sediment previously eroded and deposited in the same erosion event.

rainfall impact can also result in the breakdown of the structure of the soil, the term structure broadly covering the way in which the primary soil particles are bound together in bundles, or aggregates. In some soils this process of structural disaggregation can result in the formation of what is commonly called a surface seal. Any such deterioration in the surface structure of soil leads to a decrease in infiltration rate and thus more runoff, accentuating the danger of flow-driven erosion and reducing the amount of water stored in the soil profile which can support the growth of plants or trees.

If the rainfall rate exceeds the infiltration rate for whatever reason, then the *initial* development of sediment concentration in the surface water layer results from the difference between the rate of detachment of sediment into that layer by rainfall impact and the rate of return of sediment from the water layer by the downward process of sedimentation resulting in deposition onto the soil surface. These two processes are sketched in Figs. 8.1(a) and (b), respectively.

Figure 8.1(b) indicates that the rate of deposition will be higher for larger, rapidly settling aggregates, whereas very fine particles may move long distances before being deposited.

As the depth and flow rate of the surface water increase, the stream power, Ω of Eq. (8.1), also increases. When Ω exceeds a threshold value (which depends on the nature of the soil surface), flow-driven erosion occurs. Since the velocity of overland flow typically increases with scale, flow-driven erosion increasingly dominates rainfall-driven erosion as the scale increases. Whilst erosion at the scale of a field is typically dominated by flow-driven processes, this domination is usually complete at watershed scales.

As shown in Fig. 8.1(c), the flow-driven removal of soil from the original soil matrix, which normally possesses some strength, is often called 'entrainment'. However, the return of sediment to the soil surface in deposition quickly results in the build up of some degree of coverage of the original soil matrix by previously eroded sediment, which is then temporarily resident on the soil surface. This deposited sediment has hardly any opportunity to develop cohesive bonds with adjacent sediment before it is once again removed. Thus removal of this weak deposited material can be much more readily achieved than can entrainment of soil from the underlying cohesive soil matrix. This difference requires recognition, and this subsequent flow-driven erosion process is commonly referred to as 're-entrainment', a process sketched in Fig. 8.1(d).

As a result of the rate processes sketched in Fig. 8.1, a soil aggregate, or an individual soil particle, can be considered to make a consecutive series of hops. An aggregate will spend some short time virtually at rest on the soil bed, and then be ejected up into the flowing surface water layer either by rainfall impact or by the eddy shear stresses of overland flow (or by both of these). The ejected soil particle or aggregate will then be carried some distance downslope by the overland flow, its immersed weight in water seeking to return the sediment back to the soil surface. How far downstream the sediment moves during the period of motion depends on the velocity of overland flow and the rate of the particle's settling (called its 'settling velocity'). The settling velocity depends strongly on the size and 'wet density' of the aggregate (defined as the mass of the aggregate and included water per unit volume occupied by the aggregate). The series of hops carried out by a larger and a smaller aggregate or soil particle are sketched in Fig. 8.2. The kind of motion shown in Fig. 8.2 occurs both in shallow overland flow and in river beds. This kind of motion is commonly referred to as 'saltation'.

Figure 8.2 Illustrating a
consecutive series of short rest
periods on the soil surface for
a small and a larger soil
aggregate, followed by their
trajectories of motion after
removal from the soil surface.
The rate of return to the soil
surface increases with the
velocity with which the
particle or aggregate settles in
the fluid.

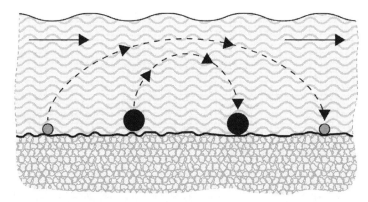

Figure 8.2 Illustrating a consecutive series of short rest periods on the soil surface for a small and a larger soil aggregate, followed by their trajectories of motion after removal from the soil surface. The rate of return to the soil surface increases with the velocity with which the particle or aggregate settles in the fluid.

Another net result of the rate processes sketched in Figs. 8.1 and 8.2 is that overland flow will carry sediment at some 'sediment concentration', c, which will vary with time and location. Sediment concentration is defined as the mass of sediment contained in a given volume of fluid, divided by that volume. Thus

$$c = \frac{\text{mass of sediment}}{\text{volume of water}} \quad (\text{kg m}^{-3}) \tag{8.2}$$

where the volume of water will contain some sediment.

The transport of sediment is intimately related to the hydrology of overland flow considered in Section 7.4, where q was defined as the volumetric flux of water per unit width of flow, or, more briefly, the unit flux. The SI unit of q is thus $\text{m}^3 \, \text{m}^{-1} \, \text{s}^{-1}$ or $\text{m}^2 \, \text{s}^{-1}$. Using Eq. (8.2), the product of q and c, expressed in terms of their definitions, is therefore given by

$$\frac{\text{volume of water}}{\text{metre width} \times \text{s}} \times \frac{\text{mass of sediment}}{\text{volume of water}} = \frac{\text{mass of sediment}}{\text{metre width} \times \text{s}}$$

Denoting the mass of sediment per second transported across unit flow width by q_s, it follows that

$$qc = q_s \tag{8.3}$$

This close relationship between the flux of water and the flux of sediment is illustrated in Fig. 8.3. The fluxes are assumed to commence on a flat section of landscape, and the flux is measured after travelling a downslope distance L. Unit flux q depends on the downslope distance; to draw attention to this, the value of q at distance L can be written $q(L)$. In Fig. 8.3, as in Fig. 7.6, the position at which the flux commences is denoted by $x = 0$.

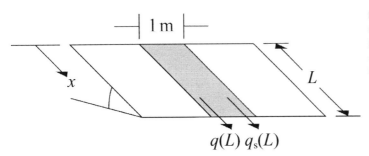

Figure 8.3 Illustrating unit flux q at $x = L$ on a plane land surface (written $q(L)$). Also shown is unit sediment flux, $q_s(L)$.

Note that L is also the area of the plane of unit strip width shown shaded in Fig. 8.3. Thus, as defined in Eq. (7.10), and using Fig. 8.3, the rate of runoff per unit area of flow, Q, is given by

$$Q = q(L)/L \qquad (8.5)$$

More generally than in Eq. (8.5), the value of q at downslope distance x from the the position of commencement of overland flow, $q(x)$, is given by

$$q(x) = Qx \qquad (8.6)$$

The loss of sediment in erosion is commonly related to the land area from which it has come. Again referring to Fig. 8.3, if the sediment flux is constant for a time period ΔT, the total loss of sediment from the shaded area of L (m^2) would be $q_s\,\Delta T$ (kg). Thus the soil loss per unit area during this time period would be

$$q_s\,\Delta T/L \quad (\text{kg m}^{-2}) \qquad (8.7)$$

The total soil loss per unit area in a complete erosion event can be obtained by summing up the expression given in Eq. (8.7).

Example 8.1

The average sediment concentration, c_{av}, and average volumetric runoff rate, $q_{av}W$, from a runoff plot of width W were measured, data being averaged over a 10-minute period as shown in Table 8.1. Calculate the total soil loss per unit area for the erosion event in units of kg m^{-2} and tonne ha^{-1} given that $W = 10$ m and the plot length is 30 m. Assume that erosion occurs from the entire plot area.

Solution
Check that $1 \text{ g l}^{-1} = 1 \text{ kg m}^{-3}$.

Table 8.1 *Data for Example 8.1 and (in the last column) calculated soil loss per unit area*

Data-averaging period ΔT (min)	c_{av} (g l^{-1})	$q_{av} W$ (m^3 min^{-1})	Soil loss/area (kg m^{-2})
10	12	0.05	0.060
10	38	0.18	0.684
10	22	0.11	0.242
10	9	0.03	0.027
Sum			0.977

Soil loss per unit area during the time period $\Delta T = q_{av} c_{av} \, \Delta T/L$ calculated from Eqs. (8.7) and (8.3) is shown in Table 8.1. The unit of kg m^{-2} for this quantity is justified, since, expressing units for the quantities in Table 8.1, and then cancelling out these units,

$$q_{av} c_{av} \, \Delta T/L = \frac{q_{av} W \, (\mathrm{m}^3)}{W \, (\mathrm{m})(\mathrm{min})} \times c_{av} \, \frac{\mathrm{kg}}{\mathrm{m}^3} \times \frac{\Delta T \, (\mathrm{min})}{L \, (\mathrm{m})} \quad \text{or kg m}^{-2}$$

where $\Delta T = 10$ min.

As shown in the summation in Table 8.1, the total soil loss per unit area

$$= 0.977 \, \mathrm{kg \, m}^{-2}$$
$$= 0.977 \left(\mathrm{kg} \times 10^{-3} \, \frac{\mathrm{tonne}}{\mathrm{kg}} \right) \times \left(\frac{\mathrm{m}^2}{\mathrm{m}^2 \times 10^4 \, \mathrm{ha}} \right)$$
$$= 9.77 \, \mathrm{tonne \, ha}^{-1}$$

(Notice that $1 \, \mathrm{kg \, m}^{-2} = 10 \, \mathrm{tonne \, ha}^{-1}$.)

8.3 Deposition characteristics of sediment

Whilst the shape of a particle can influence its settling velocity, especially if it is very non-spherical, many sedimentary particles are of a shape such that the major factors affecting their settling velocity in water of density ρ can be understood by assuming such particles to be spherical, of diameter d_p, and of density σ. Any particle settling in water rather quickly achieves a constant or 'terminal' velocity, the zero acceleration indicating that there is no net force acting on the settling particle. Thus the upward-acting drag force, resisting the particle's motion through the water, must equal the immersed weight of the particle, W. Using Eq. (3.18), the drag force is given by $C_d(\pi d_p^2/4)(\rho \, v^2/2)$. Using Archimedes' principle given in Section 3.3, the apparent weight of the sediment of mass m when immersed in water

$$= mg - \rho(m/\sigma)g$$

where $m = \sigma \pi d^3/6$, and m/σ is the particle's volume. Equating the particle's immersed weight to the drag force gives

$$v^2 = \tfrac{8}{3}g(1 - \rho/\sigma)(d_p/C_d) \qquad (8.8)$$

As discussed in Section 3.6, and as illustrated in Fig. 3.12, at low Reynolds numbers, Re, where flow is laminar, the drag coefficient $C_d = 24/Re$. Substituting this expression for C_d into Eq. (8.8) and noting that $Re = \rho v d_p/\mu$ leads to the following expression for v, known as 'Stokes' law':

$$v = [g/(18\mu)](\sigma - \rho)d_p^2 \qquad (8.9)$$

For particles of diameter less than about 70 μm (or 0.07 mm), Eq. (8.9) accurately describes the measured settling velocity of spherical particles in still water. Particles having sufficiently small values of v in still water for Eq. (8.9) to be accurate are said to be within the 'Stokes range'. However, in turbulent overland flow, or in river water, such small particles are so buffeted by turbulent eddies that they do not settle, and thus they form the 'suspended' component of the sediment load.

As shown in Fig. 3.12, the magnitude of C_d continues to fall as d_p (and hence Re) increases in magnitude. As shown by Eq. (8.8), $v \propto (d_p/C_d)^{1/2}$. Thus, as d_p increases (and C_d decreases), v can become sufficiently great compared with eddy velocities that turbulence has little effect on the velocity of settling. This is the case for the particles which form the saltation load of transported sediment.

Suppose that a size class of particles with a particular settling velocity v_i is returning to the soil bed as shown in Fig. 8.4. This process, known as 'sedimentation', is extensive in space. Thus, in considering the rate of arrival of these particles on any 1 m² of the soil or river bed, we can neglect the fact that the water is flowing, since particles lost due to horizontal motion from the imaginary cylinder defined in Fig. 8.4 will be replenished by an equivalent number due to the same lateral influent motion. Since the settling velocity of particles is v_i m s⁻¹, then in one second all the particles in the imaginary cylinder of height v_i m will have reached the river bed, or been 'deposited'. In this use of terms, deposition is the end result of the sedimentation process. The mass of sediment contained in unit volume is known as the sediment concentration, c_i. Since the cross-sectional area of the imaginary cylinder in Fig. 8.4 is 1 m², the volume of this cylinder is v_i m³. Hence the mass of sediment in the cylinder at any time is $v_i c_i$. All this sediment mass reaches the

Figure 8.4 Soil particles or aggregates of uniform settling velocity v_i m s^{-1} sedimenting towards a soil or river bed. Only particles within the imaginary cylinder are shown.

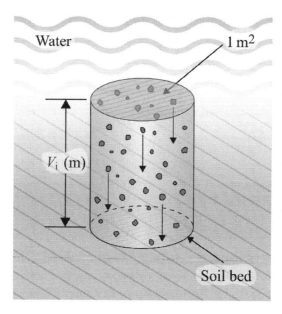

1 m^2 of soil bed at the base of the cylinder in one second, and this mass is defined as the rate of mass deposition per unit area of soil or river bed, and denoted d_i. Hence

$$d_i = v_i c_i \quad (\text{kg m}^{-2}\text{s}^{-1}) \tag{8.10}$$

Since sediment will have a range of sizes and settling velocities, of which sediment with a settling velocity v_i can be regarded as a particular class, the rate of deposition for all of the sediment, d, is given by summing over the entire range of settling-velocity classes. Thus

$$d = \Sigma d_i = \Sigma v_i c_i \tag{8.11}$$

Any naturally occurring sediment will have some particular distribution of settling velocities amongst its particles of different sizes, densities, and shapes. Of all these factors which affect the settling velocity, the effect of size is usually dominant. Figure 8.5 illustrates a common way in which the distribution of settling velocities in a particular sediment type is described or represented. Such a distribution is called the 'settling-velocity characteristic' of that particular sediment. It is like a thumbprint identifying that particular sediment. The relationship in Fig. 8.5 shows that 50% of the particular sediment represented has a settling velocity less than v_{av}, where v_{av} is the average settling velocity (0.11 m s^{-1} in this example).

The ten data points shown as small circles in Fig. 8.5 indicate ten sediment-size (or settling-velocity) ranges, each of which has the same

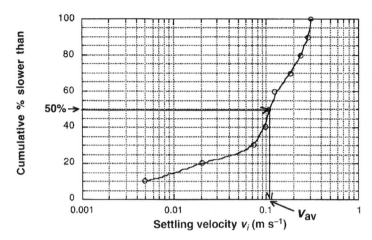

Figure 8.5 The settling velocity characteristic of a particular sediment. 50% of sediment will have a settling velocity less than the average value of v_i, denoted v_{av}.

mass, equal mass being indicated by equal steps on the ordinate, since the cumulative percentage is on a mass basis. If, in general, there are I sediment-size ranges of equal mass (where $I = 10$ in Fig. 8.5), then the sediment concentration contributed by each of the I size ranges will be c/I, where c is the total sediment concentration. Then the total rate of sediment deposition, d, from Eq. (8.11) will be given by

$$d = \Sigma v_i c / I$$
$$= c \Sigma v_i / I$$

since c is a constant, or

$$d = c v_{av} \tag{8.12}$$

where v_{av} is the mean or average settling velocity, given by

$$v_{av} = \Sigma v_i / I \tag{8.13}$$

Equation (8.12) indicates that the magnitude of v_{av} for any particular sediment concentration determines the rate at which it is deposited. The higher the rate of deposition, other things being equal, the lower the sediment concentration and so soil loss by erosion. A soil with better structural stability, which is typically associated with relatively higher levels of organic matter and an adequate clay percentage to provide cohesion, will tend to have higher values of v_{av}. Hence this mean settling velocity is an important sediment characteristic, called the sediment 'depositability' or 'depositabilty characteristic' (Marshall, Holmes and Rose, 1996).

If all particles had the same settling velocity, then the characteristic in Fig. 8.5 would be a 'step function', meaning that the distribution would go from 0 to 100% at the particular common settling velocity. Though this extreme situation is not common, natural processes can 'sort' particles into deposits of somewhat similar velocity in segments of a stream bed, at particular locations in estuaries and, most notably, on beaches.

As illustrated in Fig. 8.5, sediment normally contains particles with a wide range of settling velocities. During erosion the differently sized sediment particles all settle together in the same water layer, the faster settling particles moving through those which are slower. Thus in sedimentation there is interaction between particles of different sizes, with the result that settling velocities are different from when particles settle alone. The method of measuring settling-velocity characteristics should allow this interaction to occur, as discussed in Lovell and Rose (1988, 1991).

8.4 Soil erosion as a rate process

Now that in Eq. (8.12) we have an expression for the rate of deposition, we can commence quantitatively assessing the magnitudes of the rate processes involved in soil erosion sketched in Fig. 8.1. The various erosion processes shown in (a), (c) and (d) of Fig. 8.1 all lift or eject sediment from the soil surface into the layer of water flowing over that surface, acting to increase the sediment concentration in that layer. In opposition to these erosive processes, sedimentation towards the soil surface results in deposition, which acts to reduce sediment concentration in the overland flow. This process, shown in (b) of Fig. 8.1, is the *only* process which moderates what otherwise would be a continuous increase in sediment concentration, an increase that would become self-limiting as the concentration became so high as to approach that of a solid and so non-erodible state. Thus, it is an enigma that the continuation of soil erosion is possible only because of deposition!

Lumping together all soil-erosion processes, Fig. 8.6 shows the various rate processes which can affect the sediment concentration, c, of the overland flowing water (Eq. (8.2)). In Fig. 8.6 the difficulty of representing the exchange of sediment between the soil surface and the overlying water layer is overcome by showing the overland flow as artificially lifted up above the soil surface. Whilst this requires some imagination, it allows clear representation of the two oppositely directed processes transferring material from and to the soil surface, namely erosion and deposition (Fig. 8.6).

A tap is used to control the rate of flow of water into a hand basin. In Fig. 8.6 the convention adopted is to represent rate processes as

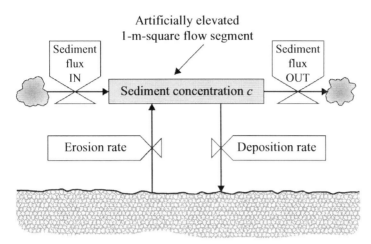

Figure 8.6 A flow or systems diagram illustrating erosion and deposition rates, and rates of sediment flux into and out of the artificially elevated segment of overland flow of sediment concentration c, shown shaded.

controlled by a tap or rate-controller symbol of the following type, where the arrow represents the flow rate:

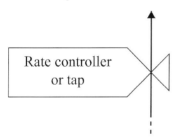

Consider the segment of (artificially elevated) overland flow of sediment concentration c, shown shaded in Fig. 8.6. This segment is fed by the flow of sediment-laden water into the segment from upslope, but also depleted by outflow from the downslope end of the flow segment. Figure 8.6 is a flow diagram of a style introduced by Forrester (1970). In this style of systems diagram, cloud symbols are used to indicate that the downslope inflow and outflow to the flow segment must come from and go to somewhere, but these locations are outside the system being represented in the diagram.

We are now in a position to assess quantitatively the typical orders of magnitude of the four fluxes shown in Fig. 8.6 which control the sediment concentration c of the flow segment. We will illustrate the possible magnitudes of these fluxes using typical feasible values of quantities involved at a field scale. However, the chief interest in this investigation lies not so much in the absolute magnitudes, but in the relative magnitudes of the fluxes, the latter information having much more generality of application than the former.

An illustration of the orders of magnitude of fluxes

The chosen values of previously introduced quantities, or assumptions adopted for the following illustration, are as follows:

- sediment concentration, $c = 20$ kg m^{-3};
- sediment depositability, $v_{av} = 0.1$ m s^{-1};
- unit flux of overland flow, q, arises from a rate of overland flow per unit area, Q, which is assumed to be 20 mm h^{-1};
- overland flow commenced at a distance $L = 30$ m upslope of the segment of overland flow shown shaded in Fig. 8.6, and the length of this flow segment is 1 m;
- a steady-rate situation is also assumed.

From Eq. (8.12), with these assumed values, the rate of sediment deposition is

$$d = cv_{av}$$
$$= 20 \times 0.1 = 2 \quad (\mathrm{kg\,m^{-2}\,s^{-1}})$$

From Eq. (8.5) the flux of water into the flow segment, denoted $q(\mathrm{IN})$, is given by

$$q(\mathrm{IN}) = QL = 20\,\mathrm{mm} \times \left(\frac{10^{-3}\,\mathrm{m}}{\mathrm{mm}}\right) \times \left(\frac{\mathrm{h}}{\mathrm{h}\,3600\,\mathrm{s}}\right) 30\,\mathrm{m}$$

or

$$q(\mathrm{IN}) = 1.67 \times 10^{-4} \quad (\mathrm{m^2\,s^{-1}})$$

Thus, from Eq. (8.3), the flux of sediment into the segment, $q_s(\mathrm{IN})$, is given by

$$q_s(\mathrm{IN}) = q(\mathrm{IN}) \times c$$
$$= 1.67 \times 10^{-4} \times 20$$
$$= 3.33 \times 10^{-3} \quad (\mathrm{kg\,m^{-1}\,s^{-1}})$$

The water and sediment fluxes out of the unit element in Fig. 8.6 will only be greater than the fluxes into the element by a factor $(L + 1)/L = 31/30 = 1.03$, or 3%. Thus the downslope fluxes IN and OUT of the element are both of order 10^{-3} kg m^{-1} s^{-1}, and the difference between these fluxes (which is what affects c), will be even smaller. Since the rate of deposition is 2 kg m^{-1} s^{-1}, the magnitudes of these downslope fluxes are three orders of magnitude smaller than the deposition rate. Since these downslope fluxes are so small in comparison, and we are considering a steady-rate situation, the erosion and deposition rates must be almost identical, since, if any difference in these rates develops, this will immediately change c.

Thus the very important conclusion which follows from this numerical investigation is that it is the relative magnitudes of the very large erosion and deposition fluxes which dominantly control the sediment concentration c, these fluxes being greater than the downslope fluxes by a factor of 10^3 in this illustration. This complete flux dominance remains, even if quite different but feasible parameter values are chosen compared with those adopted in this illustration, and whatever the mix of erosion mechanisms involved. This is not to say, however, that the downslope sediment flux is unimportant, since it is this flux which is directly responsible for the loss of sediment in erosion. However, the magnitude of this downslope flux depends on the sediment concentration, and this analysis shows that c is determined dominantly by the relative magnitudes of the two much larger or more rapid fluxes, namely those of erosion and deposition.

An analogy to this practically important insight into erosion processes might be given as follows. If the rates of deposition and erosion can be conceived as related to the rapid key depressions and their return on a keyboard operated by a typist, then these rates are very fast compared with the outflow of pages being typed, whose slower, but most important, page rate can be likened to the rate of sediment transport.

It follows that whatever slows the erosion rate relative to that of deposition will reduce net erosion. The important practical soil-conservation implications of this understanding of the relative rates of processes will be taken up in Chapter 9.

8.5 Erosion theory and soil-erodibility characteristics of sediment

Following examination of a body of soil-erosion data in which flow-driven erosion dominated, Foster (1982) and others noted that, in any given situation, there appeared to be an upper limit to the measured sediment concentration, and called it the 'transport limit'. The greatest rate of flow-driven soil erosion would be expected to occur when the soil surface is completely covered with very-low-strength deposited sediment that had previously been eroded in the same erosion event. This describes the situation in which all erosion is due to re-entrainment of deposited sediment (Fig. 8.1(d)). This would occur if all of the eroding surface were covered with a layer of deposited sediment.

Hairsine and Rose (1992) developed a theoretically based equation allowing prediction of sediment concentration at this transport limit, denoted c_t, for the equilibrium or steady-rate situation in which rates of re-entrainment and deposition are equal. This theoretically based equation uses the concept of 'excess stream power' which is defined as the difference $\Omega - \Omega_o$, where Ω_o is the threshold stream power at which re-entrainment of sediment commences. The equation for c_t was developed on the assumption that a fairly constant fraction, denoted F, of the excess stream power is effective in re-entraining sediment. Thus the assumption is that the 'effective excess stream power' is $F(\Omega - \Omega_o)$, where F has been found experimentally to be in the range 0.1–0.2, and a value of 0.1 will later be used (Proffitt, Hairsine and Rose, 1993). In situations of significant erosion Ω_o is typically quite small compared with Ω, and is often neglected.

Box 8.1 gives an approximate derivation of this theoretical expression for c_t. A more rigorous derivation of this equation for c_t, which recognises that sediment normally consists of a range of size classes, is given in Marshall, Holmes and Rose (1996). Neglecting Ω_o compared with Ω, this equation for c_t is

$$c_t = \frac{F}{v_{av}} \left(\frac{\sigma}{\sigma - \rho} \right) \rho S V \tag{8.14}$$

where v_{av} is defined in Eq. (8.13), σ is the wet density of sediment, ρ the density of water and S the slope of the land surface over which water is flowing with velocity V. Though it was derived theoretically (Box 8.1), there is good experimental support for Eq. (8.14) as the steady-rate upper limit to the sediment concentration of flow-driven eroding soil. A big advantage of Eq. (8.14) is that c_t can be predicted for any soil and land slope, provided that V can be predicted or measured.

For a given soil (i.e. particular v_{av} and σ) and for land at a particular slope S, c_t is proportional to the overland flow velocity, V (Eq. (8.14)). Now V can be related to other quantities as follows.

From Eq. (7.1),

$$q = DV \tag{8.15}$$

Also from Eq. (8.6), at distance x downslope from the commencement of overland flow (Fig. 7.6),

$$q = Qx \tag{8.16}$$

From Eqs. (8.15) and (8.16), $D = Qx/V$, which, when substituted into Manning's equation, Eq. (7.7), gives

$$V = (S^{1/2}/n)(Qx/V)^{2/3},$$

from which it follows that

$$V = (S^{0.5}/n)^{0.6}(Qx)^{0.4} \tag{8.17}$$

Thus, for a given slope and surface roughness, that is, for any given particular situation,

$$V \propto (Qx)^{0.4} \tag{8.18}$$

Box 8.1 Simplified theory for sediment concentration at the transport limit, c_t

The transport limit can be conceived as the sediment concentration achieved in the steady state of overland flow when the rate of re-entrainment of recently deposited sediment is equal to the rate of deposition. Theory development recognising that soil consists of a range of aggregate sizes is given in Marshall, Holmes and Rose (1996). In the simplified development of the theory given here, it is assumed that all the soil has a settling velocity equal to v_{av} (Eq. (8.13)).

As described in the main text, the effective excess stream power is assumed to be given by $F(\Omega - \Omega_0)$, and hereafter the threshold stream power, Ω_0, will be assumed negligible compared with Ω (a reasonable assumption for cultivated soil).

Energy is required in order to maintain the steady upward flux of re-entrained sediment particles illustrated in Fig. 8.1(d), with all particles here assumed to be identical. Since sediment is immersed in water, it only has to be lifted against its immersed weight in water. From Archimedes' principle discussed in Section 3.3, the sediment's immersed weight is $(\sigma - \rho)/\sigma$ multiplied by its un-immersed weight, where σ is is the wet density of sediment and ρ is the density of water. The work done in lifting sediment by re-entrainment against this immersed weight is equal to the immersed weight of sediment added to the water layer per second multiplied by the distance moved, taken as equal to the depth of water, D. Denote the mass of sediment re-entrained per unit area per second (the re-entrainment rate) by r.

Then the energy expended per second in re-entrainment, or the power required, is therefore given by

$$rg \left(\frac{\sigma - \rho}{\sigma} \right) D \quad (\text{W m}^{-2}) \tag{a}$$

The power requirement given in Eq. (a) is assumed to come from the fraction F of the stream power, or $F\Omega$. Equating these two expressions, then

$$r = \frac{F\Omega\sigma}{D(\sigma - \rho)g} \tag{b}$$

Now, from Eqs. (7.9) and (7.4),

$$\Omega = \rho g S D V \tag{c}$$

So, from Eqs. (b) and (c),

$$r = \frac{F\rho SV\sigma}{\sigma - \rho} \qquad (d)$$

Now, at equilibrium, when the sediment concentration is constant, the rate of re-entrainment given by Eq. (d) must equal the rate of deposition, which, from Eq. (8.12), is given by

$$d = cv_{av} \qquad (e)$$

Equating Eqs. (d) and (e), which then implies that $c = c_t$, finally gives

$$c_t = \frac{F}{v_{av}}\left(\frac{\sigma}{\sigma - \rho}\right)\rho SV \qquad (f)$$

which is Eq. (8.14) in the main text.

Hence, from Eqs. (8.18) and (8.14), for a given soil (i.e. given v_{av} and σ) and for a particular context (i.e. given slope S and roughness n),

$$c_t = k(Qx)^{0.4} \qquad (8.19)$$

where, from Eqs. (8.14), (8.17) and (8.19),

$$k = \frac{F}{v_{av}}\left(\frac{\sigma}{\sigma - \rho}\right)\left(\frac{\rho S^{1.3}}{n^{0.6}}\right) \qquad (8.20)$$

Example 8.2

Using Eq. (8.19) for c_t and Eq. (7.9) for Ω, plot the form of the relationship between c_t and Ω, using illustrative values for the parameters involved.

Solution
From Eq. (7.4) for the shear stress τ, and since $\Omega = \tau V$ (Eq. (8.1)), then

$$\Omega = \rho g SDV \qquad (8.21)$$

From Eqs. (8.15) and (8.16), $DV = Qx$, so that

$$\Omega = \rho g SQx \qquad (8.22)$$

Summarizing, for a given soil, slope and surface roughness,

$$c_t \propto (Qx)^{0.4} \quad \text{from Eq. (8.19)}$$

and

$$\Omega \propto Qx \quad \text{from Eq. (8.22)}$$

Hence it follows that there is a non-linear form of relationship between c_t and Ω (or Qx, to which it is proportional). This relationship is illustrated in Fig. 8.7 for the following particular set of values of the parameters

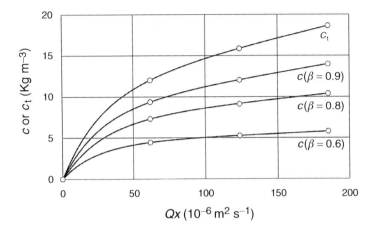

Figure 8.7 The form of the relationship between c_t and Ω (or Qx) discussed in the text. Also shown are three examples of the type of relationship that commonly provides a good fit to experimental data when measured values of c are less than c_t. This relationship is $c = c_t^\beta$, where relationships are shown for $\beta = 0.9, 0.8$ and 0.6.

involved: $Q = 20$ mm h^{-1} (or 5.56×10^{-6} m s^{-1}), $F = 0.1$, $v_{av} = 0.1$ m s^{-1}, $\rho = 1000$ kg m^{-3}, $\sigma = 2000$ kg m^{-3}, $S = 0.1$ and $n = 0.05$ m$^{-1/3}$ s. It follows from Eq. (8.20) that $k = 605$ kg m$^{-4.2}$ s$^{0.4}$. The plotted values of c_t correspond to downslope distances $x = 10, 20$ and 30 m.

The transport-limit sediment concentration, c_t, is the maximum value of c due to flow-driven erosion. However, measured values of c are often less then c_t. How much a measured value of c is less than c_t depends on a number of factors. Such factors include the soil strength, the degree and stability of aggregation of the soil, and rilling characteristics. Figure 8.7 shows that the strength of the dependence of sediment concentration c on the stream power (or on Qx) decreases as the value of a parameter β decreases.

For soil in any particular state or condition, there is experimental and theoretical support for the belief that the form of relationship between c and Ω (or Qx) is adequately, though approximately, described by simply raising c_t to a power β, where β is generally ≤ 1 (Rose, 1993). If $\beta = 1$, then $c = c_t$, corresponding to soil in its most erodible condition. As the resistance a soil offers to erosion increases, this can be described quantitatively by a decrease in the value of β. Likewise, lower resistance to erosion corresponds to higher values of β. Thus β is a parameter that provides a convenient, though approximate, description of the erodibility of bare soil, and Fig. 8.7 illustrates relationships for three particular values of β. The power, β, has been called a 'soil-erodibility factor' (Rose, 1993), and is defined by

$$c = c_t^\beta \tag{8.23}$$

so that, taking logarithms of both sides,

$$\beta = \frac{\ln c}{\ln c_t} \tag{8.24}$$

From Eq. (8.23), it follows that, if $\beta = 0$, $c = 1$ kg m^{-3}, a value so low that erosion is not a soil-degrading process.

During an erosion event both c and c_t vary with Qx as discussed earlier, and Q will vary with time. In many field soil-erosion experiments, soil loss is measured from a hydrologically defined runoff plot, such as that shown in Fig. 4.4. The unpredictable occurrence of erosion events usually means that it is not feasible to measure c as a function of time during the erosion event by manual sampling. During storms, such activity can also be unsafe. Automatic serial sediment sampling in surface runoff is difficult and not yet common, though sampling of suspended sediment in rivers is more routine.

Furthermore, in more general soil-conservation applications, the time scale of interest is usually no finer than that of a single erosion event, and may be much longer. Indeed, averaging soil-loss data over annual time periods for various land-management treatments, land slopes, etc. was developed early in soil-erosion studies, and this method of summarising data provides a very useful practical guide in erosion management (Wischmeier and Smith, 1978; Renard et al., 1991).

Thus, for practical application of knowledge about soil erosion, there is a need to be able to assess erodibility on an event-average basis or over a longer period. This need sets up a tension with the understanding of erosion as essentially a rate-driven process. This tension can be addressed as outlined in the following section.

8.6 Effective averaging over a soil-erosion event

In field erosion studies aimed at understanding erosion mechanisms it is quite feasible to measure Q (or $QL = q(L)$) as a function of time, for example using equipment illustrated in Fig. 4.5. However, since only the total soil loss for the erosion event is usually measured, only an average sediment concentration, $\bar{c}$, can be determined, where $\bar{c}$ is defined as

$$\bar{c} = \frac{\text{total soil loss during the event}}{\text{total runoff during the event}}$$

Since $q_s = qc$ (Eq. (8.3)), the total soil loss in the event is given by the sum of qc over the duration of the erosion event, which can be written as Σqc. Denoting q measured at the bottom of the plot (where $x = L$) by $q(L)$, it follows that

$$\bar{c} = \frac{\Sigma q(L)c}{\Sigma q(L)} = \frac{\Sigma Qc}{\Sigma Q} \tag{8.25}$$

Thus, from Eq. (8.25), $\bar{c}$ is really a flow-weighted average value of c.

In order to calculate an effective or average value of β for the entire erosion event, a flow-weighted average value of c_t corresponding to $\bar{c}$

is also required, and this is denoted by $\bar{c}_t$. As shown in Ciesiolka *et al.* (1995), $\bar{c}_t$ is given by

$$\bar{c}_t = k(Q_e L)^{0.4} \tag{8.26}$$

where k is given by Eq. (8.20), and Q_e is an 'effective runoff rate per unit area' for the runoff plot, which can be calculated from measurements of the rate Q as a function of time throughout the erosion event. The single value Q_e provides the appropriate event-average value of Q for interpretation or prediction of soil-erosion data, and is given by

$$Q_e = \left(\frac{\Sigma Q^{1.4}}{\Sigma Q} \right)^{2.5} \tag{8.27}$$

Using this value of Q_e in Eq. (8.26) to give $\bar{c}_t$, then the event-average value of β is given by

$$\beta = \frac{\ln \bar{c}}{\ln \bar{c}_t} \tag{8.28}$$

Thus

$$\bar{c} = \bar{c}_t^{\beta}$$

so that, from Eq. (8.26),

$$\bar{c} = k^{\beta}(Q_e L)^{0.4\beta} \tag{8.29}$$

Hence, the total soil loss per unit area in an erosion event, M (kg m^{-2}) during which the total runoff per unit land area is ΣQ is given by the product $\bar{c} \, \Sigma Q$, or

$$M = k^{\beta}(Q_e L)^{0.4\beta} \, \Sigma Q \quad (\text{kg m}^{-2}) \tag{8.30}$$

Note that k can be calculated using Eq. (8.20), which requires information on factors such as appropriate soil characteristics, land slope (assumed constant) and slope length.

Equation (8.30) shows that the total soil loss from an area of land of uniform slope depends on a combination of hydrological characteristics, soil characteristics and the soil-erodibility parameter β. It is because soil erosion is a rate-driven process that the effective-runoff-rate characteristic Q_e is involved. Equation (8.27) shows that, in order to determine Q_e, the runoff rate per unit area, Q, must be measured. While techniques illustrated in Figs. 4.4 and 4.5 provide one type of robust solution to this measurement challenge, other methods may be more appropriate at high volumetric flow rates (Ciesiolka and Rose, 1998). The total event runoff, ΣQ, is simply obtained by summing Q.

With the total soil loss, M, also measured, the erodibility parameter β can be calculated using Eq. (8.30). Alternatively, if β is known, then M can be predicted for any given hydrological event or events. The

Table 8.2 *The hydrological data for Example 8.3*

Time period (min)	Q (mm h^{-1})	$Q^{1.4}$
0 to 5	4	6.96
5 to 10	11	28.70
10 to 15	21	70.98
15 to 20	17	52.80
20 to 25	10	25.12
25 to 30	3	4.66
Sum	66	189.2

methodology outlined is described by Ciesiolka *et al.* (1995) and Rose and Yu (1998). Computer programs are helpful in such analysis (Rose and Yu, 1998), but a simple manual example of such calculation is given in Example 8.3.

Example 8.3

The runoff rate was measured, together with the total mass of eroded sediment (18.6 kg), for a rainfall event on a bare runoff plot of the type shown in Fig. 4.4. The plot slope is 8%, the downslope length of the plot is 35 m and its width is 10 m. Measurement of the relevant soil characteristics gave $v_{av} = 0.11 \, \text{m s}^{-1}$ and $\sigma = 1400 \, \text{kg m}^{-3}$, and Manning's n was found to be $0.1 \, \text{m}^{-1/3} \, \text{s}$. Assume that F in Eq. (8.14) is 0.1.

Table 8.2 shows the rate of runoff per unit area, Q, recorded in units of mm h^{-1}, averaged over 5-minute periods for convenience. Table 8.2 also shows $Q^{1.4}$ and their respective sums.

From this information and these data, calculate the effective erodibility parameter β for the erosion event using Eqs. (8.26) and (8.28).

Solution
From the data in Table 8.2, $Q_e = (\Sigma Q^{1.4}/\Sigma Q)^{2.5} = 13.9 \, \text{mm h}^{-1}$. Whilst such measurements are commonly expressed in units of mm h^{-1}, Q_e is required in SI units for the calculation of c_t. Converting units:

$$Q_e = \left(13.9 \, \frac{\text{mm}}{\text{h}}\right) \times \left(\frac{10^{-3} \, \text{m}}{\text{mm}}\right) \times \left(\frac{\text{h}}{3600 \, \text{s}}\right)$$
$$= 3.87 \times 10^{-6} \, \text{m s}^{-1}$$

The quantity k in Eq. (8.26) can be obtained by inserting values of quantities given into Eq. (8.20) to give $k = 475$ in SI units. Thus, from

Eq. (8.26) with $L = 35$ m,

$$\bar{c}_t = 13.5 \, \text{kg m}^{-3}$$

By definition

$$\bar{c} = \frac{\text{total soil loss during the event}}{\text{total runoff during the event}}$$

The total volume of runoff can be calculated as follows.
The equivalent ponded depth of runoff

$$= (\text{average value of } Q \text{ for event}) \times (\text{duration of event})$$
$$= \frac{66}{6} \left(\frac{\text{mm}}{\text{h}}\right) 0.5 \, \text{h}$$
$$= 5.5 \, \text{mm} = 5.5 \times 10^{-3} \, \text{m}$$

Thus the total volume of runoff for the event

$$= (\text{equivalent ponded depth of runoff}) \times (\text{plot area})$$
$$= 5.5 \times 10^{-3} \times 35 \times 10 \, (\text{m}^3)$$
$$= 1.925 \, \text{m}^3$$

Thus

$$\bar{c} = \frac{18.6}{1.925} = 9.14 \, \text{kg m}^{-3}$$

So

$$\beta = \frac{\ln \bar{c}}{\ln \bar{c}_t} = 0.85$$

Note that β could alternatively be evaluated using Eq. (8.30). A further example of values of β is given in Fig. 8.9 later, which follows a discussion of flow-concentration effects.

The lower the value of the erodibility, β, the lower the magnitude of soil erosion from bare soil. The major reductions in soil loss that result from effective cover protection of the soil surface will be discussed in the next chapter.

As explained in introducing the erodibility parameter β, the value of β is dominantly a soil characteristic if flow-driven erosion dominates over other erosion processes. Should such other processes make a substantial contribution to the concentration of eroded sediment, the value of β is still just as useful for describing erodibility or predicting erosion, but β is then not solely a soil characteristic. The theoretical upper limit to β in flow-driven erosion is 1, and, if β is found to be greater than 1, a contribution from some other form of erosion is indicated.

Possible causes of an increase in β for any given soil can be a reduction in soil strength from excessive mechanical cultivation, or decreasing levels of organic matter, leading to loss of aggregate structure and

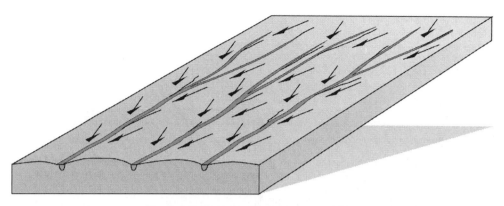

Figure 8.8 Illustrating the development of rills on a section of bare sloping hillside. Lateral flow to rills is also illustrated.

so lower values of v_{av}. Maintaining adequate levels of organic matter in a soil favours the formation and stability of soil aggregates, which also reduces soil erosion through increasing the depositability and infiltration rate. All such characteristics are strongly affected by soil texture, and in general soils high in sand or silt percentage tend to be prone to erosion.

Effects of flow concentration on soil erosion

So far in this chapter little specific attention has been paid to the observation that overland flow tends to follow preferred pathways, leading to spatial as well as temporal variation in the depth of overland flow. It is difficult to make quantitative observations of this type of irregularity if it is subtle and limited to minor unevenness in flow depth. However, especially for some types of soil, it is common for the increased erosion along preferred flow paths to develop into very visually obvious superficial channels, called 'rills'. When they have become adequately developed, rills can capture most of the downslope flow from the areas between the rills, called 'inter-rill' areas. Thus, as shown in Fig. 8.8, most of the excess rainfall and sediment developed in the inter-rill areas is likely to flow into the nearest downslope rill, especially if rills are not too far apart. If rills are close together, much of the sediment delivered laterally into rills may be due to rainfall-driven erosion. Rill development is not always obvious. However, if rills form, they are the major delivery systems for water and sediment, and it is the dynamic balance between erosion and deposition during transport in the rill which largely controls the net rate of erosion or deposition.

The existence or development of rills is itself evidence of localised enhancement of the erosion rate. Rill development can sometimes

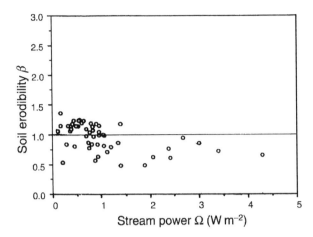

Figure 8.9 Values of β calculated for a number of erosive storm events on bare-soil erosion plots on the island of Leyte in the Philippines. The bare-soil plots developed rills, and had very high slopes of 50%, 60% and 70%. Plot weeding, hand cultivation and changes through time in soil-surface characteristics were involved in the changes in β.

resemble a miniature version of stream capture, which occurs in water-sheds, where several steams in the upper regions of a watershed can feed into or be captured by a more major stream. This type of rill development is also illustrated in Fig. 8.8.

For any given magnitude of runoff rate per unit area (Q), the type of flow concentration which occurs in rills leads to a considerable localised enhancement in stream power, as is explained in Marshall, Holmes and Rose (1996). This reference also describes how the expression for c_t given in Eq. (8.14) needs to be extended to accommodate the erosion-enhancing effect of rills. Fortunately, rills that develop during an erosion event can often remain in evidence after it, though 'drowning' of rills with sediment later in an event has been reported. The type of approach to analysis of soil-erosion data or erosion prediction developed earlier in this section which also incorporates the effectiveness of rills has been incorporated into the GUEST methodology for analysis and prediction of soil erosion (Yu *et al.*, 1997; Rose and Yu, 1998).

An example of use of the GUEST methodology is given in Fig. 8.9. Data on soil loss and runoff rate during an event as a function of time were obtained from hand-cultivated runoff plots. The plot size and hand-cultivation methods were typical of those used by subsistence farmers in the steep lands of the region. High rates of rainfall, yet soils of high infiltration rate, resulted in runoff being only a small fraction of rainfall. However, due to the very high land slopes, rates of soil erosion and sediment concentrations were high.

Figure 8.9 shows that values of β (defined in Eq. (8.28)) tend to decline as the stream power, Ω, increases, and to be greater than 1 at relatively low values of Ω. A possible reason why $\beta > 1$ at low Ω (and so shallow water depths) is that the intense rainfall could have made a significant contribution to the sediment concentration. Some of the scatter in the values of β was associated with the activities of weeding (required to keep the plot bare of vegetation) and hand cultivation, which was carried out in order to repair the results of major rill development.

The GUEST methodology referred to in this chapter (Yu *et al.*, 1997) is based on the concept illustrated in Fig. 8.6, in which erosion and deposition are conceived of as simultaneously occurring, though competing, processes. An alternative conceptual approach is developed in the WEPP methodology described by Nearing *et al.* (1990) and Toy, Foster and Renard (2002). In this methodology the term 'deposition' implies 'net deposition', which thus takes place only when the transport capacity is exceeded. Despite this conceptual difference, Yu (2003) has shown that the WEPP and GUEST modelling frameworks share a common unified mathematical structure under steady-state conditions. The difference between the two approaches thus lies chiefly in concept and so in parameter definition.

The WEPP methodology is very well documented and widely used in the USA, where parameter information required by the methodology is more readily available than it is in the rest of the world. WEPP documentation is also freely available on the world-wide web.

Much larger and deeper erosion features than rills are commonly described as 'gullies'. A gully involves cutting erosion into the subsoil, and, when gully systems are extensive, this indicates a seriously damaged landscape, often associated with inappropriate land use for the region. Just as stream formation occurs in response to the need for adequate drainage in undisturbed landscapes, so a gully can be regarded as a landscape response to accelerated overland flow initiated by a decrease in infiltration rate following unsustainable land use. In some cases gully formation can result from flow concentration caused by human construction, such as of roads and culverts.

As the scale of sediment transport increases, land-slope variation becomes significant, with a typical hillside variation in slope being shown in Fig. 8.10. When slope decreases with downslope distance, as in Fig. 8.10, a position in the landscape is reached, for any particular flow regime, type of land management and soil condition, where net erosion of soil gives way to net deposition. Siepel *et al.* (2000) have given a simple model of such hillslope erosion and net-deposition processes that also acknowledges the soil-conserving effect of vegetation or surface cover. These larger-scale issues and, in particular, the effect of land management on soil erosion and conservation at the watershed scale, will be considered in the next chapter.

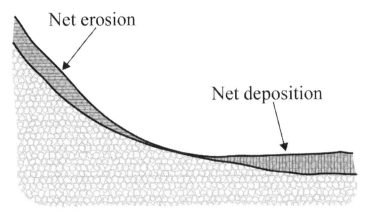

Figure 8.10 Areas of net erosion and net deposition on a typical hillslope.

Main symbols for Chapter 8

c	sediment concentration in water (Eq. (8.2))
$\bar{c}$	Average value of c during an erosion event
c_i	Sediment concentration for sediment in size class i
c_t	Sediment concentration at the transport limit (Eq. (8.14))
$\bar{c}_t$	Average value of c_t during an erosion event
d	Mass rate of deposition of sediment of all size classes per unit area of soil or river bed
d_i	Mass rate of deposition of sediment of size class i per unit area of soil or river bed
F	Fraction of excess stream power used in eroding sediment
I	Arbitrary number of size classes into which sediment is divided
k	Combined quantity defined in Eq. (8.20)
L	Downslope length of a plane land surface
n	Manning's roughness coefficient
q	Volumetric flux of water per unit width of flow (unit flux)
$q(L)$	Value of q at $x = L$
q_s	Sediment mass flux per unit width of flow
$q_s(L)$	Value of q_s at $x = L$
Q	Volumetric flux of water per unit land area
Q_e	An effective erosion-event average value of Q (Eq. (8.27))
S	Slope of land surface
ΔT	Period of time
v_{av}	Average or mean settling velocity of sediment (Eq. (8.13))
v_i	Settling velocity in water of sediment of size class i
V	Velocity of flow of water
x	Downslope distance

β Soil-erodibility coefficient defined in Eq. (8.23) or Eq. (8.24)

ρ Density of water

σ Wet density of sediment

Σ Summation symbol

τ Shear stress between overland flowing water and the soil surface

Ω Stream power

Ω_o Threshold stream power

Exercises

8.1 Describe and distinguish among the following forms of erosion:
   - mass movement of sediment as in landslides,
   - rainfall-driven erosion and
   - flow-driven erosion.

8.2 In Figs. 8.1(c) and (d) a distinction is made between removal of the original soil matrix by flow stresses, a process called entrainment, and similar removal of non-cohesive sediment deposited in the same erosion event, a process called re-entrainment. Discuss whether a similar distinction can be made when raindrop impact is the agent of sediment removal, where corresponding processes could be denoted detachment and re-detachment.

8.3 Justify that the unit flux of sediment is given by the product of the unit flux of overland flow and the sediment concentration (Eq. (8.3)).

8.4 Use Stokes' law, given in Eq. (8.9), to calculate the settling velocity in water, v, of a sedimenting particle of effective diameter $d_p = 70\,\mu m$. Find appropriate possible values for the other quantities in the equation.

8.5 In the presence of overland flow, the trajectory of a sedimenting particle is more realistically represented in Fig. 8.1(b) than in Fig. 8.4. Critically evaluate the assumption made in deriving Eq. (8.10) that lateral displacement of a particle during sedimentation, which is ignored in Fig. 8.4, does not affect the result given in Eq. (8.10).

8.6 Assuming that $v_{av} = 5\,mm\,s^{-1}$, calculate the rate of deposition of sediment, d, using Eq. (8.12) and assuming that $c = 45\,kg\,m^{-3}$. Express d in units of $kg\,m^{-2}\,s^{-1}$ and in tonne $ha^{-1}\,min^{-1}$.

8.7 In Section 8.3 an order-of-magnitude estimation of the various fluxes in Fig. 8.6 was carried out. In this illustration certain assumptions were made concerning the parameters involved. Repeat this type of flux-magnitude comparison making the same assumptions, but adopting the following different values for the parameters involved: $c = 40\,kg\,m^{-3}$, $v_{av} = 0.2\,m\,s^{-1}$, $Q = 30\,mm\,h^{-1}$ and $L = 40\,m$.

8.8 (a) Use Eq. (8.14) to calculate the transport-limit sediment concentration, c_t, assuming that $F = 0.1$, $v_{av} = 0.1\,m\,s^{-1}$,

$\rho = 1000\,\mathrm{kg\,m^{-3}}$, $\sigma = 2000\,\mathrm{kg\,m^{-3}}$, $S = \sin 7°$ and the depth of overland flow is 2 cm. In order to calculate the flow velocity V, use Manning's equation with $n = 0.03\,\mathrm{m^{-1/3}\,s}$.

(b) Calculate the flux of sediment, q_s, which crosses unit slope width at the location where the flow depth D is 2 cm. (Note that $q = DV$ (Eq. (8.15)).)

(c) Suppose that the accumulated sediment transport over a wet season can be estimated by assuming that q_s has the value calculated in part (b) and persists for a period of 40 min. Suppose also that this flux q_s flows into a large farm dam where the perimeter length of delivery to the dam is 120 m. Calculate the mass of sediment in tonnes (1 tonne $= 1000\,\mathrm{kg}$) deposited in the dam during that wet season.

8.9 Using Eq. (8.27), calculate the effective runoff rate per unit area from a runoff plot of length $L = 35\,\mathrm{m}$ and width $W = 12\,\mathrm{m}$ given the following information. The entire volumetric flux of water from the bottom of the plot was collected and measured as a function of time. Five-minute-average values of this flux in $\mathrm{m^3\,min^{-1}}$ for the runoff event are as follows:

$$0.05;\ 0.09;\ 0.15;\ 0.21;\ 0.16;\ 0.1;\ 0.04$$

Calculate Q for these data, and then calculate Q_e using Eq. (8.27). Also calculate the total volume of runoff for the event.

8.10 The average measured sediment concentration in an erosion event, $\bar{c}$, is $32\,\mathrm{kg\,m^{-3}}$. The corresponding value of $\bar{c}_t$ is $48\,\mathrm{kg\,m^{-3}}$, which was calculated from Eq. (8.14) assuming that $F = 0.1$. Using Eq. (8.28), calculate the value of the erodibility, β, of the soil in this event.

In describing the calculation of $\bar{c}_t$ (or c_t) it was noted that there was some uncertainty in the exact value of F (Eq. (8.14)). Recalculate the values of $\bar{c}_t$ and β assuming that $F = 0.2$ instead of 0.1.

References and bibliography

Ciesiolka, C. A. A., Coughlan, K. J., Rose, C. W., Escalante, M. C., Hashim, G. M., Paningbatan, E. P., and Sombatpanit, S. (1995). Methodology for a multi-country study of soil erosion management. *Soil Technol.* **8**, 179–192.

Ciesiolka, C. A. A., and Rose, C. W. (1998). The measurement of soil erosion. In *Soil Erosion at Multiple Scales – Principles and Methods for Assessing Causes and Impacts*, eds. F. W. T. Penning de Vries, F. Argus and J. Kerr. Wallingford: CABI Publishing, in association with the International Board for Soil Research and Management (IBSRAM), pp. 287–301.

Forrester, J. E. (1970). *Industrial Dynamics*. Cambridge, Massachusetts: MIT Press.

Foster, G. R. (1982). Modelling the soil erosion process. In *Hydrologic Modelling of Small Watersheds*, ed. C. T. Hann. St. Joseph, Michigan: American Society of Agricultural Engineering, pp. 297–379.

Hairsine, P. B., and Rose, C. W., (1992). Modelling water erosion due to overland flow using physical principles I. Uniform flow. *Water Resources Res.* **28**, 237–243.

Lovell, C. J., and Rose, C. W. (1988). Measurement of soil aggregate settling velocities I. A modified bottom withdrawal tube method. *Aust. J. Soil Res.* **26**, 55–71.

Lovell, C. J., and Rose, C. W. (1991). Wake capture effects observed in a comparison of methods to measure particle settling velocity beyond Stokes's range. *J. Sedimentary Petrol.* **61**, 575–582.

Marshall, T. J., Holmes, J. W., and Rose, C. W. (1996). *Soil Physics*, 3rd edn. New York: Cambridge University Press.

Nearing, M. A., Lane, L. J., Alberts, E. E., and Laflen, J. M. (1990). Prediction technology for soil erosion by water: status and research needs. *Soil Sci. Soc. Am. J.* **54**, 1702–1711.

Renard, K. G., Foster, G. R., Weesies, G. A., and Porter, J. P. (1991). RUSLE: revised universal soil loss equation. *J. Soil Water Conservation* **46**, 30–33.

Rose, C. W. (1993). Erosion and sedimentation. Chapter 14 in *Hydrology and Water Management in the Humid Tropics – Hydrological Research Issues and Strategies for Water Management*, eds. M. Bonell, M. M. Hufschmidt and J. S. Gladwell. New York: Cambridge University Press, pp. 301–343.

Rose, C. W., and Yu, B. (1998). Dynamic process modelling of hydrology and soil erosion. In *Soil Erosion at Multiple Scales – Principles and Methods for Assessing Causes and Impacts*, eds. F. W. T. Penning de Vries, F. Argus and J. Kerr. Wallingford: CABI Publishing, in association with the International Board for Soil Research and Management (IBSRAM), pp. 269–286.

Siepel, A. C., Steenhuis, T. S., Rose, C. W., Parlange, J.- Y., and McIsaac, G. F. (2002). A simplified hillslope erosion model with vegetation elements for practical applications. *J. Hydrol.* **258**, 111–121.

Toy, T. J., Foster, G. R., and Renard, K. G. (2002). *Soil Erosion: Processes, Prediction, Measurement, and Control*. New York: John Wiley and Sons.

Yu, B., Rose, C. W., Ciesiolka, C. A. A., Coughlan, K. J., and Fentie, B. (1997). Towards a framework for runoff and soil loss using GUEST technology. *Aust. J. Soil Res.* **35**, 1191–1212.

Wischmeier, W. H., and Smith, D. D. (1978). *Predicting Rainfall Erosion Losses – a Guide to Conservation Planning*. Agriculture Handbook No. 537. Washington: US Department of Agriculture.

Yu, B. (2003). A unified framework for water erosion and deposition equations. *Soil Sci. Soc. Am. J.* **67**, 251–257.

9
Watersheds and rivers

9.1 Introduction

There is general agreement that, over much of the earth's surface, there must be changes in the way that the land, rivers and oceans are used and managed in order to move towards more sustainable use and the protection of the natural processes which sustain life. The continuous growth of cities and urban populations in particular is sustained only by the supply of a diverse range of basic services, which draw on vastly greater land areas than that occupied by the cities themselves. Such services include a reliable supply of water, food, timber, energy and waste processing, all of which assumes that we have a sustained ecological resource that can provide such services indefinitely and, therefore, without degradation of the resource.

A vast number of studies and ample information provide evidence that, despite the global scale of resource capture and distribution, the current demand for resources, dominated by the 'developed' world, is being met only by a running down of environmental quality. Thus there are basic questions of concern as to the sustainability of the much-modified ecosystems which meet this demand.

In this chapter we consider some mainly physical aspects of how land use and land management affect the basic processes which have an impact on the many and linked issues of environmental quality and

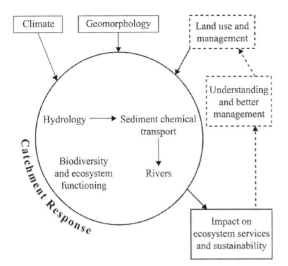

Figure 9.1 Some of the factors involved which affect catchment response, illustrating how a recognition of deleterious outcomes may lead to improvement in catchment health through the better use and management of land and water resources in the catchment.

ecological sustainability. The consequences of how land is used and managed are evident in our river systems and on the sea shores to which they ultimately deliver some of the sediment, nutrients and contaminants added to them. In Section 9.6 the concept of 'river health' is discussed as an integrated way of assessing trends in the ecological sustainability of a river basin or 'catchment'.

It is a challenge even to link together conceptually the various influences on a catchment or its component watersheds. In particular, can we understand why particular uses of land and methods of land management degrade the integrity of natural ecosystems more than others?

Figure 9.1 seeks to list some of the numerous factors and processes which affect the biophysical response of a catchment or river basin, focusing on issues of land use and management that are commonly central to the ecological integrity or biological health of a catchment.

The factors of climate and geomorphology in Fig. 9.1 were commonly assumed to be virtually uninfluenced by human activity, or it was thought that any influence would be small and long-term in nature. Whilst climate change has long been recognised to have occurred at geological time scales, significant change due to the enhanced greenhouse effect is now understood to be likely over the time scale of the next human generation. Likewise, change in geomorphological characteristics (Fig. 9.1), normally considered a slow evolutionary process, can be accelerated or modified by human activity, notably in cities and rivers, but also in rural areas where substantial alterations of land-surface characteristics occur. Satisfying the enormous needs of cities for ecosystem services and products is an important driver in such change.

The box surrounding land use and management in Fig. 9.1 is shown dashed in order to draw attention to this as a factor where change can be particularly rapid and substantial in terms of the scale of a human lifetime. Even more rapid is the mechanical clearing of timber and other vegetation using equipment such as a bulldozer.

Figure 9.1 lists just four related but distinguishable types of processes or factors involved in catchment response to changes in the inputs shown. These are the following.

- The change in hydrological response of the catchment, which in turn is closely linked to the the generation, transport and delivery to streams of sediment and associated nutrients and other chemicals. The hydrological implications of salinity will be considered later on in Chapter 12.
- Some, though by no means all, of the sediment and chemicals mobilised within a watershed reaches the streams which drain it. This delivery to streams affects the physical characteristics of the stream and stream bed, as well as the vast range of biota which is characteristic of a healthy stream.
- Both on the land and in the water bodies of a catchment, alterations in all these processes have direct consequences on diversity of life and on the great range of life-supporting processes on which life depends.

Relieving the impact on ecosystem services, such as clean water, requires an understanding of these catchment responses and a capacity to develop and implement better land-management systems.

9.2 Hydrological considerations at watershed and catchment scales

A large river basin typically contains not only a major river, but also contributing streams (Fig. 9.2). There is no universal agreement in the terminology employed, with the terms drainage basin, catchment and watershed sometimes being used interchangeably, with stream discharge via a common outlet being the common element in their description. In this book the terms 'catchment' and 'drainage basin' will be used interchangeably to denote the entire land area where overland flow of a major river system is directed towards a common outlet. The term 'watershed' will be used to denote the subareas drained by separate streams that ultimately contribute their water to the major river system (Fig. 9.2). In some literature the term watershed is used to describe the catchment boundary or divide, overland flow being shed in opposite directions on either side of this boundary (Fig. 9.2).

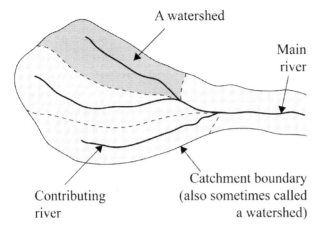

A watershed

Main river

Contributing river

Catchment boundary (also sometimes called a watershed)

Figure 9.2 The term catchment will here be used to indicate a collection of watersheds as illustrated in the figure, though there is no general agreement on terminology, the catchment boundary sometimes being referred to as a watershed or catchment divide.

The water budget for a watershed was developed in Section 4.4 using the principle of mass conservation of water, which is expressed in Eq. (4.3). Some of the ways of measuring the various component terms in Eq. (4.3) were also outlined in Chapter 4. A water budget quantitatively relates the collection by, storage within and discharge of water from a watershed or other defined region. A budget approach has the advantage that sources must be balanced by changes in storage or exports, so that there is the capacity to test the adequacy of our knowledge of the water-budget components measured or predicted separately. A budgeting approach can be applied equally to water, sediment, or nutrients at a watershed or catchment scale.

The entire volume of soil and subsoil in a watershed acts as a large sponge that buffers the commonly relatively rapid rates of rainfall, snowmelt and surface-runoff events. Change in the amount of water stored in the root zone of a watershed was addressed in Section 4.5 in terms of the one-dimensional model shown in Fig. 4.11, changes being tracked by a simple accounting or budgeting-type procedure. Use of a one-dimensional or 'bucket-type' of water-balance-accounting method has proved very useful in understanding general features of the changing hydrological balance of the soil–plant zone. This procedure is particularly useful wherever water is a limiting resource, being especially helpful in sub-humid climates, where it can be used to interpret seasonal variation in plant and crop production, for instance. However, such procedures cover only a part of watershed hydrology.

As illustrated in Fig. 4.7, water can leave a watershed by evaporation, as surface runoff, or as subsurface flow. Subsurface flowing water can enter streams or rivers, providing the continuity in stream flow following cessation of overland flow into rivers. This residual continuous stream

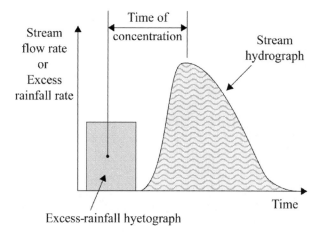

Figure 9.3 A storm hydrograph for an ephemeral headwater stream in response to an idealised block of uniform excess rainfall rate.

flow, fed by the capture of subsurface flows, is commonly referred to as 'base flow'. In contrast, it is also possible, in at least some sections of a river, for water to be lost by infiltration into regions beneath the river bed.

Particularly because of the damage that can result from river flooding, there is considerable interest in the nature of a river's flow response to storm or heavy rainfall within its watershed. For overland flow generated by excess rainfall on a simple plane land surface, Fig. 7.8, which illustrates the form of Eq. (7.19), shows that the peak in runoff rate Q occurs well after the peak in excess rainfall rate R, the magnitude of this lag depending on the parameter K_1. The magnitude of K_1 increases with slope, length and surface roughness, and decreases with slope, as discussed in Section 7.5. Whilst this role of K_1 is clear for the simple plane geometry assumed in Eq. (7.19), even in complex watersheds a similar theory is used by hydrologists and engineers, with the magnitude of K_1 being obtained as follows. A hydrograph is the measured relationship of stream flow or discharge against time. In hydrological practice the value of K_1 is obtained by fitting a model of watershed hydrology to the measured hydrograph.

Figure 9.3 shows a simple hydrograph for an ephemeral headwater stream that flows only in response to overland flow from adjacent hillslopes. The hydrograph shown is in response to an idealised hyetograph of excess rainfall of constant rate for a fixed time period (Fig. 9.3).

The flow rate of a stream, also called the 'discharge', is commonly expressed in units of $m^3 \, s^{-1}$ (cubic metres per second, or cumecs), or else in cubic feet per second (commonly abbreviated to cfs). The 'time of concentration' shown in Fig. 9.3 is one measure of the hydrological lag involved. The shape of the hydrograph given in Fig. 9.3 is typical

Hillslopes delivering overland flow to stream

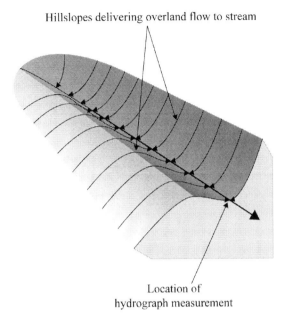

Figure 9.4 An idealised representation of a headwater watershed.

Location of
hydrograph measurement

for a short sharp rainfall event, with the rising segment or 'limb' of the hydrograph being steeper than the 'falling limb'. An idealised representation of a headwater stream watershed is given in Fig. 9.4. The theory for flow on a plane land surface given in Sections 7.5 and 7.6 provides an approximate theoretical framework for considering overland flow down either of the two hillslopes which deliver water to the stream. If the stream itself can be idealised as of constant slope, then the same theory as that used to describe flow down the hillslopes indicates that a further lag develops for flow in the stream collecting runoff from both hillslopes. This is illustrated in Fig. 9.5.

For a stream or river that is either contributed to by other streams (Fig. 9.2) or collects subsurface flow, its hydrograph will indicate some base flow prior to and following the storm hydrograph shown in Fig. 9.3. With more complex patterns of rainfall delivery the shape of the stream hydrograph is also more complex. Also rainfall over larger watersheds or catchments can no longer be assumed spatially uniform, which is a reasonably accurate approximation for small watersheds. Rainfall may cover only part of a catchment, and furthermore, the way in which the rainfall pattern moves over the component watersheds will strongly affect the hydrographs of all streams, including the main river to which the streams contribute.

There is ample evidence to show that, in humid climates, water contents are typically higher in valley bottoms, and, given suitable

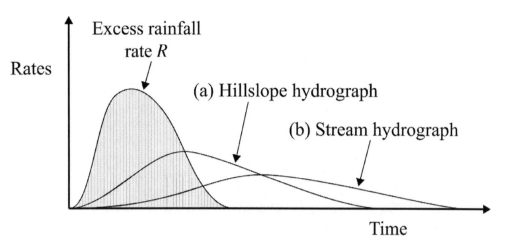

Figure 9.5 Two hydrographs resulting from an excess-rainfall storm event on the watershed. (a) The hillslope hydrograph measured at the base of either hillslope in Fig. 9.4. (b) The stream hydrograph measured at the stream exit shown in Fig. 9.4.

topography, zones of saturation can be produced as illustrated in Fig. 6.8. During rainfall, overland flow is produced on such saturated zones of negligible infiltration (Section 6.3). Since an increase in rainfall rate will lead to an increase in infiltration outside the saturated zone, then the average infiltration rate, $\bar{I}$, for the watershed as a whole will also increase. This type of positive response of $\bar{I}$ to P is illustrated in Eq. (6.5). In Sections 6.4 and 6.5 experimental justification is given for this type of response in situations in which overland flow is largely generated by rainfall rate exceeding infiltration rate (so-called infiltration-excess, or Hortonian, overland flow).

Equation (6.5) approximately encapsulated the typical effect of a spatially varying infiltration rate. Such variation, which is related to variability in soil properties, pore space and voids and to other factors such as spatial variation in vegetation, significantly affects the distribution of moisture in soil in the root zone. Such recognition of spatial variability in infiltration rate is also related to the well-established partial-area concept in hydrology. This concept recognises that the regions of a watershed which contribute to overland flow (and thus to sediment transport) change dynamically during a runoff event.

In humid zones where precipitation exceeds evapotranspiration for significant periods of time, and where also some less-permeable restrictive layer exists in the soil profile, water can infiltrate and then move laterally in the soil from hillslope to valley bottom (Fig. 9.6(a)). In the absence of such a restrictive layer, infiltrating water can move down

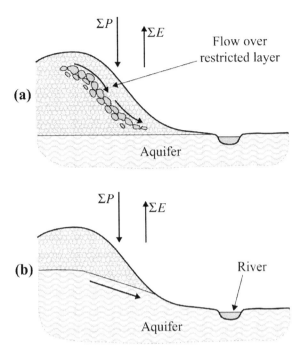

Figure 9.6 The cross-section of a humid-zone watershed where rainfall over the period of consideration (ΣP) exceeds evapotranspiration (ΣE). (a) The presence of a flow-restricting layer in the soil profile leads to saturation and lateral flow towards the aquifer and river. (b) With no impeding layer, infiltrating water builds up the aquifer, and lateral aquifer flow can be an even greater contributor to river flow.

through the soil profile and build up the depth of the water table; the slope in this water table can also generate lateral flow (Fig. 9.6(b)). In all such circumstances the topography of the watershed plays an important role in its hydrology (O'Loughlin, 1986; Grayson and Western, 2001). When overland flow occurs, this is likely to be saturation overland flow on saturated areas.

However, in drier climates, there is evidence that it is commonly a useful simplification to consider a watershed as consisting of two almost separate components: (a) the hillslopes and (b) the valley bottoms which collect and convey water, thereby being moister than the hillslopes. During dry periods, movement of water in the hillslopes is dominantly vertical, and can be considered independently of change in the valley bottom (Fig. 9.7). It is during less-frequent runoff-producing rainfall that the connection between hillslopes and valley bottom is established, with infiltration-excess, or Hortonian, overland flow.

As mentioned in Exercise 6.4, a widely used method of estimating total runoff (ΣQ) from total effective rainfall ($\Sigma P - F_o$) on watersheds is the SCS method developed by the US Soil Conservation Service (USDA Soil Conservation Service, 1985). The term F_o is interpreted as the depth of rainfall required to generate runoff, a term incorporating the effects of interception by vegetation and pre-wetting of the soil surface. It is

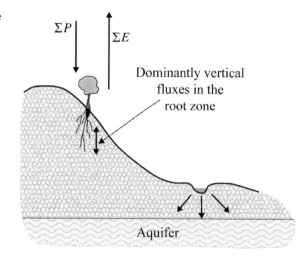

Figure 9.7 A semi-arid zone or a dry period in a humid zone. ΣP and ΣE are as in Fig. 9.6, but with $\Sigma P < \Sigma E$. Water fluxes on the hillslope are dominantly vertical and restricted to the root zone of vegetation.

sometimes referred to as an 'initial abstraction' or an 'initial infiltration amount'. As shown by Yu (1998), the SCS method is compatible with the exponential infiltration equation (6.5), provided that the rainfall rate is also exponentially distributed. This description of the rainfall-rate distribution has general experimental support. This form of distribution implies that lower rainfall rates are more common than high rates, following a pattern similar to that illustrated in Fig. 6.15(b) or Fig. 6.18.

Using the symbol I_m from Eq. (6.5) to indicate the maximum mean infiltration rate for the watershed and T to denote the effective duration of a runoff event, the water-storage term S_t (usually denoted S in the SCS method's documentation) is given by

$$S_t = I_m T \tag{9.1}$$

It then follows that the equation used in the SCS method for the runoff accumulated over a storm event of effective duration T is

$$\Sigma Q = \frac{(\Sigma P - F_o)^2}{(\Sigma P - F_o) + I_m T} \tag{9.2}$$

In using this equation, when information on the magnitude of F_o is lacking, it is commonly approximated by $0.2 S_t$ or $0.2 I_m T$. With this approximation, Eq. (9.2) becomes

$$\Sigma Q = \frac{(\Sigma P - 0.2 I_m T)^2}{\Sigma P + 0.8 I_m T} \tag{9.3}$$

Figure 9.8 presents Eq. (9.3) in graphical form for a range of values of $I_m T$.

If $I_m T = 0$, then, from Eq. (9.3), $\Sigma Q = \Sigma P$, which is shown in Fig. 9.8 as a 1 : 1 line. $I_m T = 0$ is appropriate for describing runoff from roofs or

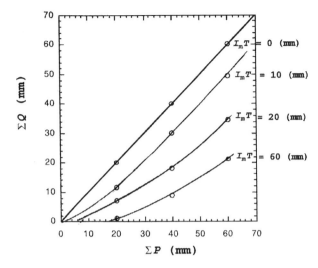

Figure 9.8 Relationships between event total runoff ΣQ and total rainfall ΣP for a range of values of the storage term $I_m T$ in Eq. (9.3). (Note that $I_m T$ is denoted S in the SCS procedure's documentation (USDA Soil Conservation Service, 1985).)

pavement. For $I_m T > 0$, Fig. 9.8 shows that the relationship between ΣQ and ΣP becomes more and more curved as $I_m T$ increases. The typically curved nature of this relationship is well supported by results from a very wide range of watershed studies at scales to which the SCS method could be expected to apply. As the storage term $I_m T$ increases, ΣQ decreases for any given ΣP, this being due to the increase in infiltration described by this storage term.

Whilst the approximation made by assuming $F_o = 0.2 I_m T$ is useful in understanding the general effect on total runoff of an increase in storage of water in the soil profile, F_o depends on factors in addition to I_m and T. As shown by Ward and Elliott (1995) and Yu et al. (2000), the magnitude of F_o depends on the amount of water stored in the soil prior to the runoff event. Hence F_o can be related to antecedent rainfall, and Fig. 9.9 illustrates one such relationship. However, the details of such a relationship may vary with season, location and other factors, as well as with $I_m T$.

In the documentation for the SCS method (USDA Soil Conservation Service, 1985), the storage parameter $I_m T$ (written S in this documentation) is inversely related to a 'curve number'. This curve number has been used to summarise experience on how storage depends on soil type and other factors. Choice of the SCS curve number is guided by any available information on the infiltration characteristics of the soil, the type of land use and other factors. Ward and Elliott (1995) provide a useful summary of the SCS methodology using English units widely used in the USA, rather than the metric units used elsewhere.

Figure 9.9 The relationship
between the amount of prior
10-d rainfall and the average
value of F_o (the initial
infiltration amount), at a site
in Khon Kaen, Thailand. Error
bars represent one standard
deviation. (From Yu *et al.*
(2000).)

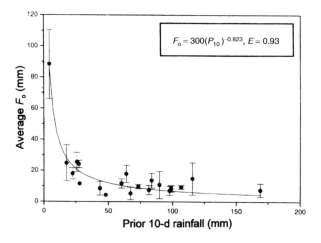

Figure 9.9 The relationship between the amount of prior 10-d rainfall and the average value of F_o (the initial infiltration amount), at a site in Khon Kaen, Thailand. Error bars represent one standard deviation. (From Yu *et al.* (2000).)

Example 9.1

Use the value of $I_m = 50\,\mathrm{mm\,h^{-1}}$ derived to fit the data in Figs. 6.14 and 6.17, assuming an effective duration of runoff, T, of 0.5 h. Using the relationship for F_o given in Fig. 9.9, calculate the form of the relationship between ΣQ and ΣP for the following amounts of prior 10-day rainfall (P_{10}, in mm): 25, 75 and 150.

Solution
See Fig. 9.10.

The methodology increasingly being used to model the hydrology of watersheds and the aggregates of watersheds in a catchment or river basin involves computer-based geographical information systems (GIS). GIS-based methods greatly facilitate spatial coordination of described characteristics such as surface topography, rainfall rate, soil characteristics, land use, land cover and the geometry of stream systems. De Vantier and Feldman (1993) have reviewed the application of GIS to hydrological modelling. By efficiently handling the spatial complexity involved in modelling processes at the catchment scale, such techniques hold out considerable promise, though gathering the basic data and parameters needed to ensure realism in prediction provides very substantial challenges.

9.3 Sediment transport in watersheds

This section extends the interpretation of processes involved in erosion and deposition by water given in Chapter 8 to the larger scale of a watershed. Furthermore, a brief review of why some of the erosion-reduction

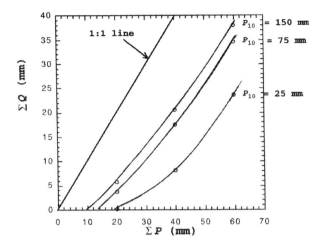

Figure 9.10 The relationship between ΣQ and ΣP obtained using Eq. (9.2) when $I_m T = 25$ mm with a range of values of initial infiltration, F_o, taken from Fig. 9.9.

management opportunities or soil-conservation practices employed at that scale are effective will be given.

At watershed scales the mass movement of soil and rock, as in landslides, can play a substantial if spasmodic role in sediment transport. As discussed in Chapter 2, soil strength interacting with slope and climatic variables plays a major role in initiating such events, which can be massive and devastating in their effect.

Extensive and erosive overland flows of water can occur at the time of melting of extensive snow fields. The freezing and thawing of soil involved in snow-covered regions leaves the soil in a weak and erodible condition, though these processes can be advantageous in breaking down large clods formed by cultivation.

When overland flow is generated by rainfall rather than snowmelt, at watershed scales runoff commonly has the opportunity to reach such velocities and depths of flow that soil erosion is dominantly driven by flow rather than rainfall impact (processes discussed in Section 8.2). However, especially when rain falls on bare soil, it can break down the structural integrity of soil aggregates, and the resulting fine sediment can infill and clog the passages through the soil, reducing the rate at which infiltration can occur (Section 8.2). In this way rainfall impact can result in an increase in overland flow, and enlarge the areas which generate such flow.

The erosion model developed by the Water Erosion Prediction Project of the US Department of Agriculture's Agriculture Research Service (referred to as the WEPP model), which is based on extensive experimentation carried out in the USA, allows prediction of the effects of different agricultural practices on soil erosion (Lane *et al.*, 1992;

Laflen *et al.*, 1997). Erosion between rills is treated as driven by rainfall impact, and erosion within rills as flow-driven. Different versions of WEPP have been developed for application at the hillslope and field scales.

Equation (8.3) shows that the sediment flux depends directly on the product of the unit volumetric flux of water, q, and the sediment concentration, c. For bare soil the sediment concentration at the transport limit, c_t, provides an upper limit to flow-driven sediment concentration, and, from Eq. (8.14),

$$c_t \propto SV \qquad (9.4)$$

where S is the land slope and V the velocity of overland flow. From Manning's equation (Eq. (7.7)), V also depends on S. In the solution to Example 7.3, we found that, for steady sheet flow down a uniform hillslope with excess rainfall rate R,

$$V = (S^{0.3}/n^{0.6})(RL)^{0.4} \qquad (9.5)$$

indicating that V increases with downslope flow distance in proportion to $L^{0.4}$. In Eq. (9.5) Manning's n is a measure of the roughness of the soil surface to overland flow. The formation of rills, which is quite common with rather bare soil and illustrated in Fig. 8.8, collects and channels water, thereby increasing V and thus c_t (Eq. (9.4)).

Equations (9.4) and (9.5) also show that c_t will decrease with a reduction in slope. Even if the sediment concentration c is less than c_t, due to the erodibility factor β of Eq. (8.28) being less than unity, in general c will still decline with S. Thus, in the typical segment of hillslope shown in Fig. 8.10, regions of net erosion on a hillslope will usually be followed by regions of net deposition. Thus, even if an entire hillslope segment is laid bare by cultivation, much of the sediment eroded from the higher slope segment of the hillslope will be deposited on the slope segment below it. In Fig. 9.11 this segment of net deposition is adjacent to a stream.

Thus, by no means all the sediment mobilised in a watershed by erosion will generally reach a stream. In some watersheds it is estimated that about 5% to 10% of sediment mobilised is delivered to the stream within it. However, this sediment-delivery fraction depends substantially on the topography and other land characteristics of the watershed, as well as on the nature of the runoff-generating event. As will be further discussed in Sections 9.5 and 9.6, it is common, especially when cultivated land is extensive in a watershed, for enough sediment (and chemicals attached or sorbed to it) to reach streams for stream-water quality and its biological suitability to be degraded. Sediment contributing to this degradation in river habitat can also come from gullies and erosion of the stream

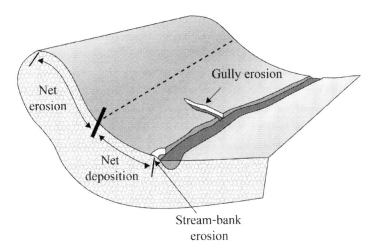

Figure 9.11 Illustrating that, in a typical watershed hillslope, regions of increasing downhill slope and potential net erosion are commonly followed by regions of gentler slope where net deposition will occur. The possibility of erosion of gullies and stream banks is also indicated.

bank (Fig. 9.11). In some particular climatic and soil situations these particular spatially concentrated forms of erosion can be the dominant contributor to sediment in streams.

It was demonstrated towards the end of Section 8.3 that sediment concentration is dominantly controlled by the balance between the two fast processes of erosion and deposition. These two processes remove sediment from, and return sediment to, the soil surface, respectively (Fig. 8.6). No change in sediment concentration implies that these two oppositely directed fluxes are of equal magnitude. This understanding helps us appreciate why a decrease in slope (and so in flow velocity) leads to net deposition, as discussed in association with Fig. 9.11.

However, even when the slope does not decrease, there are several important and commonly employed soil-conservation methods that considerably decrease the downhill flux of eroding sediment. One of the simplest and most effective of such methods is to leave or grow a strip of reasonably dense vegetation along the contour so that it intercepts and slows down overland flow. This 'buffer strip' or 'contour hedgerow' of vegetation could be grass or any other reasonably ground-hugging vegetation the land user may be interested in growing, and can include vegetation with harvestable products (Fig. 9.12). It is a very important objective that such a protective strip of vegetation is left in place or reinstated on all stream and river banks, because this location is effectively the last line of defence of the stream from polluting sediments generated in the watershed it drains. This 'riparian', or river-bank, strip of vegetation is also shown in Fig. 9.12.

In the presence of overland flow the hydraulic resistance provided by buffer strips slows down the flow even before it reaches the strip. The

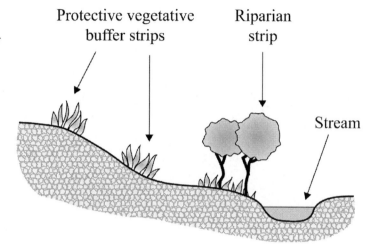

Figure 9.12 Illustrating possible locations of sediment-flux-reducing buffer strips in association with cultivated land. Also shown is a well-vegetated stream-protecting riparian strip, which preferably would include trees as well as ground-hugging plants and litter.

reduction in velocity in this upstream region of decelerated flow implies a reduction in erosion rate, so that the deposition process wins and net deposition of sediment occurs. Because larger sediment or aggregates settle more quickly than does smaller or finer sediment, the layer of deposited sediment formed just upslope of the buffer strip tends to be richer than the eroding sediment in larger aggregates. Figure 9.13(a), which represents an early stage in sediment trapping, seeks to display the feature of size-selective deposition, coarser sediment being deposited first. Some of the fine sediment might be neither deposited upstream of the buffer strip nor be trapped by it, thus moving through and beyond the buffer strip with the onflowing water.

Figure 9.13(b) illustrates a later stage in deposition upslope of a vegetative buffer strip, where the deposited sediment is sufficient to alter substantially the slope of the land surface over which it flows. The resultant reduced slope will also favour deposition over re-entrainment of the sediment. If an erosion event is of sufficient severity, or the distance between buffer strips is too great, the deposit of sediment illustrated in Fig. 9.13(b) may overwhelm the soil-conserving capacity of the buffer strip.

Soil accumulation, illustrated in Fig. 9.13(b), together with downslope soil displacement due to cultivation can ultimately result in the slow formation of terrace-like steps on the hillside. Even though there are very labour-intensive to construct, some 13 million hectares of constructed terraces are reported to have been built in recent decades on sloping agricultural land in the People's Republic of China to overcome devastating erosion problems in that country (Tang Ya et al., 2002). However, these authors have also reported on distinct soil-fertility advantages over

(a)

Incoming
sediment-laden water

Vegetated buffer
strip

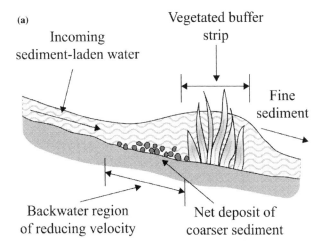

Fine
sediment

Backwater region
of reducing velocity

Net deposit of
coarser sediment

(b)

Incoming
sediment-laden water

Vegetated buffer strip

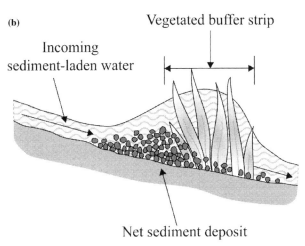

Net sediment deposit

Figure 9.13 (a) Illustrating
the early consequences of the
decelerated flow region of
hydraulic adjustment formed
upslope of a vegetated buffer
strip or contour hedgerow.
The increase in flow depth,
and corresponding reduction
in velocity in the backwater
region, reduces the rate of
erosion, allowing net
deposition of sediment to
occur. (b) A possible later
stage than that shown in
(a) of accumulation of trapped
sediment upslope of a
vegetated buffer strip or
contour hedgerow.

terracing of using contour hedgerows of perennial nitrogen-fixing shrubs
or trees. Such vegetative barriers are periodically pruned back to restrict
their competition with the crop grown between the hedgerows, which is
mulched with the hedgerow prunings.

Meyer, Dabney and Harmon (1995) examined the effectiveness for
trapping sediment of buffer strips or hedges, 0.2 m wide in the direction of
flowing sediment, consisting of the stiff grasses vetiver and switch grass.
Nearly all sediment of sand size and above was deposited upslope of the
hedge (as shown in Figs. 9.13(a) and (b)), but the finer the sediment the
more likely it was to pass right through the hedge. Thus the sediment-size
distribution was found to be the dominant factor governing the efficiency
of the buffer strip for trapping sediment.

Such soil-conserving methods are helpful at all scales, and, if there is flow in the watershed stream, it is the final integrator of the effects of the variety of land uses and management practices within it. Recognising that the use of land for agriculture, and also urban, roadway, and industrial uses, commonly leaves soil exposed at some time in the relevant cycle of operations, the timely application of soil-conserving methods can minimise both on-site and off-site damage by net erosion or by net deposition.

Soil on lands modified by human action can be most effectively conserved by seeking to mimic as closely as possible the type of soil-surface protection and strengthening that occurs in most (though not all) natural catchments. For example, it has been found possible in cropping systems to reduce the amount and the degree of aggression in tillage operations. Such 'minimum-tillage' systems of operation have often been found to lead to acceptable or even improved agricultural productivity, as well as contributing to the long-term sustainability of the farming system and reducing ecological damage.

Furthermore, keeping in place an effective protective cover on the soil surface has commonly been found to reduce soil loss to levels similar to natural rates in all but extreme events. Perhaps the most widely practised soil-conservation measure is the return to the soil surface of stems and other parts of crops that are not commonly used commercially. Though this practice is commonly effective at reducing soil erosion, even full cover by such material can be swept away in extreme rainfall events, resulting in severe soil loss. Soil is weakened by cultivation, although the root systems of growing crops or retained stubble of a harvested crop can provide some of the soil strengthening provided by natural vegetation. In parts of Africa and elsewhere, vegetation unsuitable for human consumption is often in great demand for animal feed, so its retention for soil protection is seen as a loss, which it may indeed be in the short term.

As explained in texts covering soil conservation (Hudson, 1981; Morgan, 1986; Toy, Foster and Renard, 2002), there is a very wide variety of soil-conserving methods that can be assessed for suitability and acceptability in different contexts. The main soil-conservation methods have been classified into those of overall land-management methods (such as changing land use), agronomic/soil-management methods, vegetative measures (such as vegetative buffer strips) and structures (such as constructed terraces of various kinds). Because of the multitude of factors involved, implementation of sustainable soil-conserving practices requires ingenuity and local knowledge, informed by an understanding of hydrological and soil-erosion processes, to which Chapter 8 and this chapter seek to contribute.

In practice the calculation or prediction of sediment transport in watersheds commonly involves the use of computer-implemented models of the processes involved. At a small-watershed or hillslope scale the WEPP methodology mentioned earlier is increasingly being used in the USA, the aim being for it to replace earlier methodologies. The use of WEPP outside the USA is currently limited by the availability of the type of information required for its use.

As scale is increased, there is considerable evaluation of alternative methodologies for predicting sediment transport and its water-quality implications (Ghadiri and Rose, 1992). For reasons mentioned in Section 9.2, in recent times methodologies have increasingly made use of computer-based GIS platforms. Savabi *et al.* (1995) and de Roo, Wesseling and Ritsema (1996) reported significant examples of this trend.

Evidence of erosion and deposition at watershed and catchment scales can be obtained using isotope tracers of natural or human origin. One tracer that has widely been used for assessing net erosion or deposition over the last 50 years or so is caesium-137 (or ^{137}Cs). This isotope is a product of nuclear explosions in the atmosphere, which has spread the radioactive isotope world wide. It is brought down from the atmosphere by rainfall and is very strongly absorbed at the soil surface. Net deposition is indicated by an excess accumulation of ^{137}Cs, and net erosion by a deficit in its concentration compared with sites where no erosion or deposition has occurred. McHenry and Bubenzer (1985) and Walling (1988) give examples of use of this methodology.

9.4 The transport of nutrients and other chemicals in watersheds

The manner in which any chemical associated with soil moves in watersheds depends greatly on its solubility in water. Soil-related chemicals can be divided approximately into two groups:

(a) chemicals that are rather insoluble in water, and which are so strongly 'sorbed' to the soil that they effectively become part of it (the term 'sorbed' implies a substantial attachment between the chemical and the soil, without specifying the particular bonding mechanism involved); and

(b) water-soluble chemicals, which are free to move with flowing water (either overland or subsurface).

Understanding the transport of group (a) chemicals presents the same enormous challenge as that of interpreting sediment transport at the watershed scale, the topic reviewed in Section 9.3. A similar comment can be made in respect of the soluble group (b) chemicals, for which the corresponding challenge is to understand surface and subsurface hydrology as discussed in Section 9.2 and Chapter 12, with more consideration of groundwater being given in Chapter 10.

Transport of these two broad classes of chemicals will now be considered in that order.

Transport of sorbed chemicals

The transport of sediment discussed in Section 9.3 is directly linked to the transport of any chemicals that are sorbed to or are part of the sediment. Such chemicals can be plant nutrients, pesticides, organic matter and heavy metals, as well as micro-organisms and oxygen-demanding material. The transport of sorbed chemicals can be directly inferred from the transport of sediment, given the concentration of the chemical in the sediment. However, the concentration of any sorbed chemical generally varies with the size of the fundamental component of the sediment, as given by its mechanical analysis. This size dependence is partly because the surface area per unit mass of particles increases as particle size decreases. For example, the surface area per unit mass of the clay fraction (particles of diameter <0.002 mm) can be thousands of times greater than that for the coarse sand fraction (particles of diameter 0.2–2 mm), thus allowing much greater opportunity for sorption of chemicals by clay particles than by sand. During soil erosion the finer fractions of soil settle slowly, if at all, and thus move faster downslope than do coarser fractions, which settle more rapidly, their progress being by a series of short hops (Fig. 8.2). This differential rate of movement is more pronounced for rainfall-driven than for flow-driven erosion, and also in situations in which net deposition occurs.

Now, if the coarser fractions of soil consist of sand, which sorbs but little amounts of chemicals, eroding sediment can be higher both in fine sediment and in sorbed chemicals than was the original complete soil. Thus eroding sediment is commonly more highly enriched in sorbed chemicals than was the original uneroded soil. The degree of this chemical enrichment is measured by the 'enrichment ratio', E_R, defined for any particular chemical by

$$E_R = \frac{\text{concentration of chemical in eroded sediment}}{\text{concentration of chemical in original surface soil}} \qquad (9.6)$$

The phrase 'original surface soil' in Eq. (9.6) implies soil in the surface layer which is eroded. E_R is typically greater than unity, and, if so, then greater amounts of sorbed chemicals are transported in eroded sediment than would be expected from the amount of soil eroded.

Thus the sorbed-chemical loss in eroded sediment is given by the product

Sediment loss $\times$ concentration of sorbed chemical in original soil $\times E_R$

In soils of higher clay content, larger sedimentary units are typically complex soil aggregates, not just a coarser sand fraction as we assumed earlier. Such aggregates have the large clay content typical of the original soil, so the concentration of sorbed chemicals in these aggregates may be no lower than that in the finer fractions which are preferentially transported during erosion. Hence, for well-aggregated soils of high clay content, chemical enrichment can be negligible, with values of E_R close to unity.

At watershed scales, chemical enrichment is likely to be most strongly associated with the sorting of sediment by size which occurs in locations of net deposition, though this is an issue requiring further research.

Any soil-conservation practice that reduces soil erosion and, in particular, which limits delivery of sediment to streams is also effective at reducing the transport and delivery of sorbed chemicals. The major plant nutrients phosphorus and nitrogen are of particular significance in causing 'eutrophication' of streams by promoting excessive growth of algae. Both nutrients can be present in soil in sorbed or well-incorporated forms. However, especially at higher levels of fertiliser application, they can be present also in more soluble forms that can move with overland flowing waters.

The vital plant nutrient nitrogen (N) can be present in soil in many forms. Substantial amounts of N are insoluble and often locked up in organic matter in ways rendering it not readily available for uptake by growing plants or trees until the compounds concerned are broken down into other forms. From the point of view of erosion and chemical transport, such insoluble and sorbed chemicals behave similarly.

Transport of soluble chemicals

Soluble chemicals move with water, during infiltration into the soil profile, movement to and within groundwater and overland flow. The movement of saline salts and other soluble contaminants is discussed in Chapter 12.

Whilst much transported nitrogen (N) is insoluble, nitrogen compounds can be converted into soluble forms such as ammonium (NH_4^+) and nitrate (NO_3^-), which also come from applied fertilisers. The process of conversion of other forms of N into NH_4^+ is called 'mineralisation', and further transformation to NO_3^- is denoted 'nitrification'. 'Denitrification' is the conversion of nitrate to gaseous nitrogen, which is then removed to the atmosphere. All these processes are mediated by soil organisms and bacteria. Soil-science texts such as Brady and Weil (1999) describe the wide range of compounds in which the element N can exist, the movement from one form to another being best described in terms of cycles (the 'nitrogen cycle' in this case).

The proportion of nitrogen which is in soluble mineral forms is typically less than 2% of total N in the soil, and less than 10% in eroded sediment. Though erosive loss of N in soluble forms is generally much less than that in organic forms, it is an important short-term loss to the majority of plants which can take up N only in its soluble forms. Furthermore, if mineral forms of N reach water bodies, it is directly available to promote algal growth and eutrophication.

The disposal of sewage following limited treatment and the development of fertilised agricultural operations adjacent to streams, rivers and estuaries have all contributed to well-publicised environmental problems of eutrophication and excessive algal growth of various forms. Whilst mineral forms of N play a vital role in basic ecosystem functioning, environmental problems arise from excessive inputs from the variety of sources mentioned earlier. 'Point sources', such as those resulting from urban or city sewage systems, are more amenable to technical treatment than is the diffuse or non-point-source delivery of excess nutrients and other chemicals arising from erosion and runoff from agricultural and some other extensive land-based operations.

Because of their strategic location with respect to streams, riparian strips or zones (illustrated in Fig. 9.12) can help provide a protective buffer between streams and the chemical flow arising from adjacent upslope activities such as agriculture, dairying, forestry, etc. Of course, the effectiveness of a riparian buffer strip depends on the characteristics of that strip and on the rate and character of the fluxes impinging on it en route to the stream. Whilst soluble chemicals can reach a stream in overland flow, such flow is typically much more intermittent than subsurface flow. Studies in the USA, Europe and New Zealand in particular have shown that zones of riparian vegetation can lead to the removal of a substantial fraction of nitrate (NO_3^-) from subsurface flow passing through the riparian root zone prior to discharge into the stream. Research on just how such nitrate removal occurs is currently under way: how much is taken up by roots of the riparian vegetation, and how much by microbial populations in the soil, either by uptake or by conversion into nitrogen gas in denitrification?

Thus the protection provided by a riparian buffer zone is particularly important for the entire network of small streams in watersheds. This network is the major collector of surface and subsurface flow from extensive land areas, and the provider of much of the flow for downstream rivers.

Pesticides and herbicides vary widely in their water-solubility and sorption characteristics, the nature of these characteristics substantially controlling whether loss is in water or sediment-bound phases.

9.5 Rivers

A catchment approach is increasingly being adopted as a most useful framework in which to consider the interplay between human activity and the characteristics of natural land and water resources. Rivers in a catchment are the natural veins that collect and transport water and its associated constituents, which are the essential lifeblood of river ecosystems. The form of the complex patterns and paths of the prior contributors to major river systems can depend on the resistance offered to flow by bedrock, which depends on the form, structure and geological characteristics of the underlying rock material.

Major river systems develop over geological time scales, during which a good deal of sediment eroded from higher in a drainage basin is commonly deposited lower in the basin on gentler slopes to form 'alluvial deposits' of sediment. These deposits can be substantially deeper than the current depth of incision of the major river into it. The surface slope of land formed in this way is necessarily low, and a sinuous or meandering form of river when viewed in plan is typical of such 'alluvial rivers'.

The two dominant processes of erosion and deposition in rivers can be strongly influenced by changes in land use and vegetative cover within a river basin (Schumm, Mosley and Weaver, 1987). The impact of human activity modifying river flow or altering land characteristics in river basins has received increasing study. Examples of such modification include constructing dams on rivers, abstraction of water for irrigation or urban water supply and changes to river morphology designed to improve navigation or reduce flood levels. Such activities modify flow characteristics of rivers and the amount and size characteristics of transported and deposited sediment, thus altering the form of the river bed. Even if there is no such direct modification of river flow, significant changes in vegetation or land use in a river basin alter the characteristics of delivery to, and transport by, the river of water, sediment and nutrients. These changes in the river can disrupt the availability of habitats suitable for the reproduction and growth (and thus the survival) of many in-stream biota.

Erosion and stream capture of sediment in the upper reaches of river basins, where river slopes and velocities are relatively high, can lead to very substantial deposition of much of this sediment in lower river reaches, where slopes and flow velocities are gentler. In its lower reaches the river bed can be substantially raised by such depositional infilling, with the common consequence of over-bank flow and flooding of adjacent low-lying land. Such processes can be exaggerated by the extraction or diversion of river water for irrigation or urban use. The construction of dams, which is ironically partly justified for flood mitigation, can also increase downstream deposition through reducing river flow, hence exacerbating the downstream flooding problems they are intended to overcome. This issue is dramatically illustrated by the severe flooding in the lower reaches of the Yangtse and Yellow rivers in China, where enormous loss of life has followed from failure or over-topping of levees constructed to reduce such risk. Similar examples of this type of behavioural response of rivers to human diversion of water and dam construction occurred with the Rio Grande and Green rivers in the USA.

An important characteristic of rivers is how fast they flow (on average) past a cross-section, recognising that this speed will vary spatially, even when the volumetric flow rate is steady. Although a river bed is irregular in form, a general framework for understanding river flow is provided by simplified models. These models have their origin in the hydraulics and fluid mechanics of flow in open channels and in the characteristics and behaviour of liquids considered in Chapters 3 and 7.

The overall velocity of river flow basically results from the balance between the rate at which energy becomes available in downslope flow and the resistance to flow offered by the bed and bank walls of the river; this resistance is enhanced by other objects in the river, such as fallen trees and vegetation.

Figures 7.4 and 7.5 in Section 7.4 show the downslope and upslope forces acting on an element of flow, these forces being equal in steady, non-accelerated flow. Let us now carry out exactly the same kind of force-balance analysis, except applying it to the element of steady flow in a uniform river reach shown in Fig. 9.14. Here the shear stress, τ, acting between the water and the wetted perimeter of the river (of length p_r) is assumed uniform. Then the upstream force on the river element will be $\tau p_r \Delta x$.

Generalising Eq. (7.2), the downstream force on the river element is given by $\rho g A \Delta x S$, where A is the cross-sectional area of the river (Fig. 9.14) and S is the slope of the river bed. Equating these two forces gives an equation similar to Eq. (7.4), namely

$$\tau = \rho g R S \qquad (9.7)$$

where $R = A/p_r$ is called the 'hydraulic radius' of the flow. In a wide shallow river, the hydraulic radius R is not greatly different from the average

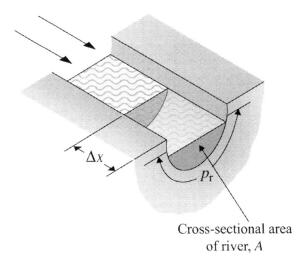

Figure 9.14 An element of a steadily flowing uniform river of wetted perimeter p_r and cross-sectional area A.

Cross-sectional area
of river, A

depth of the river, and the wetted perimeter p_r will be approximately equal to the width of the river.

As discussed in Section 7.4, τ is commonly related to Manning's resistance coefficient, n. Equation (9.7) leads to a form of Manning's equation suitable for describing a river's flow velocity, V. This equation, which is similar to Eq. (7.7) for overland flow, is

$$V = (S^{1/2}/n)R^{2/3} \qquad (9.8)$$

River beds can take a wide variety of forms, just two possibilities being larger fluctuations, such as 'dunes', and smaller periodic fluctuations or 'ripples'. Such bed forms increase the resistance to flow offered by the river bed by a resistance component that is often called a 'form resistance'. In the absence of such bed forms the bed is said to offer a 'grain resistance'. It has been found that the magnitude of Manning's n can be approximately estimated from a description of the nature of river-bed materials. As a rough guide, the grain resistance of a river bed consisting of sand is given by $n \approx 0.02$; for a bed of gravel, $n \approx 0.03$; for a bed of cobbles, $n \approx 0.04$; and for boulders, $n \approx 0.05$. (Values of n are often given without units, as here, though strictly its unit is $\mathrm{m}^{-1/3}$ s). Any variation in bed slope or river cross-section, or the presence of bed forms, river meandering, or flow obstructions such as vegetation or fallen trees, will lead to an increase beyond the basic grain resistance of a straight uniform river (see, for example, Dingman (1984) for more detail). Furthermore, if the depth of the stream falls to less than several times the magnitude of bed-roughness elements, this can also lead to an increase in Manning's n.

Example 9.2

A section of a steep mountain stream has a slope $S = 0.1$, a cross-sectional area of flow, A, of $9.5\,\text{m}^2$ and a wetted perimeter $p_r = 5\,\text{m}$. Calculate the volumetric discharge across this stream section, in $\text{m}^3\,\text{s}^{-1}$ assuming that Manning's n can be calculated (see Dingman (1984)) from

$$n = (n_0 + n_1)m$$

where n_0 is the basic value of n for a straight and regular stream section, n_1 is the additional flow resistance due to irregularity in the section and m is the resistance contribution due to changes in direction of the stream bed. Assume that $n_0 = 0.04\,\text{m}^{-1/3}\,\text{s}$ (cobble-covered stream bottom), $n_1 = 0.01$ (moderate irregularity) and $m = 1.15$ (appreciable meandering).

Solution
Manning's

$$n = (0.04 + 0.01) \times 1.15$$
$$= 0.058\,\text{m}^{-1/3}\,\text{s}$$

The hydraulic radius is

$$R = A/p_r = 1.9\,\text{m}$$

Thus, from Eq. (9.8), the average flow velocity is

$$V = 9.4\,\text{m s}^{-1}$$

Thus the volumetric discharge is

$$AV = 80\,\text{m}^3\,\text{s}^{-1} \quad \text{(or cumecs)}$$

Stream location with respect to the entire stream network in the drainage basin is an important factor affecting the character of materials forming the beds of rivers. 'Higher-order' or tributary streams may flow extensively on bedrock surfaces, or perhaps cut into superficial materials, in which case deposits of the resultant erosion products may cover much of the stream bed. Lower down the drainage basin, deep alluvial deposits of material eroded by the river higher in its basin commonly cover the entire river bed.

Some sediment that moves downstream, such as boulders and cobbles, are so large that their movement is by rolling along the bed or by mass movement. Sediment that is too large to be carried up into the flow by turbulent eddies is commonly referred to as bedload, and its movement can be most noticeable during periods of high discharge.

Another component of sediment on river beds can be picked up by turbulent eddies and bounce along the channel floor, returning rather frequently to the river bed because of its relatively high settling velocity, which is the speed with which the sediment component settles in water. This type of intermittent motion, involving a sequence of removal from the bed and return to it, is described as 'saltation'. The material carried by it is called the 'saltation load'. This type of sediment transport also occurs in shallow overland flow as described in Chapter 8.

The dominant importance of the settling velocity is again emphasised by another component of sediment transport in rivers known as the 'suspended load'. Either due to its small size or due to its low density, the settling velocity of this sediment component is sufficiently small that it is of similar magnitude to the velocity of eddies in turbulent river flow. This situation was discussed in Section 3.4. Since, on average, any downward settling is counteracted by buffeting by eddies, this fine load component remains suspended, and thus is continuously transported with the flow of the river. This suspended material (sometimes referred to as 'wash load') can be distinguished from the 'dissolved load', wherein chemicals originating from whatever sources are in true solution.

The description given in Section 8.3 of the process of deposition and the discussion of sediment-deposition characteristics are equally applicable to the deposition of sediment in rivers. In river flow, eddies of a scale larger than is possible in shallow overland flow can develop. Also larger-scale rotational flows are a characteristic of meandering streams (Gordon, McMahon and Findlayson, 1994).

From the earlier description of bedload, saltation load and suspended load, it follows that somewhat different types of physical processes are involved in each of these three components. Since the bedload component doesn't lose contact with the river bed, and suspended components don't return to the river bed in active river flow, it is only the saltation load that involves both removal from and return to the river bed. This range of components contributing to sediment transport, combined with the spatial and temporal complexity in river geomorphology, provides a considerable challenge to physical modelling of sediment transport in river systems.

The flux of suspended sediment is in principle the most readily calculated of the three load components, being given by the product of the concentration of suspended sediment and the flow velocity. However, both these quantities vary with time and with position in the river's cross-section, so sampling that would allow accurate measurement even of this suspended load is still a challenging task (Richards, 1982).

Figure 9.5 above shows as hydrograph (a) the arrival of flow at an upland stream bank from a contributing hillslope. The rate of delivery of

Figure 9.15 Showing the relative time relationship commonly found between the sediment concentration of the suspended load and stream flow during a runoff event.

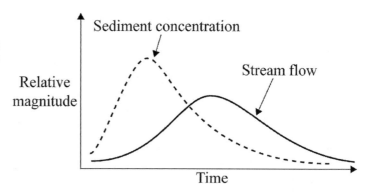

Figure 9.15 Showing the relative time relationship commonly found between the sediment concentration of the suspended load and stream flow during a runoff event.

sediment and its sorbed chemicals to the stream is closely related to this hydrograph. This is one reason why the peak in sediment concentration measured in a stream during a runoff event is found to precede the peak in the stream hydrograph as illustrated in Fig. 9.15. In contrast, the rates of movement of the stream bedload and saltation load are more closely related to the stream hydrograph shown as (b) in Fig. 9.5, though often with a time lag. The concept of stream power, defined in Eq. (9.9), is invoked in some types of deterministic models used to interpret or predict the transport of sediment as bedload or saltation load (Richards, 1982).

9.6 River health

The ecological roles and values of river systems are increasingly being recognised and appreciated. Indeed, rivers have been described as the arteries of ecosystems, their delivery of water being just as important to ecosystems as blood is to human life. The criteria used in assessing the value of rivers are diverse, but can include the diversity and abundance of in-stream and riparian communities (e.g. fish, flora and bird life), which usually depend on how little the river system is disturbed. How representative or rare the life depending on the river may be can also contribute to the value ascribed to a river. The state of 'river health' is a term commonly used to summarise information on trends in the ecological diversity and abundance of the multiple component populations of life which depend on the river for their existence.

There are direct relationships between river health and human health, and rivers can sustain a close relationship between human and other biological populations. The intimate dependence of indigenous and river-dwelling peoples on fish and other food sources from rivers and lakes

made them sensitive to such disturbance, natural or otherwise. It is estimated that 1.2 billion people world wide lack access to safe drinking
water. In 50 states of the USA members of the public are warned that
the level of contaminants in freshwater fish is such as to pose a potential
health risk. The gross modifications to river health resulting from persistent pollution by discharges of chemicals or sewage, and the catastrophic
damage caused by accidental major releases of chemicals into mighty
rivers such as the Rhine, have succeeded in alerting most civilisations
to the need to understand the range of factors affecting the ecological
health of rivers and to institute changes necessary to reduce or avoid
such degradation.

Whilst human-induced change to river morphology, and the consequent threat to biophysical integrity, appears to have been a consistent
feature of human civilisations, there is growing recognition that much
of the serious degradation occurring in rivers can be avoided by improving industrial processes and implementing conscious river management.
This has led to a rapidly growing area of knowledge commonly referred
to as 'river management', which brings together biophysical, geomorphic
and ecological knowledge of river systems in a management framework.
An objective of such management can be to maintain or restore variety
in ecological niches in terms of such factors as composition and form of
the river bed, flow depth and maintenance of underwater vegetation. In
rivers from which substantial amounts of water are diverted, a specific
objective can be to manage water abstractions in such a way as to satisfy minimum environmentally desirable flow criteria, such flows being
referred to as 'environmental flows'.

Thus it is clear that rivers cannot be managed in isolation from their
watersheds, which is why this chapter brings together the two topics
of watersheds and rivers. Watersheds include the important elements of
riparian and river-bank areas, flood plains over which the river spreads
during high flows, bringing and gaining life during this expansion of
flow. Integral to maintaining a healthy river is allowing rivers the opportunity to expand over their natural flood plains and wetlands, which
form a single ecological unit. This opportunity is often denied by allowing inappropriately located and constructed housing and other human
developments on these river subsystems, then requiring the construction of massive levee structures to reduce flooding of these naturally
flood-prone areas.

From the point of view of river health, the water flow in many
rivers has been over-allocated to extraction for irrigation or other
human-related consumption. Experience in some regions of the world
indicates that, where more than about a third of median river flow is

extracted, there is likely to be a serious decline in river health. The efficiency of use of water in irrigation is typically poor, and, when this is so, a decrease in use of irrigation need not reduce primary production assisted by irrigation, and sometimes results in benefits to landscape as well as river health.

When it is intact, the riparian zone can protect stream banks from excessive erosion during high flows, and encourages the net deposition of sediment from overland flow, as described in connection with the buffer strips shown in Fig. 9.12. Without such protection, the sediment and associated nutrients and other chemicals not deposited within the watershed can enter the stream or river associated with it.

The impacts of excessive sediment entering a stream can be broadly separated into two classes: effects due to fine sediment (which tends to move downstream with the river flow) and those due to coarser sediment, which quickly settles to the stream bed, so adding to the bedload and saltation load.

The fine sediment increases the turbidity of the river water, reducing light and oxygen levels, and carries nutrients and other chemicals with it. Such movement can be over long distances, eventually reaching estuaries and the sea. In some slowly flowing rivers, such as in inland Australia, the nitrogen and phosphorus carried by fine sediment is responsible for the choking blooms of blue–green algae in warm seasons, some of which can be toxic to humans and animals. In coastal areas, important habitats such as wetlands, estuaries, sea-grass beds and coral suffer degradation due to fine sediment and its associated chemicals. Near-shore parts of the Great Barrier Reef located off Australia's north-eastern coastline, one of the great coral reefs of world-heritage stature, have suffered degradation due to the blanketing of coral with excessive algal growth supported by fine sediment and nutrients delivered by rivers. It is estimated that the amount of sediment delivered by rivers adjacent to this reef has increased by at least a factor of four, due mainly to agricultural development in the river catchments.

Let us now consider the effect on river health of the coarser fractions of sediment which enter rivers, the sand and gravel size fractions in particular. These coarser fractions are rather quickly deposited on and smother the stream bed, destroying plant and animal habitats. Deeper pools are a common and biologically significant feature of many streams, and these tend to be lost due to infilling with coarser sediment.

Periods of particularly great delivery of sediment to streams can arise, for example, from a combination of unusually intense rainfall and landscape disturbance, such as clearing of vegetation. The resultant large input of sediment to the stream can lead to the formation of a band of coarse bedload, often referred to as a 'sand slug' or 'sediment slug'.

Figure 9.16 Slugs of sandy sediment forming a braided appearance to flow rather than a single channel in the Rigarooma River, Tasmania, Australia. (From Bartley (2001).)

Such slugs cause degradation of river health as they slowly work their way downstream (Fig. 9.16). Movement of the slug is slow since most displacement occurs during floods or greater flows, and, especially if river flow is regulated to minimise flooding, movement of the sand slug is likewise curtailed. Sediment slugs move like waves down a stream, but may remain in a large river system for many hundreds of years.

The spatial complexity of habitats in natural streams and rivers provides important spawning and breeding sites for a wide range of stream-dwelling organisms, including fish. Both coarse and fine sediment can reduce the suitability and extent of such ecologically important habitats.

Clearly streams and rivers have limits to their ability to cope without damage with the multiple physical, chemical and biological impacts due to excessive entry of sediment into them. Maintaining river health depends on management of the watersheds they drain and of all the activities which take place within them.

Main symbols for Chapter 9

A Cross-sectional area of a river

c_t Sediment concentration at the transport limit

F_o Initial abstraction of rainfall prior to runoff generation (expressed as equivalent ponded depth of water)

I_m Maximum possible value of the spatial mean infiltration rate for a field

L Downslope length of a field

n Manning's roughness coefficient

p_r Wetted perimeter of a river

P Rainfall rate

Q Volumetric runoff rate per unit area

R Hydraulic radius of a river $(= A/p_r)$

S Land slope

S_t Equivalent ponded depth of water stored in a watershed

T Effective duration of a runoff event

V Velocity of flow

ρ Density of water

Σ Summation symbol

τ Shear stress between flowing water and the river bed

Exercises

9.1 Rework Example 9.2 after having converted all data given in metric units into English-system units.

9.2 At a particular location on a river the average flow velocity is $1.2\,\mathrm{m\,s}^{-1}$. If the hydraulic radius is 40 m and the wetted perimeter is 95 m, calculate the river's volumetric flow rate at this location. Express the result in units of $\mathrm{m^3\,s^{-1}}$ and $\mathrm{ft^3\,min^{-1}}$.

9.3 In a study of sediment transport in an upland agricultural catchment in West Java, Indonesia, a simplified 'Meyer–Peter'-type equation was used to estimate bedload transport in streams. The simplified form of equation employed gave the rate of bedload transport, G_s $(\mathrm{kg\,m^{-1}\,s^{-1}})$, as

$$G_s^2 = 250SG^{2/3} - 42.5d_{50}$$

where G is the volumetric discharge $(\mathrm{m^3\,s^{-1}})$, S is the stream slope $(\mathrm{mm^{-1}})$ and d_{50} is the median diameter of the bedload material (m). Calculate G_s for $G = 500$ and $3000\,\mathrm{l\,s^{-1}}$, given $S = 0.00673$ and $d_{50} = 1.8\,\mathrm{mm}$.

(*Note.* The larger the value of d_{50}, the greater the rate of deposition of sediment. The negative sign for the term including d_{50} in the equation for G_s recognises that deposition reduces the concentration of sediment and thus the rate of sediment transport.)

9.4 Consider a section or reach of length L of a steadily flowing river of constant cross-section. Since in this context the terms p and $\rho V^2/2$ in Bernoulli's equation, Eq. (3.17), are constant, it follows that the decrease in potential energy $\rho g(z_1 - z_2)$ for water flowing from height z_1 down to height z_2 over the reach must equal the energy loss per unit river volume, ΔE_{loss}. The term ΔE_{loss} will be the work done against the force arising from the shear stress τ given by Eq. (9.7), this force per unit length of river being given by τp_r, where p_r is the wetted perimeter of the river in cross-section (see Fig. 9.13).

Remembering that work done is the product of force and distance, and that ΔE_{loss} is energy lost per unit volume of river water, show that the energy-loss term, ΔE_{loss}, over the reach of length L is equal to $\rho g(z_1 - z_2)$. This expression is exactly equal to the decrease (or negative change) in potential energy of water over the section, thus satisfying Eq. (3.17). (Note that the hydraulic radius is $R = A/p_r$, and that the slope, S, is given by $S = (z_1 - z_2)/L$.)

9.5 Consider a reach of river where bedload transport is not occurring, but flow is sufficiently rapid for re-entrainment of sediment to occur. Suppose that processes are taking place at a steady rate in this reach, so that there is equilibrium in the exchanges between the saltating load and bed sediments. If the rate of re-entrainment per unit area of bed sediment, r_i, is the same irrespective of the size of bed sediment that takes part in saltation, then r_i can be written as

$$r_i = kc_{bi} \tag{i}$$

where k is a constant not dependent on size class i and c_{bi} is the concentration of particles of size class i in the bed sediment.

Summing Equation (i) over all the size classes i,

$$r = kc_b$$

where r is the total rate of re-entrainment of all size classes and $c_b = \Sigma c_{bi}$. Denote the rate of deposition of particles of size class i by d_i, with $d = \Sigma d_i$. In the steady-rate system assumed to exist between re-entrainment and deposition, it follows that

$$r_i/r = d_i/d$$

Then use Eqs. (8.10) and (8.11) to show that

$$c_{bi}/c_b = v_i c_i / \Sigma v_i c_i \tag{ii}$$

What does Equation (ii) tell us about the relationship between the size characteristics of the bed and those of the saltating load?

(*Note*. Equation (ii) commonly gives a good description of measured data for rivers in which the saltation load dominates the bedload.)

9.6 If you are familiar with a particular stream, river, or lake, use the knowledge you have gained from this familiarity to consider the issues of river health introduced in Section 9.6.

References and bibliography

Bartley, R. (2001). Australia is on the move . . . down the creek. *Aust. Landcare.* December issue, 6–10. Moonee Ponds, Victoria: Agricultural Publishers.

Brady, N. C., and Weil, R. R (1999). *The Nature and Properties of Soil*, 12th edn. Englewood Cliffs, New Jersey: Prentice-Hall International, Inc.

de Roo, A. P. J., Wesseling, C. G., and Ritsema, C. J. (1996). LISEM: a single-event physically based hydrological and soil erosion model for drainage basins, 1. Theory, input and output. *Hydrol. Processes* **10**, 1107–1117.

de Vantier, B. A., and Feldman, A. D. (1993). Review of GIS applications in hydrological modelling. *J. Water Resources Planning Management* **119**, 246–261.

Dingman, S. L. (1984). *Fluvial Hydrology*. New York: W. H. Freeman and Company.

Ghadiri, H., and Rose, C. W. (1992). *Modelling Chemical Transport in Soils; Natural and Applied Contaminants*. London: Lewis Publishers.

Gordon, N. D., McMahon, T. A., and Findlayson, B. L. (1994). *Stream Hydrology: An Introduction for Ecologists*. Chichester: John Wiley and Sons.

Grayson, R. B., and Western, A. W. (2001). Terrain and the distribution of soil moisture. *Hydrol. Processes* **15**, 2689–2690.

Hudson, N. (1981). *Soil Conservation*, 2nd edn. Ithaca, New York: Cornell University Press.

Laflen, J. M., Elliott, W. J., Flanagan, D. C., Meyer, C. R., and Nearing, M. A. (1997). WEPP – predicting water erosion using a process-based model. *J. Soil Water Conservation* **52**, 96–102.

Lane, L. J., Renard, K. G., Foster, G. R., and Laflen, J. M. (1992). Development and application of modern soil erosion prediction technology. *Aust. J. Soil Res.* **30**, 893–912.

McHenry, J. R., and Bubenzer, G. D. (1985). Field erosion estimated from ^{137}Cs activity measurements. *Trans. Am. Soc. Agric. Eng.* **28**, 480–483.

Meyer, L. P., Dabney, S. M., and Harmon, W. C. (1995). Sediment-trapping effectiveness of stiff-grass hedges. *Trans. ASAE* **38**, 809–815.

Morgan, R. P. C. (1986). *Soil Erosion and Conservation*. Harlow: Longman.

O'Loughlin, E. M. (1986). Prediction of surface saturation zones in natural catchments by topographic analysis. *Water Resources Res.* **22**, 794–804.

Richards, K. (1982). *Rivers: Form and Process in Alluvial Channels*. London and New York: Methuen.

Savabi, M. R., Flanagan, D. C., Habel, B., and Engel, B. A. (1995). Applications of WEPP and GIS-GRASS to a small catchment in Indiana. *J. Soil Water Conservation* **50**, 477–483.

Schumm, S. A., Mosley, M. P., and Weaver, W. E. (1987). *Experimental Fluvial Geomorphology.* New York: John Wiley and Sons.

Tang Ya, Sun Hui, Xie Jiasui, and Cheng Jianzhong (2002). Soil conservation and sustainable management of sloping agricultural lands in China. *Proceedings of the 12th International Soil Conservation Organisation Conference, Beijing, Volume III.* Beijing: Tsingua University Press, pp. 1–5.

Toy, T. J., Foster, G. R., and Renard, K. G. (2002). *Soil Erosion: Processes, Prediction, Measurement, and Control.* New York: John Wiley and Sons.

USDA Soil Conservation Service (1985). SCS *National Engineering Handbook,* Section 4, Hydrology. Washington: USDA.

Ward, A. D., and Elliot, W. J. (1995). *Environmental Hydrology.* New York: CRC Press Inc.

Walling, D. E. (1988). Erosion and sediment yield research – some recent perspectives. *J. Hydrol.* **100**, 113–141.

Yu, B. (1998). Theoretical justification of the SCS method for runoff estimation. *J. Irrigation Drainage Eng.* **124**, 306–310.

Yu, B., Sombatpanit, S, Rose, C. W., Ciesiolka, C. A. A., and Coughlan, K. J. (2000). Characteristics and modelling of runoff hydrographs for different tillage treatments. *Soil Sci. Soc. Am. J.* **64**, 1763–1770.

10

Movement of water through the groundwater zone

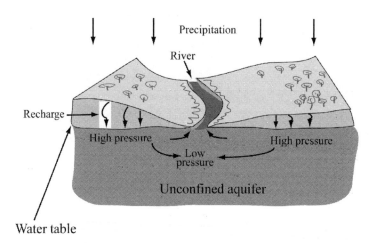

10.1 Introduction

In Chapters 4–9 we have followed the fate and some of the consequences of water falling in precipitation, evaporating, infiltrating into the soil, flowing over the land surface and contributing either to lakes or to rivers, which mostly discharge ultimately into the ocean. In this chapter we turn to the fate and environmental consequences of water that infiltrates into the land surface, some of which moves through the soil profile and contributes to the substantial volumes of water stored within the earth's crust. Water that fills or saturates all the available pore space within soil or rock materials is referred to as 'groundwater' (Fig. 4.1). The subsurface region where groundwater is stored and is free to move is called an 'aquifer'.

This subsurface water is not only a large fraction of fresh water on the planet (about 22%), but is also a growing source of drinking water for the earth's population. Thus stores of groundwater are considerably greater than the total fresh water stored in lakes, rivers, water storages and other surface water sources, which amounts only to some 6% of planetary fresh water. On earth, most fresh water is stored in frozen forms, such as ice sheets and glaciers. Only a tiny percentage of water at any time occurs in the atmosphere or biosphere, though this very small fraction is vitally important.

Figure 4.7 shows that an important component of the water balance of a watershed is the flows which occur beneath the soil surface due to the presence of aquifers. The presence of aquifers is not restricted to humid regions. Groundwater can travel, even though slowly, over hundreds of kilometres from intake areas and extend, through such movement, into more arid regions, where it becomes a most important source of water.

Protecting the quality of groundwater so that it remains fit for human or animal use without the need for substantial decontamination is a major environmental objective. However, the quantity and quality of ground-water is under threat in a substantial number of major population centres around the world, as is also the case for more dispersed populations.

Especially if the upper surface of the groundwater (the 'water table') is sufficiently deep to be beneath the root zone of vegetation, the loss of groundwater by evaporation or evapotranspiration is negligible. This makes it a more efficient way of storing water than surface storage, where loss by evaporation can remove a substantial fraction of the water stored. Sometimes groundwater is naturally rich in salts dissolved from the surrounding rock, or delivered in solution from the soil surface. Especially when the quality of groundwater is such that it is suitable for human consumption (perhaps with minor processing), and if surface water is limited, then groundwater becomes a most important source of potable water.

The quantity of groundwater is reduced when groundwater is removed at a greater rate than it is replaced by natural replenishment processes. Because of such 'over-pumping' of groundwater for irriga-tion in some areas of Texas in the USA, use of irrigation has had to be reduced. Where groundwater is a major source of potable water, such as in Perth, Western Australia, and coastal areas of Florida, good water management is required in order to ensure that over-pumping, which could lead to intrusion of salt water from the adjacent sea, is avoided. Thus the continued supply of fresh groundwater in such contexts involves a fine balance between its rates of removal and recharge.

Overland flows are generally ephemeral in nature. Thus groundwa-ter is commonly not only a major contributor to stream flow, but also provides the characteristic continuity in stream flow between surface runoff events. If this continuity of flow is reduced or removed by deple-tion of groundwater, then the complex food webs on which stream life depends are disrupted. Waste and inefficient use of groundwater has been documented widely.

Groundwater quality is under threat in locations where there is the opportunity for contamination by materials applied to the surface, or

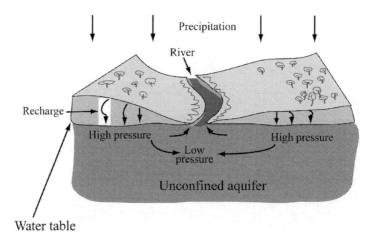

Water table

Figure 10.1 An element of a watershed with a river intersecting an 'unconfined aquifer', this term indicating that the groundwater is not under pressure due to geological confinement. The water table is shown as being affected by the shape of the land surface. The unconfined aquifer is being fed by recharge from infiltrating water, and groundwater flow from regions of higher pressure to regions of lower pressure is feeding water into the river.

from buried landfills, which yield contaminants transported by water flowing through the soil profile. There is a vast range of natural and manufactured contaminants that have been found to have this mobility, including sewage, herbicides and insecticides used in agriculture, fertilisers, eroded naturally occurring excess nutrients and products regarded as waste produced during industrial activity, radioactive and otherwise. The major plant nutrient, nitrogen, can be present in soil in the form of the nitrate ion, NO_3^-, which is mobile and can move down or 'leach' through the soil profile and thereby accumulate in groundwater. If groundwater is high in nitrate, and is accessed for drinking, this can present health problems, especially for young children. Industrial and municipal wastes also pollute groundwater. A quite different and complex pollutant is acid drainage, which can result from disturbance by mining in some situations. Further consideration of the movement of contaminants in groundwater is given in Section 12.4.

The depth beneath the earth's surface at which groundwater is encountered varies greatly. Typically the depth to groundwater increases with the (current) aridity of the climate. Some extensive bodies of groundwater in currently arid areas have been built up during past, less-arid, geological eras. Especially in areas of substantial rainfall, the shape of the groundwater surface (or water table) is strongly influenced by the shape of the earth's surface, or 'topography' (Fig. 10.1). In such moist

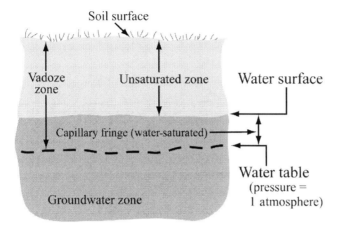

Figure 10.2 The water table is the surface where water pressure is atmospheric, dividing the vadoze zone above from the groundwater below it. The 'capillary fringe' above the watertable is water-saturated, though at a pressure less than atmospheric. The medium above the capillary fringe is the 'unsaturated zone'. Since atmospheric pressure is taken as a datum, where the water pressure is defined as zero, water pressure below the water table is positive, and the pressure in the capillary fringe above it is less than zero, or negative.

climatic contexts, groundwater can also be found at shallow depths, the incisions made by streams commonly intersecting the surface of groundwater (Fig. 10.1).

10.2 Groundwater at equilibrium

At the scale implied in Fig. 10.1, saturation of the pore space is indicated as occurring at the water-table surface. More exactly, the water table is the surface where the pressure in the groundwater is equal to the local atmospheric pressure. As shown in Fig. 10.2, the water table is commonly located a little below the surface of the water-saturated zone. The water-table surface, where pore water is at atmospheric pressure, divides the groundwater zone below it from the 'vadoze zone' above it (Fig. 10.2).

The saturated zone immediately above the water table (shown in Fig. 10.2) is called the capillary fringe since its existence depends on the phenomenon of capillarity discussed in Section 2.7 and illustrated in Fig. 2.18. Using Fig. 2.18(b), it was shown that the pore water pressure is less than that of the atmosphere, and the water is then described as being in suction. This suction acts as a weak glue on the soil, for example holding sand grains together and giving sufficient strength to beach sand for shapes to be modelled in it. This suction can develop in the pore water of any porous material, such as soil, even when, as in the capillary

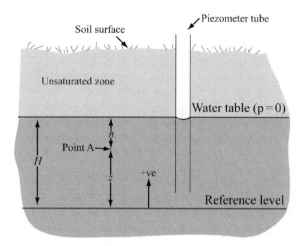

Figure 10.3 For an unconfined aquifer the level of the water table is indicated by the level of the water surface in a piezometer tube. At this level in the soil, the pore water pressure is equal to the local atmospheric pressure, which is regarded as a pressure datum, being taken as zero pressure. The text considers the pore water pressure at point A, a distance h below the water table and at height z above an arbitrarily located reference level. The water table is assumed to be stationary. (The capillary fringe is omitted because it is irrelevant to the issues considered in connection with this figure.)

fringe, water saturates the entire pore space (Fig. 10.2). The height of the capillary fringe depends strongly on the mechanical analysis of the soil in that zone (see Section 2.3). The fringe height increases with the fineness of the soil material, and so can be substantial in clay soils, though it is negligible in gravel or other coarse material with relatively large pores. Water pressure in the capillary fringe will be examined in Section 11.2, but will be ignored from here on in this chapter.

Groundwater can be under pressure due to geological confinement as discussed in Section 10.4. Such groundwater is referred to as being in a 'confined aquifer'. However, for an 'unconfined aquifer' or 'water-table aquifer' the depth of the water table can be determined by drilling or coring a hole or piezometer well into the earth and measuring the depth of the free-water surface (Fig. 10.3). A tube let into the earth's subsurface, sealed along its length, open to entry of water at its bottom and to the atmosphere at its top is called a 'piezometer'.

Bernoulli's equation (Eq. (3.17)) described the various energy components for a streamline in flowing water. Let us apply this equation to water in a soil matrix for the special situation in which the water is at rest and not moving relative to the matrix, so that all forces are at equilibrium. Thus the fluid velocity v in Eq. (3.17) will be set to zero.

Since the component term E_{loss} in this equation arises from energy lost due to motion, then, since $v = 0$, $E_{\text{loss}} = 0$ also. In this equilibrium or steady-state situation, Bernoulli's equation reduces to

$$p + \rho g z = \text{constant} \tag{10.1}$$

where p is the water pressure and z is the height measured above an arbitrary reference plane (Fig. 10.3). On dividing by ρg, Eq. (10.1) becomes

$$p/(\rho g) + z = \text{constant} \tag{10.2}$$

The component term $p/(\rho g)$ has the dimensions of length, and is commonly referred to as the 'pressure head', being the height of a column of water that would exert the pressure p. The soil water pressure at the water table is atmospheric, which is taken as the zero datum pressure. Thus, at point A beneath the water table in Fig. 10.3, it follows from Eq. (3.2) that the pressure $p = \rho g h$, where the pressure head is h. On replacing $p/(\rho g)$ by h, Eq. (10.2) becomes

$$h + z = \text{constant}$$

and, from Fig. 10.3, this constant is shown to be H. Thus it follows that

$$H = h + z \tag{10.3}$$

where z is called the 'elevation head', since it is the elevation above the datum of the point under consideration (Fig. 10.3). This sum of the pressure and elevation heads is defined as the 'hydraulic head', denoted by H.

It is found, quite generally, that water moves in response to any spatial change in hydraulic head, water moving from where H is higher to where it is lower. This is discussed further in Section 10.3. The spatial consistency of H in the situation shown in Fig. 10.3 is more clearly evident if the elevation, pressure and hydraulic head are all plotted on a horizontal axis as a function of height above the arbitrarily chosen reference level used as the datum for height measurement. This is done in Fig. 10.4, where the spatial constancy in H indicates that no movement of soil water will occur.

In the static situation depicted in Fig. 10.4, it is clear that, wherever the point A in Fig. 10.3 is located beneath the water table, the variation in h and z is such that $h + z = H$ is constant, being the same for all positions within this zone. Thus, as previously assumed in writing Eq. (10.1), no movement of water will occur in this equilibrium situation.

In Eq. (10.3) the elevation head, z, is determined assuming that z is taken to be positive in the upward direction (Fig. 10.4). With this

Figure 10.4 The spatial variation with height above the reference level of the elevation head, z (zero at reference level), pressure head, h (zero at water-table height) and hydraulic head, $H = h + z$ (Eq. (10.3)).

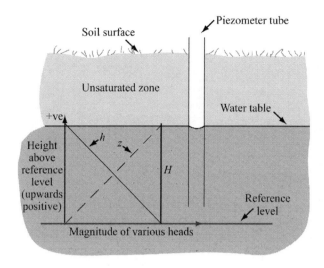

assumption, the magnitude of z at any location is positive if the reference level from which it is measured is selected to be below the location of interest. It is sometimes convenient to measure z in a positive-downward sense, say from an arbitrary datum taken as the soil surface; in this case Eq. (10.3) would become $H = h - z$ (z positive downwards).

Let us now consider the principles involved in describing or predicting the movement of groundwater.

10.3 Movement of groundwater

So far we have considered groundwater in a static or equilibrium situation. Let us now consider how groundwater moves if the hydraulic head, H, varies in space, and if the subsurface material has pores through which water can move. Whilst water may be unable to move appreciably through some dense unfractured rocks, quite often rocks have fissures, joints or cracks in them, and, nearer the earth's surface, rocks can be broken down or undergo weathering, a process involved in the formation of soil (Chapter 1). As illustrated in Fig. 2.2, soil commonly has some degree of aggregation, leaving relatively large pores through which water can move, even if it follows a somewhat complex pathway (Fig. 10.5). Even soil less well aggregated than that in Figs. 2.2 and 10.5(a) commonly has small pores which allow movement of water. The smaller the pore size, and the more tortuous the path water has to take through such pores, the slower the flow. Figure 10.5 shows possible steps in replacing the more complex reality hinted at in Fig. 10.5(a) by hydraulically equivalent flow tubes (Fig. 10.5(b)), which may then be straightened out

(a) (b) (c)

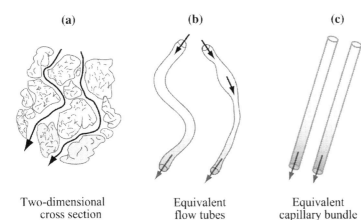

Two-dimensional cross section   Equivalent flow tubes   Equivalent capillary bundle

Figure 10.5 Illustrating in (a) the tortuous pathways taken by water as it moves through a complex water-filled pore space in saturated soil or porous rock. Successively simplified models of the flow pathways are shown in (b) and (c).

into an equivalent bundle of capillary tubes to provide a crude but simple conceptual model.

Poiseuille's equation, Eq. (6.2), shows flow in a tube to be proportional to the fourth power of its radius, emphasising the extremely strong role pore size plays in controlling movement of water.

Example 10.1

Adopting the capillary-bundle model of movement of water through a porous medium (Fig. 10.5(c)), how is the actual mean flow velocity, v, related to the volumetric flux of water per unit cross-sectional area of soil, q? (In physical sciences a flux per unit area normal to the flux is called a flux density.)

Solution
In any area A (m^2) normal to the capillary bundle, an area A_s will be occupied by solids, so that fluid can move only through the cross-sectional area $A - A_s$. Thus mass continuity in any flow through the capillary bundle requires that

$$v(A - A_s) = Aq$$

so that

$$v = \frac{q}{(A - A_s)/A}$$

Note that $(A - A_s)/A$ is the porosity, ε, of the capillary bundle, where porosity was defined in Section 2.3 and Eq. (2.6).

Thus

$$v = q/\varepsilon \qquad (10.4)$$

Figure 10.6 The change in
hydraulic head in the direction
s from H_1, at distance s_1, to
H_2, at distance s_2.

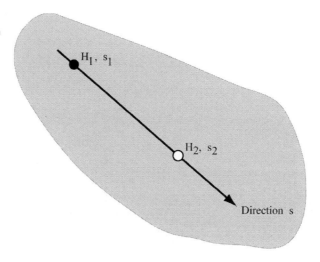

Figure 10.6 The change in hydraulic head in the direction s from H_1, at distance s_1, to H_2, at distance s_2.

where a typical value of ε in soils might be 0.3. If so, $v = 3.3q$, the actual mean velocity being greater than the flux density.

Darcy's law

It was the need to design a sand-filtration system to improve the water quality for the city of Dijon that led the French hydraulic engineer Henry Darcy to address the question of how water moves in a saturated porous medium. Darcy published his results in 1856, and the equation generalising his findings has come to be known as 'Darcy's law'. His experiments showed that the discharge of water through any given type of sand bed was proportional to the spatial rate of change in hydraulic head, referred to as the 'hydraulic gradient'. The hydraulic gradient in any particular direction s is the ratio of the change in hydraulic head measured over some distance interval in the direction s and the magnitude of that distance interval. More explicitly, using Fig. 10.6, the hydraulic gradient is given by

$$\text{hydraulic gradient in direction } s = (H_2 - H_1)/(s_2 - s_1) \qquad (10.5)$$

Groundwater flows in the direction of decreasing hydraulic head, just as overland water flows downhill in the direction of decreasing elevation head, since in the latter context the pressure head varies little, if at all. Since in Fig. 10.6 $H_1 > H_2$, then $H_2 - H_1$ in Eq. (10.5) is a negative quantity. Since the other term in this equation, $s_2 - s_1$, is positive, it follows that the hydraulic gradient is negative in the direction of flow. This simply indicates that flow takes place in the direction of decreasing hydraulic head.

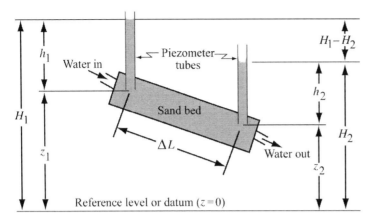

Figure 10.7 Illustrating the change in hydraulic head from H_1 to H_2 as water flows through a filter consisting of a sand bed packed into a container. Head differences are indicated over a length interval ΔL. Flow is in the direction of decreasing H, a direction in which the hydraulic gradient is negative.

The essence of Darcy's experiments is illustrated in Fig. 10.7, in which water for human consumption was filtered by passing the flow through a filter consisting of uniformly packed sand.

Darcy's equation can be written in terms of the flux density (or specific discharge) of water flowing through the sand filter, q ($m^3\,m^{-2}\,s^{-1}$ or $m\,s^{-1}$) as follows:

$$q \propto \frac{H_1 - H_2}{\Delta L}$$

but the hydraulic gradient is $(H_2 - H_1)/\Delta L$, so

$$q \propto -\frac{H_2 - H_1}{\Delta L}$$

or

$$q = -K\frac{\Delta H}{\Delta L} \qquad (10.6)$$

where the constant of proportionality, K, in Eq. (10.6) is called the 'hydraulic conductivity' of the sand or other porous medium in the column. Equation (10.6) has been found to provide a most useful description of flow through any relatively stable type of porous medium, including soil, and is not restricted to the sand of Darcy's experiments. Equation (10.6) is the commonly used form of Darcy's law, and an alternative indication of its plausibility using previous material in this book is given in Box 10.1.

The hydraulic conductivity, K, of a water-saturated porous medium is a characteristic of that particular medium. If the hydraulic conductivity of a porous medium such as the sand bed in Fig. 10.7 varies with distance down the container, then knowledge of the measured hydraulic gradient, $\Delta H/\Delta L$, and the specific discharge, q, allows calculation of the average value of K over the measurement interval.

Box 10.1 An approximate justification of Darcy's law

Though Darcy's law (Eq. (10.6)) was originally based solely on experimental evidence, there have been attempts to justify the law in terms of basic theory of viscous fluid flow. Whilst these attempts involve use of the calculus, the following comments based on earlier material in this book at least indicate the plausibility of Darcy's law for fluid flow through a porous medium such as soil or rock, without providing a rigorous proof.

Despite the complexity of the movement of water through porous materials such as sand or soil, we will seek, rather boldly, to apply Bernoulli's energy-conservation equation to water flowing through such media, as illustrated in Fig. 10.5. Some of the assumptions made in deriving Bernoulli's equation in Section 3.5 might not be satisfied in such an application, which is one reason for describing the following argument as only an approximate justification of Darcy's law.

It would be expected that a good deal of energy would be lost by water as it flows through the small and tortuous passages in porous media such as soil and rock. Thus the energy-loss term E_{loss} in Bernoulli's equation, Eq. (3.17), would be expected to be a dominant term. In practice a loss in energy results in a loss in hydraulic head, ΔH. Thus E_{loss} will be written as ΔH.

Also, movement of water through such porous media is slow. Thus the kinetic-energy term $\frac{1}{2}\rho v^2$ in Eq. (3.17) would be expected to be negligible.

Thus, a suitable form of Eq. (3.17) for flow through a saturated porous medium from a location denoted by subscript 1 to a location 2 of lower hydraulic head would be

$$h_1 + z_1 = h_2 + z_2 - \Delta H$$

or, using Eq. (10.3),

$$\Delta H = H_2 - H_1$$

Equation (3.18) showed that the drag force per unit area, D/A, experienced by flow around an isolated object was proportional to $C_d v^2$, where C_d is a non-dimensional drag coefficient and v is the flow velocity. In a porous medium, fluid is flowing around many objects, not just one. However, since similar physical processes are involved, we will extrapolate Eq. (3.18) to flow through porous media. Flow through such media will generally be at very low Reynolds numbers.

For such flow over a spherical object, Fig. 3.12 shows that

$$C_d \propto 1/Re \propto 1/v$$

where v is the flow velocity. Assuming that this relationship will also apply to porous media flow, then the force per unit cross-sectional area,

$$D/A \propto C_d v^2$$

$$\propto v^2/v$$

or

$$\propto v$$

where v would be the mean velocity defined in Eq. (10.4).

The work done by this force (per unit area) as the flow moves a distance Δs will be given by the product of force and distance moved, and so will be proportional to $v \, \Delta s$. This work done will be reflected in the loss of energy represented by the drop in hydraulic head. Hence

$$v \, \Delta s \propto -\Delta H$$

with the negative sign indicating that the head H decreases in the direction of the flow velocity v. Also, as shown in Eq. (10.4), $v \propto q$, the flux density or unit discharge. Thus

$$q \propto -\Delta H/\Delta s$$

or

$$q = -K \frac{\Delta H}{\Delta s}$$

which is Darcy's equation, Eq. (10.6).

The form of Eq. (10.6) also indicates that the hydraulic head has the nature of what in physics is called a 'potential' or 'potential function' as discussed in Rose (1979).

Since $\Delta H/\Delta L$ is non-dimensional, q and K must have the same dimensions and units. Since q is a volume flux per unit area, its SI unit is m s^{-1}, the unit of velocity. However, the flux density q is not really a velocity, and q is sometimes called the 'Darcy flux' or 'unit discharge'. Perhaps somewhat confusingly, it is also referred to as the 'Darcy velocity'. When the pore space is saturated, q is equal to the product of the real mean velocity, v, and the porosity, ε, as shown in Eq. (10.4).

As Darcy found in his experiments, the magnitude of the hydraulic conductivity K depends chiefly on the nature of the porous medium

Table 10.1 *Order-of-magnitude value ranges of K (m s^{-1}) for various natural earth materials (adapted from Freeze and Cherry (1979))*

Material	Typical range of K (m s^{-1})
Gravel	$10^{-3}-10^{-1}$
Clean sand	$10^{-6}-10^{-2}$
Silty sand	$10^{-7}-10^{-3}$
Silt, loess	$10^{-9}-10^{-5}$
Clay	$10^{-10}-10^{-8}$
Unweathered marine clay	$10^{-12}-10^{-9}$
Shale	$10^{-13}-10^{-9}$
Fractured rock	$10^{-8}-10^{-4}$

transmitting the water. The hydraulic conductivity of a clean sand may be six orders of magnitude higher than that of dense clay, reflecting the enormous variation in the ability of these two soil materials to transmit water. Similar enormous variation exists in rock materials, and fractured rocks can be just as 'permeable' to water as sand. (A permeable material is one that has a high hydraulic conductivity.)

Characteristics of soil or rock material that play important roles in determining its hydraulic conductivity are its bulk density (defined in Eq. (2.4)), porosity (defined in Eq. (2.6)) and the degree of continuity of pore space. A typical order-of-magnitude range for values of K is given in Table 10.1 for a range of saturated porous natural earth materials.

Example 10.2

(Adapted from Bouwer (1978)).

Effluent from a sewage-treatment plant is spread on a hillside using sprinkler irrigation. The hillside has a slope of 3% with relatively shallow (6-m deep) soil draining to a stream as shown in Fig. 10.8. The rate of application of treated effluent (dominantly water) is 25 mm d^{-1}, over a width of 20 m (Fig. 10.8). After an approximately steady state had been achieved, the water table was found to have the geometry shown in Fig. 10.8, where the water table is initially at the soil surface, but declines in depth towards the stream. Calculate what the average hydraulic conductivity of the soil layer must be.

Solution

If z_1 and h_1, and z_2 and h_2 are the elevations and pressure heads at two locations distance L apart, then Darcy's equation, Eq. (10.6), can be

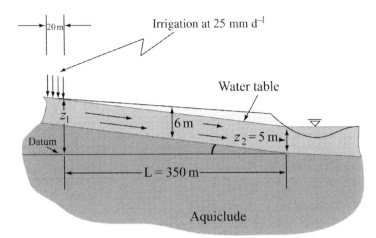

Figure 10.8 The subsurface flow field resulting from the sprinkler application of treated sewage effluent at a distance $L = 350$ m upslope from a stream (not to scale).

written

$$q = -K \frac{(z_2 + h_2) - (z_1 + h_1)}{L} \qquad (10.7)$$

On applying Eq. (10.7) to Fig. 10.8, h_1 and h_2 at the water table are both zero by definition. Thus the height of the water table gives the hydraulic head, which is constant with depth in the water at any particular location (cf. Fig. 10.4). Thus surfaces of constant hydraulic head are vertical, so the direction of flow, which is always in the direction of decreasing H and normal to the surface of constant H, will be approximately horizontal in this case (Fig. 10.8).

From Fig. 10.8, adopting the arbitrary horizontal datum shown, $z_1 = 350 \sin 0.03 + 6$ (m) and $z_2 = 5$ (m). Thus

$$q = -K \frac{(5 + 0) - (350 \sin 0.03 + 6 + 0)}{350} \qquad (i)$$
$$= 3.38 \times 10^{-3} K$$

Let q and K be given in units of m d^{-1} (the SI unit being m s^{-1}). Then, by definition of q, the downslope flow rate of water per unit width of land will, to a good approximation, be

$$q(6 + 5)/2 \, \text{m}^2 \, \text{d}^{-1}$$

and the rate from input of water will be

$$20(25 \times 10^{-3}) \, \text{m}^2 \, \text{d}^{-1}$$

Equating these two expressions for the downslope flow rate gives

$$q = 9.09 \times 10^{-2} \, \text{m} \, \text{d}^{-1}$$

Thus, from (i),

$$K = 2.69 \, \mathrm{m \, d^{-1}}$$

Knowing K, the expected outcomes of alternative irrigation regimes for disposal of treated sewage can be predicted. Combined with information on improvement in water quality as the effluent moves through the soil, this information could aid in assessing risk to water quality in the stream of Fig. 10.8.

Example 10.2 shows that, provided that the flow is in a dominantly horizontal direction, flow in an unconfined aquifer can be estimated quite well using the slope of the water table as the hydraulic gradient (see Fig. 10.1). This is called the 'Dupuit–Forschheimer' assumption, which can give accurate solutions for flow with mildly sloping water tables. Exercises 10.2 and 10.3 at the end of this chapter give further examples to which this assumption can be applied.

In Fig. 10.8 the lines of constant hydraulic head are essentially vertical, since the sum of pressure and hydraulic head is the same at all heights in all vertical planes, which are surfaces of constant hydraulic head. Since Fig. 10.8 is a two-dimensional cross-section, the vertical lines of constant hydraulic head are the edges of vertical plane surfaces on which H is constant. Surfaces of constant hydraulic head are called 'equipotential surfaces' for reasons explained in Rose (1979). In Fig. 10.8, to a close approximation, the direction of flow is horizontal and thus is perpendicular to the equipotential surfaces provided by the series of vertical planes of equal head or potential. This is illustrated in the following sketch.

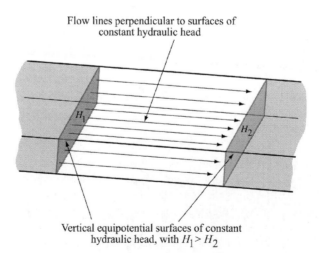

Flow lines perpendicular to surfaces of
constant hydraulic head

Vertical equipotential surfaces of constant
hydraulic head, with $H_1 > H_2$

(a)

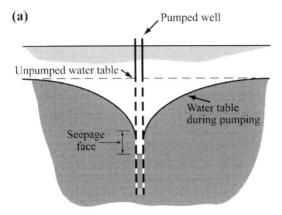

Figure 10.9 (a) The draw down of a free water table due to extraction of water by pumping from a well in an unconfined aquifer. (b) A view in plan showing equipotential lines (dashed circles) and flow lines (solid) for radial flow to the water-extraction zone in the lower section of the well.

(b)

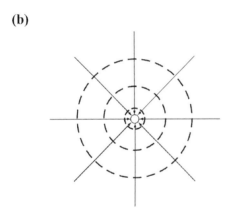

In fact, the form of Eq. (10.6) implies that the direction of flow (indicated by flow lines) must be normal (or perpendicular) to equipotential surfaces in a hydraulically 'isotropic' medium. (An isotropic medium has the same hydraulic conductivity in all directions.) In contrast to Fig. 10.8, when the flow is vertical, as can occur during infiltration into the vadoze zone of a horizontal land surface, the equipotential lines (or surfaces) are horizontal.

Equipotential lines and flow lines must always intersect each other at right angles, or be 'orthogonal' to each other. Any set of orthogonal equipotential lines and flow lines is called a 'flow net'. In general flow and equipotential and flow lines are curved, as is illustrated in Fig. 10.9.

Figure 10.9(a) illustrates in cross-section the typical shape developed in a water table as water is pumped from some depth below a free water table. In this context the flow net in plan consists of circular equipotential lines with radially inward flow lines (Fig. 10.9(b)). Such a flow net is used in the analysis of flow. When the saturated aquifer becomes exposed to the atmosphere during pumping, a 'seepage face' develops as in Fig. 10.9(a). The water table of the aquifer intersects the slotted wall of the well at a distance above the free-water level in the well. Water then moves out of the aquifer and down the seepage surface to the free-water surface.

Though the situation in Fig. 10.9 can be analysed analytically (Bouwer, 1978), in general such analysis is carried out using numerical methods implemented with computing techniques (see, for example, Freeze and Cherry (1979) and Hornberger *et al.* (1998)). A commonly used model to simulate groundwater flow, named MODFLOW, was developed by the United States Geological Survey (McDonald and Harbaugh, 1988).

Just as in Chapter 6 the rate of infiltration into the earth's surface was reported as being commonly spatially variable, the same can be true of hydraulic conductivity, both laterally and with depth. Some reasons for this will be evident from material in the next section.

10.4 Groundwater in natural subsurface formations

The general complexity of subsurface geological structures implies comparable complexity in their hydraulic behaviour. The investigation of 'hydrogeological' structures or formations at significant depths commonly involves the drilling of wells or taking cores as explained in texts such as Bouwer (1978) and Freeze and Cherry (1979). A large body of expertise has been built up in this area of study, driven by interest not only in water, but also in fuels such as oil, gas and coal.

A region of the subsurface that allows significant movement of groundwater through it is called an aquifer. As shown in Fig. 10.10, an aquifer may be unconfined, implying that it is not subject to pressure arising from overburden. However, an aquifer bounded above and below by a layer of very low hydraulic conductivity (an 'aquitard'), or a layer of essentially zero permeability (an 'aquiclude'), is called a 'confined aquifer'.

The level of water achieved in the piezometer measures the pressure head (and thereby the hydraulic head) at the point of measurement, which is the entry point for water at the base of the piezometer. The equilibrium level of water achieved in a nest of piezometers defines a 'piezometeric surface' (Fig. 10.10). The slope of the piezometeric surface gives the hydraulic gradient for horizontal flow in an approximately horizontal aquifer.

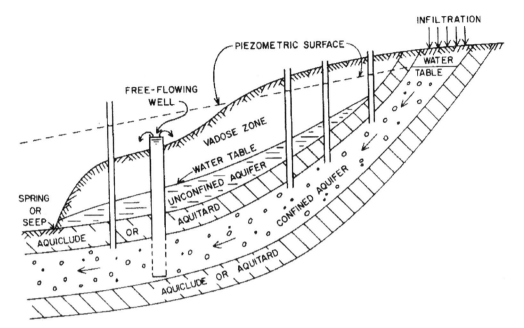

Figure 10.10 A schematic cross-section of an unconfined and confined aquifer. The water level in the piezometers which open into the confined aquifer defines the piezometric surface. (After Bouwer (1978).)

If a tube is drilled into a confined aquifer, water will rise above the confined aquifer if the aquifer is under pressure, as sometimes occurs. If the pressure is sufficient, then water can rise above the earth's surface, making it available for use without pumping, providing a 'flowing artesian well' (Fig. 10.10).

Figure 10.10 illustrates the gross generalisation that movement of water tends to be more vertical in the vadoze zone and more lateral in aquifers.

A water-filled layer of sand or gravel between layers of dense clay or shale would be an example of a confined aquifer. The Great Artesian Basin in Australia (Fig. 10.11) is an extensive example of such an aquifer, underlying some 1.7 million km^2, more than a fifth of the continent. Water from this basin, reputed to be the earth's largest artesian supply, has been tapped for about 100 years. Flow from the artesian bores is used chiefly for cattle and sheep grazing in the currently arid and semi-arid lands above the basin. As early as 1914 a report documented a diminution of flow from hundreds of bores, yet many artesian bores in this basin still flow quite uncontrolled. Whilst the rate of recharge of this aquifer is unknown, reports of decreasing water pressure and of some bores failing

Figure 10.11 The Great Artesian Basin in Australia (shown shaded).

after flowing for many years suggest that the rate of extraction of water from the basin exceeds its current rate of replenishment. The federal government has constitutional limits on its ability to act on this matter, and the three states and one territory which the artesian basin overlaps have yet to implement plans adequate to ensure sustainable use of this water resource.

Main symbols for Chapter 10

A Area

C_d Drag coefficient

g Acceleration due to gravity

h Pressure head

H Hydraulic head ($= h + z$, Eq. (10.3))

K Hydraulic conductivity of a porous medium

L Length

q Volumetric water flux per unit cross-sectional area of flow

s Distance in a particular direction

v Mean velocity of flow of water through a porous medium

x Horizontal distance

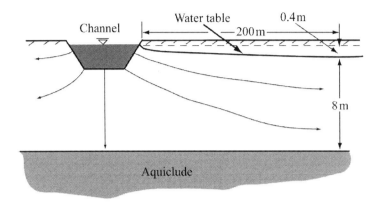

Figure 10.12 Water supply at a steady rate from an irrigation-water-conveying channel to form an unconfined aquifer bounded beneath by an impermeable layer (not to scale).

z Height measured above an arbitrary reference plane(Fig. 10.3), also used as a symbol for elevation head (Eq. 10.3).

Δ Finite difference of any quantity

ε Porosity of a porous medium

ρ Density of water

Exercises

10.1 A sand-filtration system is needed to supply drinking water to a community that requires a supply of 5000 gallons (US) a day. The sand to be used has been measured as having a hydraulic conductivity of at least $4 \, \text{m d}^{-1}$. The horizontal bed of the sand filter is to be 4 m in depth and of square shape in plan view. Flow of water through the saturated sand bed will be driven by gravity with negligible ponding on the top of the bed and with free flow at its outlet. (Thus the pressure heads at the top and bottom of the filter bed can both be assumed to be zero.)

(a) Calculate the necessary size of the sand bed. (Table 1.2 shows that gal (US) $\times$ 0.003 785 = volume in m^3.)

(b) Why would decreasing the depth of the sand bed not increase the rate of flow of water through it?

10.2 Wade (1999) has reported that, prior to lining, a conveying channel for irrigation water constructed in highly permeable sandy soils in Senegal, West Africa, lost 75% of water entering a monitored section of channel to seepage through the walls of the channel. As shown in Fig. 10.12, such seepage from the conveyed water can form an unconfined aquifer above a shallow impermeable layer.

Figure 10.13 A segment of an aquifer in which water flows across an aquiclude with an angle of slope α. Any capillary-fringe effects are assumed negligible.

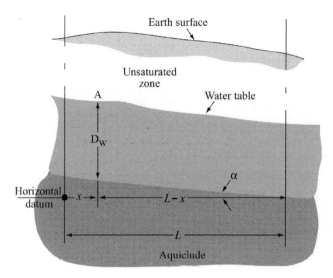

Such seepage not only represents a loss of potentially valuable water for irrigation, but also can lead to waterlogging and add to the potential salinization of areas adjacent to the channel.

The extent of the capillary fringe is negligible in this sandy soil whose hydraulic conductivity, K, is rather uniform and found to be about $5\,\mathrm{m\,d^{-1}}$. The water level in a piezometer located 200 m from the channel is 0.4 m below the (constant) water level in the conveyancing channel (Fig. 10.12).

Ignoring the fact that flow lines for water leaking from the channel are initially curved, and making the Dupuit–Forscheimer assumptions of uniform almost horizontal flow, calculate the rate of loss of water by seepage from the channel indicated by these observations.

10.3 Figure 10.13 illustrates a section of length L of an unconfined or water-table aquifer above an impervious rock aquiclude whose surface slope is α. This slope is assumed sufficiently small that the Dupuit–Forscheimer assumptions lead to a solution of sufficient accuracy. Thus the depth of water D_w can be taken to be essentially the same irrespective of whether it is measured vertically or normal to the aquiclude surface.

Assume that there is no net flow to the water table from the unsaturated zone above it, there being a balance between infiltration and evaporation of water in this zone.

(a) Show that the hydraulic gradient in the horizontal direction is given by

$$\frac{\Delta H}{\Delta x} = \frac{\Delta D_w}{\Delta x} - S$$

where D_w is the depth of water and $S = \sin \alpha$. (Note that Eq. (i) in Example 10.2 is a particular example of this more general expression.)

Then, using Darcy's equation, show that the flux of water per unit width of flow is given by

$$KD_w \left(S - \frac{\Delta D_w}{\Delta x} \right) \quad (m^2 \, s^{-1})$$

where K is the hydraulic conductivity of the aquifer (assumed spatially uniform). Comment on the origin of each of the two terms in this expression. (Note that the product KD_w is called the 'transmissivity' of the aquifer).

(b) Use the expression $KD_w(S - \Delta D_w/\Delta x)$ for an approximately horizontal flux derived in (a) to calculate the seepage loss in Exercise 10.2.

10.4 There are quite a few different techniques used to obtain the magnitude of the hydraulic conductivity, K, of aquifers. It is sometimes possible to obtain a relatively undisturbed sample or core of aquifer material, and, if so, the hydraulic conductivity of the saturated core can be obtained by encasing the core in laboratory equipment called a 'permeameter'. One type, called a 'falling-head permeameter', operates with the principles shown in Fig. 10.14.

Figure 10.14 shows a cylindrical core sample of aquifer matrix of depth L and cross-sectional area A encased and fitted so that a water head of height h is above the sample. To increase measurement sensitivity the cross-sectional area of the water-supply tube, a, is made smaller than A.

In the measurement of the core sample's hydraulic conductivity, water is allowed to flow through the core, the pressure head at outlet from the core being zero in the arrangement shown in Fig. 10.14. Observations are then made as the water level declines (whence the name used for this type of permeameter). Suppose that, at time t_1, the water-column height is h_1, and that, at time t_2, it has fallen to h_2. It may be shown that the following expression gives the hydraulic conductivity of the saturated core sample from the aquifer:

$$K = \frac{aL}{A(t_2 - t_1)} \ln \left(\frac{L + h_1}{L + h_2} \right) \tag{10.8}$$

Figure 10.14 The principle of the falling-head permeameter.

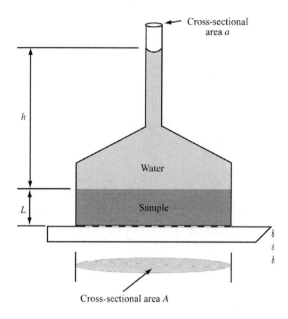

Cross-sectional area a

h

Water

L

Sample

Cross-sectional area A

Suppose that $a/A = 0.05$, $L = 10\,\text{cm}$, $h_1 = 35\,\text{cm}$, $h_2 = 15\,\text{cm}$ and $t_2 - t_1 = 15\,\text{min}$. Calculate the value of K for the sample.

10.5 At the sea edges of offshore sand islands it is possible for there to be an interface between salt groundwater under the sea surface and fresh groundwater under the sand (Fig. 10.15). Because of the shape of this interface, the fresh groundwater is said to form a 'freshwater lens'. This lens of fresh water is effectively floating above the salt water, and, because the fresh water has a lower density than the salt water, there can be restricted mixing between the two groundwater bodies.

Pressure at the point A (Fig. 10.15) on the fresh/salt interface must be the same whether it is calculated from the depth h_2 below sea level, or from the depth $h_1 + h_2$ below the water table in fresh water. Denote the freshwater density by ρ_f and that of sea water by ρ_s. Using Eq. (3.2), but recognising the difference in fluid densities, show that

$$h_2 = \frac{h_1}{\rho_s/\rho_f - 1}$$

Note that, taking the appropriate ratio of 1.025 for ρ_s/ρ_f, $h_2 = 40h_1$.

Note 1.   It follows that the depth of the groundwater lens at any position where the assumptions are satisfied can be calculated from a measurement of h_1.

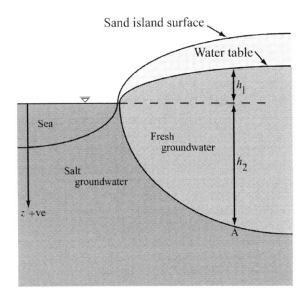

Figure 10.15 A freshwater lens floating over salt groundwater at the sea edge of an oceanic sand island.

Note 2. If fresh groundwater is extracted from a well on land then the salt groundwater can intrude into the space previously occupied by fresh groundwater. This is referred to as 'salt-water intrusion'. Such intrusion can lead to a well yielding salt water despite having previously yielded fresh.

10.6 The bottom of a lake is covered by a sediment layer of depth z_s, overlying a somewhat impermeable confining layer of thickness z_c. An intact core of the sediment layer was extracted, and its saturated hydraulic conductivity, K_s, determined.

Two piezometers were then installed. The open lower end of piezometer A (Fig. 10.16) was located at the bottom of the sediment layer, just above the confining layer. Piezometer B was installed with its open end just below the confining layer (Fig. 10.16). Steady water heights in the piezometers were recorded.

(a) Derive an expression for the vertically downward Darcy flux, q, in the sediment layer involving the height difference ΔH_1 shown in Fig. 10.16, the depth z_s and the hydraulic conductivity K_s.

(b) Steady piezometer readings indicate a steady rate of flow, so the Darcy flux through the confining layer must also be q. The head difference ΔH_2 is shown in Fig. 10.16. Derive an expression for the 'hydraulic resistance' of the confining layer, defined as z_c/K_c.

Figure 10.16

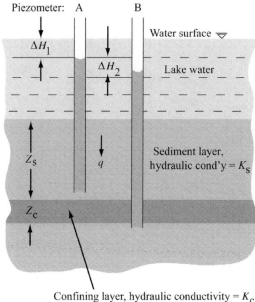

Confining layer, hydraulic conductivity = K_c

(c) How might the analysis of parts (a) and (b) be applied in an investigation of the hydrology of the lake?

10.7 A unit for water volume that is used in some groundwater-hydrology literature is the 'acre foot'. An acre foot is the volume of water enclosed in an acre surface area of water that is uniformly one foot deep.

Suppose that the average equivalent ponded depth of water stored in the Great Artesian Basin in Australia (shown in Fig. 10.11) is 125 m. The area of the basin is given as 1.7 million km². What is the volume of water stored in this artesian basin expressed in acre-foot units?

References and bibliography

Bouwer, H. (1978). *Groundwater Hydrology*. New York: McGraw-Hill Book Company.

De Smedt, F., and Wierenga, P. J. (1978). Approximate analytical solution for solute flow during infiltration and redistribution. *Soil Sci. Soc. Am. J.* **42**, 407–412.

Freeze, R. A., and Cherry, J. A. (1979). *Groundwater*. Englewood Cliffs, New Jersey: Prentice-Hall International, Inc.

Hornberger, G. M., Raffensperger, J. P., Wiberg, P. L., and Eshleman, K. N. (1998). *Elements of Physical Hydrology*. Baltimore, Maryland: The Johns Hopkins University Press Ltd.

McDonald, M. G., and Harbaugh, A. W. (1988). A modular three-dimensional finite-difference groundwater flow model. USGS Open File Report 83-875.

Rose, C. W. (1979). *Agricultural Physics*. London: Pergamon Press.

Wade, M. (1999). Improvement of surface irrigation in the sandy zones of the Senegal River. *International Programme for Technology and Research in Irrigation and Drainage, IPTRID Network Magazine*. Issue 14, pp. 8–9.

11

Movement of water through the unsaturated zone

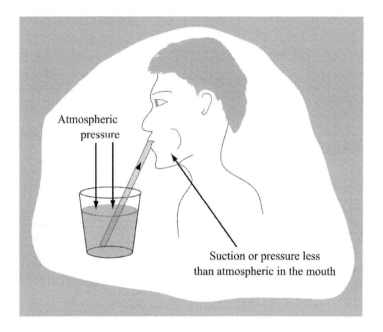

11.1 Introduction

There can be small volumes of air trapped as bubbles in the groundwater and capillary fringe, but through time these can dissolve or escape to the air above. Thus the groundwater zone is dominantly water-saturated. In contrast, in the unsaturated zone some of the pore space within the porous solid (such as soil or rock) is filled with gas rather than water. The common gas in the pore space of the unsaturated subsurface zones is air, usually of rather different composition from that of the atmosphere, with concentrations of the gases carbon dioxide and water vapour usually being considerably enhanced. It is of great importance to plant life that the carbon dioxide given off by the activities of plant roots and soil organisms is able to be exchanged with atmospheric air. Figures 2.2 and 2.4 show how liquid and gaseous components share the pore space in the solid matrix.

The growth of most plant and tree roots occurs in the unsaturated zone of the soil profile. In fact, with important exceptions such as rice, many plants will not survive too long if their complete root zone is saturated with water. Thus movement of water in the unsaturated zone is of fundamental importance to life on earth. Even the transmission of water and solutes to groundwater usually occurs without complete saturation of the vadoze zone, such transport commonly taking place dominantly through preferred pathways that offer less resistance to flow than does the bulk of the soil matrix. Channels formed from decayed roots, soil cracks and connected pores due to biotic activity are a few sources of such highly conductive channels. (The spatial variability in infiltration rate considered in Chapter 6 is one consequence of this soil-profile characteristic.)

The movement of chemicals with water that is infiltrating to the groundwater is one reason for strong environmental interest in the unsaturated zone and its transmission properties. The excessive use of fertilizers in some agricultural systems has led to considerable research on their movement in soils, in particular for the very mobile nitrate ion. The uptake of nutrients and other chemicals by food crops also takes place dominantly in the unsaturated zone.

The increase in concentration of naturally occurring saline salts in the unsaturated zone resulting from changes in land use is considered in Chapter 12. Such mobilisation of salt can lead to sufficiently high salt concentrations to cause the death or stunting of most plant life. This scourge of 'salinity' has affected civilisations in the past, and is an encroaching and major environmental problem in parts of many countries today.

Compared with the movement of water in the saturated zone considered in Chapter 10, such movement in the unsaturated zone has the added complexity that the degree of unsaturation can vary both in space and with time. The degree of unsaturation is expressed in terms of soil water content, which can be defined in alternative ways as shown in Eqs. (2.7) and (2.8). However, soil water content is not the only important variable related to partial saturation in the unsaturated zone. Even with the same soil water content, the difficulty that a plant root has in removing water from unsaturated soil can vary enormously with the texture and structure of the soil which is holding the water. Understanding this tenacity of water retention in unsaturated soils involves the concept of capillarity, which is discussed in Section 2.7 and reviewed here.

In Section 2.7 the phenomenon of capillarity is found to explain how the addition of some moisture to dry sand sufficiently increases its strength that it can be used to form sculptured shapes. In Fig. 2.18(a) this concept was related to the rise of water in a capillary tube. The pressure

Figure 11.1 The use of suction to create a pressure difference between the two ends of the straw and thus an upward flow of liquid.

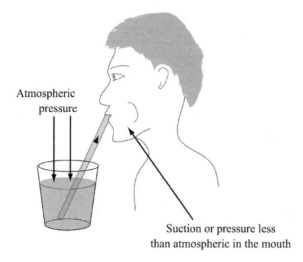

Figure 11.1 The use of suction to create a pressure difference between the two ends of the straw and thus an upward flow of liquid.

Atmospheric pressure

Suction or pressure less than atmospheric in the mouth

less than atmospheric in the capillary water column is called a suction. The concept of water experiencing a positive suction is possibly easier to grasp than the concept of a negative pressure, or pressure less than atmospheric, although they are only alternative ways of describing the same condition.

Possibly our most common experience of a suction, or suction pressure, is invoked whenever we drink a liquid from a container using a straw. The straw is not only a capillary tube, but also a means by which one can easily apply a suction or pressure less than that of the local atmosphere produced inside one's mouth. (This is an innate ability, right from birth!) Then, since the surface of the liquid in the container is subject to the local atmospheric pressure, the fluid in the container is driven up the straw into one's mouth by the pressure difference which has been set up (Fig. 11.1).

11.2 The capillary fringe – a common start to the unsaturated zone

As shown in Fig. 10.2, the vadoze zone extends a little below the unsaturated zone to include the capillary fringe. Though the fringe is saturated, water in the capillary fringe is at a pressure less than that exerted by the local atmosphere. The reason for this can be explained using Fig. 11.2, which depicts a static system in which water is in equilibrium. Denote by p_A the pressure at point A at the water table, and by p_B the pressure at point B located at height z above the water table (Fig. 11.2). Then it

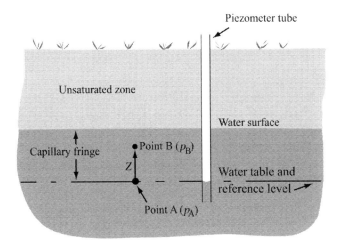

Figure 11.2 Investigating the variation of water pressure in the capillary fringe.

follows from Fig. 11.2 that, since the water column between points A and B will exert a pressure ρgz (Eq. (3.2)),

$$p_A = p_B + \rho gz \qquad (11.1)$$

However, as shown by the height of water in the piezometer tube, the pressure at the water table is that of the local atmosphere (Fig. 11.2). It is convenient and conventional to take this local atmospheric pressure as an arbitrary datum, and so regard this pressure as zero. (Since it is only pressure differences that are significant, it is really unimportant which pressure is taken as a datum.) Thus, taking $p_A = 0$, it follows from Eq. (11.1) that

$$p_B + \rho gz = 0$$

or

$$p_B/(\rho g) = -z \qquad (11.2)$$

so the pressure p_B is negative, or less than the atmospheric pressure. Thus, although the soil water pressure, and thus the pressure head, is positive beneath the water table, it is negative above it, and given by $-z$, assuming that z is measured in a positive-upwards direction from the water table as reference level. This is illustrated in Fig. 11.3.

It is a commonly used convention to represent the pressure head by the Greek symbol ψ when the pressure head is negative, as it is in the capillary fringe (Fig. 11.3). In the unsaturated zone above the capillary fringe the pressure head ψ (pronounced psi) is even more negative than in the fringe. This symbol, ψ, was introduced in 1907 by an early soil scientist named Buckingham. Apart from its historical origin, using

Figure 11.3 Variation above
and below the water table in
the saturated zone of the
component pressure and
elevation heads. The origin is
at A. The pressure head is
positive beneath the water
table and negative above it. In
contrast, the elevation head is
positive above the water table
and negative below it. H is the
hydraulic head, the sum of the
two component heads.

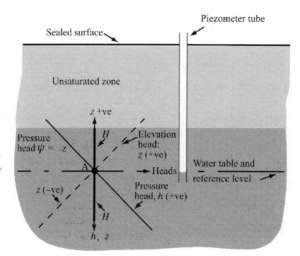

the symbol ψ can remind us of its essentially negative character. This
(negative) pressure head, ψ, equal to $-z$, is shown in Fig. 11.3.

Thus the pressure head is continuous in its magnitude from positive
values beneath the water table, through zero at A on the water table, to
negative values above the water table (Fig. 11.3). In fact, if equilibrium
(that is a zero-flux condition) also applied up into the unsaturated zone,
then this linear decrease in pressure head would also continue there.
For there to be no water flux in the unsaturated zone up to the soil
surface, however, the soil surface would have to be sealed by a barrier
impermeable to water, as indicated in Fig. 11.3.

The hydraulic head when the pore water pressure is negative

The hydraulic head, H, was defined in Eq. (10.3) as the sum of the pressure and elevation heads. This definition still holds whatever the context,
and so also applies to the unsaturated zone and the capillary fringe where
the pressure head, ψ, is negative. Thus

$$H = \psi + z \quad (z \text{ positive upwards}) \tag{11.3}$$

assuming that z is chosen to be positive in the upward direction.

From Fig. 11.3, it follows from Eq. (11.3) that, above the water table,

$$H = -z + z = 0$$

Also, since below the water table z is negative but h is positive and
equal to z, H is zero in this region also. As discussed in Section 10.2

for groundwater, so also quite generally in the unsaturated zone, no change, or zero gradient in the hydraulic head H, is the requirement for equilibrium or no flux of water, as is assumed in Fig. 11.3.

Beneath the water table in Fig. 11.3 the (positive) pressure head is denoted h, an essentially positive quantity here indicating the vertical distance beneath the water table to the point where the pressure head is being defined. In contrast to this, above the water table, and in the unsaturated zone quite generally, the symbol h is commonly (though not universally) used in such a way as to retain its essentially positive character. This use relates to the concept of suction introduced in Fig. 11.1 and in Section 2.7 to describe pore water whose pressure is less than atmospheric. This use of h as a suction head is now described.

The concept of a suction head, h

The suction, or extent to which the pore water pressure is less than atmospheric, can also be represented by an equivalent 'head', just as was done when the water pressure was positive (as it is below the water table). In the context of the capillary fringe above the water table shown in Fig. 11.2, let us now use the symbol h to represent the height of a water column that would exert a positive pressure equal in magnitude but opposite in sign to the negative pressure p_B. Then

$$p_B = -\rho g h$$

Substituting this expression for p_B into Eq. (11.2) gives

$$-h = -z$$

or

$$h = z = -\psi \qquad (11.4)$$

where, used in this way, h is commonly called a 'suction head'. Equation (11.4) indicates that, at equilibrium, the suction head is equal to the height above the water table. This statement is illustrated in Fig. 11.4.

It follows from Eqs. (11.4) and (11.3) that, either in the unsaturated zone or in the capillary fringe,

$$H = \psi + z = -h + z \quad (z \text{ positive upwards}) \qquad (11.5)$$

where in this equation h is the suction head and z is measured, positive upwards, from any chosen reference level.

In working with data on unsaturated soil, location is often conveniently given as a depth beneath the soil surface. In this context it is more convenient to take z to be positive downwards from the soil surface as a datum. This change in directional sign convention has no effect on

Figure 11.4 Water pressures above the water table are negative (i.e. less than local atmospheric pressure) and denoted by ψ, an essentially negative quantity. Alternatively, this zone can be considered to be in suction, represented by an essentially positive suction head h. At equilibrium, $h = z$, the height above the water table.

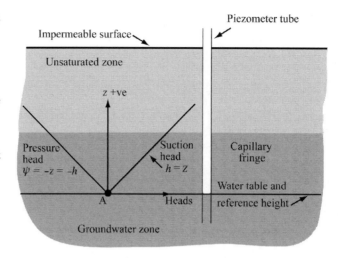

the pressure head, but changes the sign of the elevation head. With this sign convention for z, then,

$$H = \psi - z = -h - z \quad (z \text{ positive downwards}) \tag{11.6}$$

This concept of a suction head is directly related to the concept of soil water suction, introduced in Section 2.7, where it was denoted τ_s. The suction pressure τ_s, an essentially positive quantity, defines how much the pore water pressure is less than atmospheric. Thus, in terms of Fig. 11.2, at point B

$$\tau_s = -p_B$$

Also, as shown in Eq. (2.18),

$$\tau_s = \rho g h \tag{11.7}$$

with h providing the suction head equivalent to the suction pressure τ_s.

11.3 Water and soils in equilibrium

Whilst the extent of the capillary fringe can be significant in soils of high clay content, the attention given to it in Section 11.2 might be regarded as out of proportion to its spatial extent. This may well be so, but consideration of the spatial variation at equilibrium in the pressure head, ψ, and the corresponding suction head $h \,(= -\psi)$ in that zone provides a useful introduction to these concepts which are of quite general significance

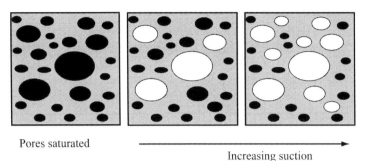

Pores saturated

Increasing suction

Figure 11.5 A cross-section of soil showing soil pores stylised as circular in shape. The progressive draining of water (shown in black) from soil pores of a variety of sizes as soil water suction increases is illustrated. The space vacated by the removal of water is filled by air.

in the unsaturated zone, and are by no means restricted to the capillary fringe.

As water is withdrawn there can be more than one mechanism involved in the development of a suction or water pressure less than atmospheric in the matrix of unsaturated soils. In soils that do not shrink or swell this suction can be largely interpreted in terms of the concept of capillarity discussed in Section 2.7 and illustrated in Figs. 2.2 and 2.18. Soil can be conceived as having a wide range of pore sizes. On draining from saturation, quite large pores will drain first, being emptied of water at very low suction values. As the soil continues to dry or lose water, then suction increases, and progressively smaller pores will lose the water they hold. Indeed, from Eq. (2.18), the suction τ_s is inversely proportional to the effective radius, r, of the soil pore which just retains water at that suction. An infinitesimal increase in τ_s, and the pore of radius r will be emptied of water. Thus, with a rigid soil matrix, the higher the applied suction the smaller the size of pore which remains water-filled. Thus the relation between suction and size of water-filled pore is inverse (Eq. (2.18)). This progressive pore emptying with increasing suction is illustrated in Fig. 11.5.

In contrast to non-swelling soils, for clay soils, which shrink with withdrawal of water and swell with addition of water, the interpretation of the effect of increasing suction is quite different from that just given for rigid, or dimensionally stable, soils. Observation of the surface of shrinking and swelling soils shows the development of cracks on drying, and closer observation shows that the clay particles pack ever closer together as suction increases. The resulting changes in the structural arrangement and properties of clay minerals are complex and not fully understood. It is forces between the clay plates or other structures of clay minerals as they are brought together on drying, rather than capillarity, which are involved in the changes in pressure head accompanying a decrease in water content. The mechanisms involved in water retention in such soils are also dependent on the 'polar' nature of the water molecule. Polar molecules are aligned in the short-range electric field arising from the net electrical charge present on clay surfaces (see Marshall, Holmes and Rose (1996) for a fuller description).

Whatever the cause of soil water suction, in order for plant or tree roots to be able to extract water from the unsaturated zone, the greater the suction, the larger the amount of work that needs to be done per unit of water extracted. Thus the energy state of water in the unsaturated zone, as indicated by the magnitude of the suction, plays an important role in movement of water through the soil–plant–atmosphere continuum. In this important sense of enabling, and yet commonly limiting, the growth of terrestrial vegetation, water in the unsaturated zone is of absolutely fundamental environmental significance.

Measurement of the soil water pressure head, ψ (or suction head, h)

The measurement of ψ for unsaturated media requires techniques different from the piezometer measurement of positive (or zero) pressure head (Figs. 10.3 and 10.10). In principle, ψ can be measured if water in a measurement system can be maintained in contact with the pore water and its pressure reduced until there is no exchange of water between that in the measurement system and that in the soil. The water pressure in the measurement system will then be ψ. The measurement problem is that of how to maintain effective contact between water in the measurement system and in the soil when the pressure in the measurement system is reduced, since such contact would normally be broken as pressure is reduced. This problem can be overcome by locating a fine-pored membrane between the two water bodies, namely the water in the soil and that in the measurement system. The smaller the pores in this interfacing membrane, the lower the pressure in the measurement system can be before air bursts through the membrane's pores and the measurement system becomes ineffective. A measuring instrument with such an interfacing membrane is called a 'tensiometer'. Tensiometers can be made in a variety of forms, the principle of a form commonly used for field soil measurements being sketched in Fig. 11.6.

After water in the instrument has achieved pressure equilibrium with that in the soil (this being achieved through exchange of water), the pressure (less than atmospheric) is read from a manual recording pressure gauge, or more conveniently can be electronically logged with a fitted pressure transducer (Fig. 11.6). The porous cup providing the membrane interface referred to earlier can be unglazed ceramic. Equipment of the type illustrated in Fig. 11.6 operates over only a modest, even

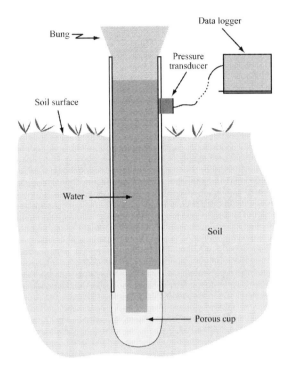

Bung

Data logger

Pressure
transducer

Soil surface

Water

Soil

Porous cup

Figure 11.6 The principle
of a common form of
tensiometer for field use.

if important, range of suctions. Marshall, Holmes and Rose (1996) give
greater detail on methods of measuring the pressure head, ψ, and the
instrument types involved which can extend measurement beyond the
tensiometer range.

The moisture characteristic

For any given porous medium such as soil there is a relationship between
the pressure head ψ and soil water content (expressed either volumetri-
cally or gravimetrically). This relationship is called the 'moisture characteristic' of the porous medium. This relationship is not unique, since
it depends to some extent on whether the medium is taking up water
(adsorption), or water is being withdrawn (desorption). The difference
between adsorption and desorption characteristics is said to represent
'hysteresis' in the relationship.

The moisture characteristic of soil at a particular location can be
determined in the field by measuring at the same time both the suction
head (e.g. with a tensiometer) and the soil water content at as near as
possible the same location. *In situ* soil water content can be measured

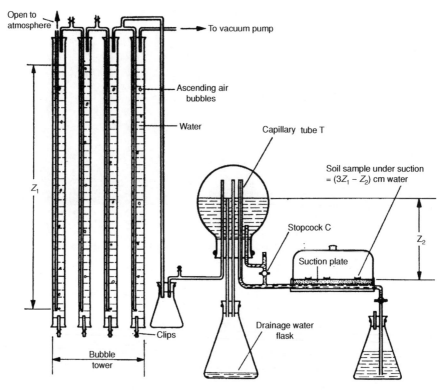

Figure 11.7 A suction plate with ancillary suction-control equipment (not to scale). (After Rose (1979).)

non-destructively, for example with a 'neutron moisture meter' (described in Marshall, Holmes and Rose (1996)), or a 'capacitance meter' (Paltineau and Starr, 1997). Both these measurement techniques yield the volumetric water content θ (Eq. (2.8)). Field investigation of the moisture characteristic at higher water contents can be achieved by irrigation if necessary; however, it may take a long time (and consistently dry weather) for water content to decline in order for one to investigate the relationship adequately at low water contents.

Hence, in order to speed up their determination, moisture characteristics are often measured by bringing soil samples from the field to laboratory equipment. For suction heads of less than a few metres of water, a convenient type of laboratory equipment for moisture-characteristic determination is the 'suction plate'. The suction head applied to the suction plate, and thus to the soil samples placed upon it, can be controlled as indicated schematically in Fig. 11.7. As with a tensiometer bulb, the suction plate is commonly constructed from fine-pored unglazed ceramic

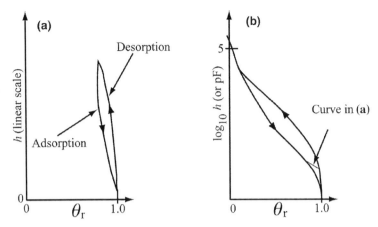

Figure 11.8 Soil moisture characteristics showing hysteresis in the relationships between the suction head, h, and the saturation ratio, θ_r. Arrows on the characteristics indicate the direction of change in water content. (a) Both h and θ_r are plotted on a linear scale. (b) The suction head is alternatively plotted as $\log_{10} h$ (termed pF when h is expressed in cm head of water). pF $= 5$ is very dry.

material, whose pores are sufficiently fine to remain water-filled up to a limit set by the air-entry value. Figure 11.7 illustrates one convenient method for controlling and varying the suction head applied to the plate by using bubble towers connected to a vacuum pump. Achievement of equilibrium between suction in the plate and that in the soil can be confirmed by sequential weighing of soil samples temporarily removed from the suction plate. Weighing wet, and then oven-drying, gives the soil's gravimetric water content for the suction applied. Conversion to volumetric water content requires field measurement of the bulk density (Eq. (2.9)).

To obtain measurement of moisture characteristics up to much higher suctions than can be sustained by a suction plate, equipment called a 'pressure-membrane apparatus' is commonly used (Marshall, Holmes and Rose, 1996; Rose, 1979).

Hysteresis in the moisture characteristic can be investigated by measuring the characteristic obtained by draining an initially saturated sample, and, in contrast, that obtained by wetting an initially very dry sample. Such hysteresis in the moisture characteristic is schematically illustrated in Fig. 11.8, where the volumetric water content is expressed as a fraction of the saturated water content. This water-content ratio is called the 'saturation ratio', θ_r. The suction head, h, can cover a very wide range of values, so that a plot using a logarithmic scale for h (as in Fig. 11.8(b)) is often preferred over the use of a linear scale (Fig. 11.8(a)). An advantage of the use of a logarithmic type of plot is that reasonable detail of the important relationship at the wetter end of the relationship can be displayed, whilst allowing coverage of the entire feasible range of h. It is conventional in some soil-science literature to use $\log_{10} h$ with h expressed in cm of water column height, and this is termed pF.

Units of head or pressure

The suction head h (or the pressure head $\psi = -h$) is expressed in a variety of units in the literature. The SI unit for either head is m of water (being negative in the case of ψ: see Eq. (11.4)). A head of water is just a convenient way of expressing pressure, and a head h can be converted into a pressure, p, using $p = \rho g h$, where ρ is the density of water. The SI unit of pressure is the pascal (Pa). Thus the water head h corresponding to 1 Pa $= 1/(10^3 \times 9.8) = 0.102 \times 10^{-3}$ m H_2O, or close to 0.1 mm H_2O. Hence the common use of the much larger pressure unit, the megapascal, or MPa, 1 MPa $= 10^6$ Pa $= 102$ m H_2O. The kilopascal, 1 kPa $= 0.102$ m H_2O, is also used. Yet another pressure unit used in related literature is the bar, 1 bar $= 100$ kPa. The bar is close in magnitude to sea-level atmospheric pressure.

Moisture characteristics

Further examples of moisture characteristics are given in Fig. 11.9, for a range of soil types described in terms of their mechanical analysis (discussed in Section 2.3). Note that, in Fig. 11.9, it is recognised that ψ (expressed in MPa) is a negative quantity, and it is plotted as an abscissa rather than as an ordinate as is used in Fig. 11.8.

Both Figs. 11.8 and 11.9 illustrate the very non-linear nature of moisture characteristics. Especially for coarse-textured soils (those high in sand content and lower in clay), it is noticeable from Fig. 11.9 that ψ does not have to decrease much below zero before most water is removed from the soil, and the rate of decrease in water content with $-\psi$ thereafter is quite small. This is one reason why, when coarse-textured and well-structured surface soils are allowed to drain from saturation, after a day or two they reach an approximately reproducible water content, which is called the 'field-capacity' value. To be more exact, the water content of such top soil on drainage from near saturation by rainfall or irrigation falls continuously; however, after a day or so the rate of fall of water content usually declines noticeably. Hence the concept of field capacity for well-draining soils is only an approximate concept, but is useful, and thus widely used, especially in agricultural and irrigation applications.

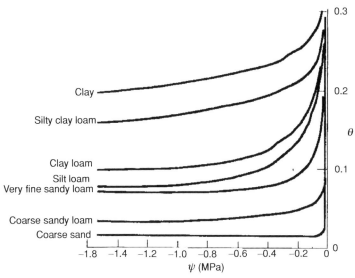

Figure 11.9 Desorption moisture characteristics for a range of soil types in the USA. θ is the volumetric water content. (After Hanks (1992).) (Note that 1 MPa = 102 m water head.)

Because drainage to this field capacity is rapid for such soils, the water content at field capacity is effectively an upper limit to the amount of stored water that is available for the growth of relatively shallow-rooted plants, such as most agricultural crops. The pressure head in soil at field capacity is commonly in the range of -1 to -2 m (or -10 to -20 kPa), but can vary with soil type, and is best experimentally determined in the field situation for which it is to be applied.

The lower limit to the amount of water available for extraction by plants corresponds to the low pressure head or high soil water suction at which plants permanently wilt, this limit being called the 'permanent wilting point'. Whilst this lower limit varies a little among different types of plants, it corresponds to the soil water content when ψ is approximately -150 m (or -1.5 MPa, as indicated in Fig. 11.9). (This is one reason why the greatest height of trees is less than 150 m, being about 100 m.)

It is the difference between the field capacity and the wilting-point water content which is of most significance to agriculture and to plant growth generally. This is because it is this water-content difference which is the relevant indication of the water-content range which is available for uptake by plants, rather than the magnitude of either the upper or the lower limit to availability.

Referring to Fig. 11.9, it can be seen that, for any given value of the pressure head, ψ, the water content θ is considerably higher for fine-textured than for coarse-textured soils. However, this does not necessarily imply that fine-textured soils can store a larger amount of water available for use by plants than can coarse-textured soils. It is usually soils of intermediate texture (e.g. loamy sands) that store the greatest amount of plant-available water.

One of a number of empirical equations that provide an approximate fit to the shape of soil moisture characteristics is the following, proposed by Campbell (1974):

$$\frac{\theta}{\theta_s} = \left(\frac{\psi}{\psi_e}\right)^{-1/b} \tag{11.8}$$

where θ_s is the volumetric water content at saturation ($= \varepsilon$, the porosity of the medium), ψ_e is interpreted as the pressure head at which air commences entry into the initially saturated soil, and b is a parameter evaluated by fitting Eq. (11.8) to the experimental data.

Example 11.1

Parameters for use in Eq. (11.8) were determined for a forested volcanic soil of low bulk density near Rotorua, New Zealand (Tomer, 1999). These soils are highly permeable and of high porosity. This was part of a study on the effects of sprinkler irrigation with tertiary treated municipal effluent. The values of parameters in the top (A) horizon of one soil profile for the desorption characteristic were $\theta_s = 0.68$, $\psi_e = -0.68$ kPa and $b = 8.65$. Calculate the moisture characteristic and compare your results with those in Fig. 11.9.

Solution

The moisture characteristic for this soil obtained using Eq. (11.8) is given in Fig. 11.10. Notice that the pressure head ψ (or psi) is given in kPa, not MPa as in Fig. 11.9, with -100 kPa corresponding to -0.1 MPa. Recognising this plotting difference, and comparing as best one can the moisture characteristic for this soil with those given in Fig. 11.9 for soil types in the USA, it can be seen that this New Zealand soil of recent volcanic origin appears to hold considerably more water over the limited pressure-measurement range than does even the clay soil in Fig. 11.9. Whilst the data summarised in Fig. 11.9 were in no way intended to cover all world soils, this comparison emphasises the need for measurement of moisture characteristics to be made for soil at the location of interest.

11.4 Movement of water in the unsaturated zone

It is found that Darcy's law (Eq. (10.6)) also provides a very good description of the movement of water in unsaturated porous media such as soil,

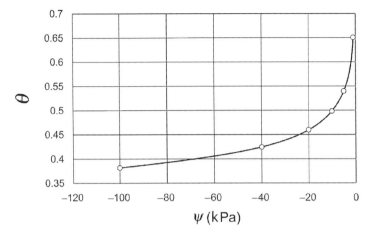

Figure 11.10 The fit using Eq. (11.8) to the desorption moisture-characteristic data obtained by Tomer (1999) for a soil of recent volcanic origin in New Zealand. The fit is given by $\theta = 0.68$ $(\psi/-0.68)^{-1/8.65}$.

provided that it is recognised that the hydraulic conductivity, K, is no longer a constant for any given medium, but is very strongly dependent on the water content or pore-water suction. The reason for this strong dependence can be understood from Fig. 11.5, which shows soil pores emptying of water as soil suction increases, commencing with the larger pores. Furthermore, Poiseuille's equation, Eq. (6.2), shows the flow through a cylindrical pore to be proportional to the fourth power of its radius. It follows that, as the radius of pores which are still water-filled, and hence conductive, is reduced by a factor of ten (say from 2 mm to 0.2 mm), then the water flow, other things being equal, will be reduced by a factor of 1/10 000!

Thus a very rapid decrease in hydraulic conductivity with decreasing water content or increasing suction would be expected – and this is experimentally found to be the case.

As noted in Section 2.3, the water content of porous media can be alternatively expressed on a gravimetric (w) or volumetric (θ) basis. The water-content dependence of K in unsaturated media can be recognised by writing K as $K(\theta)$. Thus Darcy's law for flow in unsaturated porous media in the direction s is written

$$q = -K(\theta)\,\Delta H/\Delta s \qquad (11.9)$$

where, as in saturated-medium flow, q is the volumetric flux density (m s^{-1}), or Darcy flux, and $\Delta H/\Delta s$ is the hydraulic gradient in the direction s (non-dimensional if the hydraulic head H is expressed in terms of the head of water).

Figure 11.11 The
relationship between $K(\theta)$ and
θ for the particular volcanic
soil described by Tomer
(1999). $K(\theta) = 46(\theta/0.68)^{20.3}$.

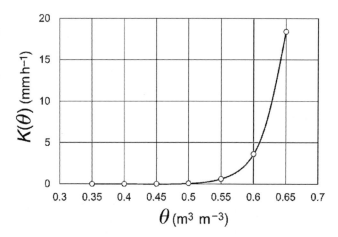

Figure 11.11 The relationship between $K(\theta)$ and θ for the particular volcanic soil described by Tomer (1999). $K(\theta) = 46(\theta/0.68)^{20.3}$.

Example 11.2

An empirical relationship used by Ross (1990) to fit experimental data on the water-content variation in $K(\theta)$ is

$$K(\theta) = K_s(\theta/\theta_s)^{2b+3} \qquad (11.10)$$

where K_s is the hydraulic conductivity at pore saturation, when the volumetric water content is θ_s, and b is the same fitting parameter as that used in Eq. (11.8). Using the same values of θ_s and b as for the New Zealand soil, which were given in Example 11.1, calculate and plot $K(\theta)$ as a function of theta (θ). The corresponding value of K_s is 46 mm h^{-1}.

Solution
Inserting the quantities appropriate for this particular soil over the experimental data range investigated into Eq. (11.10) gives

$$K(\theta) = 46(\theta/0.68)^{20.3}$$

in the units of K_s, namely mm h^{-1}. The very large value of the power, 20.3, indicates that there is a very strong dependence of K on θ. This implies the very non-linear form of the relationship shown in Fig. 11.11, which gives the result of the calculations. The form of Eq. (11.10) indicates that $\ln K$ will be linearly related to $\ln \theta$. (You might like to confirm this numerically.)

From Eqs. (11.5) and (11.6) it may be remembered that, in the equation for H, the sign given to the gravitational-head term, z, naturally depends on the choice of either upwards or downwards as the positive direction. In numerical calculation, of course, the same result is obtained whatever choice is made, the choice being simply a matter of convenience. In examples given later in this chapter it is convenient to use the soil surface as the datum level, and thus for the positive direction of z to be taken as downwards from that surface. Adopting this sign convention of downwards positive, then, from Eq. (11.6), $H = \psi - z$, since an increase in z will decrease the hydraulic head.

Using this expression for H, it follows from Eq. (11.9), taking the direction s to be z, that

$$q = -K(\theta)\frac{\Delta}{\Delta z}(\psi - z)$$

or

$$q = -K(\theta)\frac{\Delta \psi}{\Delta z} + K(\theta) \quad (z \text{ positive downwards}) \qquad (11.11)$$

The right-hand side of Eq. (11.11) shows that water moves because of two different kinds of force. The second flux component $K(\theta)$ indicates a downward flux whose origin is the gravitational force, expressed through dependence on the gravitational head, z. The magnitude of this term depends only on $K(\theta)$, and so, for any given soil, is a function of water content.

The first term on the right-hand side of Eq. (11.11) is the flux component due to the gradient in pressure head, ψ, though the magnitude of this flux component also depends, as would be expected, on $K(\theta)$.

In what follows in this chapter, $K(\theta)$ will be written simply as K, the dependence of K on θ in all unsaturated media being implicitly acknowledged.

Some people find it easier to think in terms of suction head, h, instead of pressure head, ψ, just because h is a positive quantity and ψ is negative. This common preference arises from the fact that, as water is lost from soil, h increases, whereas ψ becomes more negative (i.e. a bigger negative number). In terms of suction head, and with the same sign convention for z, Eq. (11.11) becomes

$$q = K\frac{\Delta h}{\Delta z} + K \quad (z \text{ positive downwards}) \qquad (11.12)$$

Equation (11.12) indicates that the downwardly directed flux q increases with gravity (the term K), and also if the suction head increases with z (i.e. if a uniform soil becomes drier with depth). As given in Eq. (11.6), in terms of h, the hydraulic head is

$$H = -h - z \quad (z \text{ positive downwards}) \qquad (11.13)$$

Thus, with this sign convention for z, and since h is an essentially positive quantity in unsaturated soils, H is always a negative quantity.

Infiltration into a uniform soil at a particular location

As discussed in Chapter 6, at field and watershed scales infiltration is strongly dominated by spatial variation in infiltration rate. In that chapter some of the causes of such spatial variation, which included macropores and large voids, were outlined. However, restricting consideration to a particular location or a small area free of such features, some of the characteristics of infiltration were described in Section 6.2. For a soil whose hydraulic characteristics are uniform with depth, Fig. 6.2 illustrates possible profile variations in water content. Also Fig. 6.4 shows how the infiltration rate typically declines with time, with that rate depending on the initial water content in the soil profile.

The rapid drop in $K(\theta)$ as the soil water content declines is illustrated in Fig. 11.11. This very non-linear characteristic of $K(\theta)$ helps in understanding the rather sudden spatial decline in θ at the wetting front illustrated in Fig. 6.2. Whilst Fig. 11.11 illustrates a rather dramatic variation in $K(\theta)$ with θ, a strong non-linear dependence of this general type is typical of all soils. The resulting rapid decrease in θ with depth at the wetting front illustrated in Fig. 6.2 provides support for the simple 'bucket-type' water-balance methodology described in Section 4.5 and illustrated in Fig. 4.11.

Figure 6.4 shows the typical decline in infiltration rate following the ponding of water in a ring infiltrometer (Fig. 6.3). Though the mathematical analysis of infiltration is not given in this text, some features can be understood in terms of concepts that have already been presented. For example, it can be appreciated that there will be a difference in soil water suction between the near-saturated soil surface and the drier soil at and beneath the wetting front. Thus the suction gradient $\Delta h / \Delta z$ in the term $K \Delta h / \Delta z$ of Eq. (11.12) can be understood to play a significant role in increasing the rate of infiltration beyond the flux of magnitude K which is due to the gravitational gradient $\Delta z / \Delta z$, or unity. The wetter the soil profile prior to the commencement of infiltration the less the downward suction gradient will be. This gives at least a qualitative explanation of the difference in infiltration-rate behaviour between infiltration into initially wet soil and infiltration into initially dry soil (Fig. 6.4). The positive contribution to the infiltration rate associated with the suction gradient has been called a 'capillary pull'.

As infiltration continues, Fig. 6.4 indicates that the rate of infiltration slows and approaches an approximately constant value, which is commonly somewhat close to the hydraulic conductivity of the saturated soil. As described for example in Marshall, Holmes and Rose (1996), the influence of the downward capillary pull or suction-gradient term continues to diminish with time, or alternatively with the accumulated

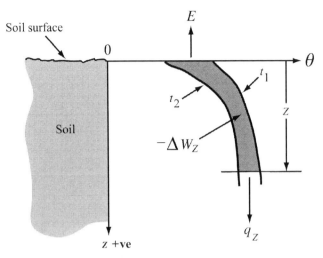

Figure 11.12 Plots of water-content (θ) profiles at an earlier time, t_1, and a somewhat later time, t_2. Owing to the combined loss of water at average rate E by evaporation from the soil surface and an average drainage flux q_z at depth z, water has been lost over the time interval $\Delta t = t_1 - t_2$. Let ΔW_z represent a positive increase in storage of water from the soil surface to depth z over the time interval Δt. Then, because water is lost, the increase is negative, and so is denoted $-\Delta W_z$.

amount of infiltrated water. The inverse dependence of I on ΣI in the Green–Ampt type of infiltration, Eq. (6.1), describes this behaviour.

Mass conservation of water to depth z in soil

Imagine a cylindrical column of soil of unit cross-sectional area, commencing at the bare soil surface and extending downwards into the soil to a depth z. This column is separated from the surrounding soil profile by an imaginary or conceptual surface only in order to apply the principle of mass conservation of water to it. During a period of no rainfall or other input of water, profiles of the volumetric water content, θ, to depth z were measured at two different times, t_1 and t_2 (Fig. 11.12).

Since the soil has no vegetation growing in it, loss of water from this cylindrical volume of soil must occur either by evaporation to the atmosphere or by drainage from the volume to soil below depth z. Over the time interval $\Delta t = t_1 - t_2$ the loss from the soil surface by evaporation will be $E \Delta t$, where E is here interpreted as the average evaporation rate over the time interval Δt. Likewise, if q_z is the average downward flux of water at depth z in the soil over the time interval Δt, then the drainage loss over this period is $q_z \Delta t$. Let ΔW_z represent a positive increase in storage of water from the soil surface to depth z over the time interval Δt. Then, because of the loss of water from the soil volume, the increase is negative, and so is denoted $-\Delta W_z$.

Mass conservation of water for the defined column of soil can be given in words by

outflux at depth z + outflux at the soil surface

= decrease in water stored in the column

or, in symbols:

$$q_z \, \Delta t + E \, \Delta t = -\Delta W_z \qquad (11.14)$$

where the decrease in amount of stored water is written as a negative increase, or $-\Delta W_z$.

Whilst the term ΔW_z can readily be calculated from sequentially measured water-content profiles, the difficulty in employing Eq. (11.14) is that, in general, both q_z and E are unknown. If either of these can be determined, then the other can be calculated using the mass-conservation equation (11.14). In the following subsection, one method of determining the average evaporation rate over a time period Δt is explained.

Determination of the rate of evaporation from the soil surface

Especially if a soil profile is drying after having initially been wet, there is an upper zone, where movement of water is upwardly directed in response to evaporation from the soil surface, and a lower zone, in which water is draining downwards, perhaps dominantly due to gravity. At the interface between these two zones water does not move at all. So, at this interface, the gradient in the hydraulic head, H, must be zero. With z taken positive downwards, from Eq. (11.13) the gradient of H in the downward direction is given by

$$\frac{\Delta H}{\Delta z} = -\frac{\Delta h}{\Delta z} - \frac{\Delta z}{\Delta z}$$
$$= -\frac{\Delta h}{\Delta z} - 1$$

Hence, if $\Delta H/\Delta z = 0$, then $\Delta h/\Delta z = 1$, or the slope of the suction-head profile is unity. Figure 11.12 shows the profiles of θ at two successive times, t_1 and t_2. The time-average profile of θ over this time interval will be somewhere between the two profiles shown in Fig. 11.12. The suction-head profile corresponding to this time-average profile of θ is shown as h on the right-hand side of Fig. 11.13. The suction head h can be measured directly or inferred from measured θ via the appropriate

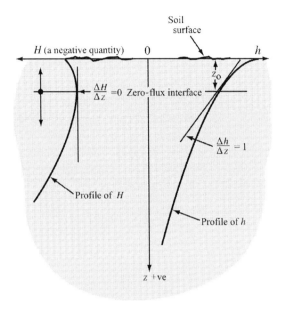

Figure 11.13 Particular profiles of time-averaged values of the suction head, h (on the right of the figure), and hydraulic head, $H = -h - z$ (on the left). The water flux is upwards above the zero-flux interface, at depth z_0 beneath the soil surface, and downwards below it.

moisture characteristic. The slope of the suction-head profile, $\Delta h / \Delta z$, is shown to have the value unity at a depth z_0 beneath the soil surface. As indicated earlier, this is the depth at which $\Delta H / \Delta z$ is zero, indicating that there is zero water flux.

With the profile of h given on the right-hand side of Fig. 11.13, and since $H = -h -z$, H can readily be calculated. The calculated profile of H (an essentially negative quantity) is shown plotted on the left-hand side of Fig. 11.13. Note that H initially increases (becomes less negative) with depth z beneath the soil surface. Since water moves from where H is higher to where it is lower, this segment of the H profile indicates an upward movement of water to be lost by evaporation at or near the soil surface. The slope $\Delta H / \Delta z$ is positive in this upper surface zone. However, as shown in Fig. 11.13, the slope $\Delta H / \Delta z$ decreases with depth, becoming zero, as it must, at depth z_0, where $\Delta h / \Delta z = 1$. It is where the tangent to the profile of H is vertical in Fig. 11.13 that $\Delta H / \Delta z = 0$ is located. Below depth z_0, H decreases (or becomes more negative), indicating a downward movement of water. In this lower zone $\Delta H / \Delta z$ is negative, since H decreases with increasing z, and this is consistent with a downward (positive) direction of water flux because of the negative sign in Darcy's equation, Eq. (11.9).

Figure 11.12 shows how the decrease in amount of water stored to depth z, namely $-\Delta W_z$, can be determined from θ profiles at two successive times, t_1 and t_2. Thus, once the depth, z_0, of the zero-flux plane has been determined for a particular time interval, the water-loss term for that depth can be determined as follows.

With the term $q_z \, \Delta t = 0$ at depth z_0 (the zero-flux interface), it follows from Eq. (11.14) that

$$E \, \Delta t = -\Delta W_z \quad (z = z_0) \tag{11.15}$$

where the quantity $-\Delta W_z$ is now measured to depth z_0 (Figs. 11.12 and 11.13).

With the evaporation $E \, \Delta t$ over the time interval Δt of successive measurements calculated in this way, then, using Eq. (11.14), the water flux $q \, \Delta t$ can be determined for any desired depth z in the soil profile, with $-\Delta W_z$ being determined from the soil surface down to that particular depth z.

One of the important uses of this information on surface evaporation is to determine the hydraulic-conductivity characteristics of the soil profile, on the basis of *in situ* measurements of soil water content.

Such basic hydraulic characteristics of soil provide the basis for predicting what movement of water is expected to occur in any given situation. One field method used to calculate such characteristics that is based on successive water-content-profile measurements is given in the next subsection, and more fully in Olsson and Rose (1978). Since this is most commonly carried out during the drainage of a wetted profile, this method is sometimes called the 'internal-drainage method' of determining hydraulic-conductivity characteristics in the field.

Field determination of hydraulic-conductivity characteristics of soil

From Eq. (11.14) the flux at depth z over the time period Δt is given by

$$q_z \Delta t = -E \Delta t - \Delta W_z \tag{11.16}$$

However, from Eq. (11.12), at depth z,

$$q_z = K_z \left(\frac{\Delta h}{\Delta z} \right)_z + K_z$$

where $(\Delta h / \Delta z)_z$ is the slope of the suction-head profile at depth z. Thus

$$q_z \, \Delta t = K_z \left[\left(\frac{\Delta h}{\Delta z} \right)_z + 1 \right] \Delta t \tag{11.17}$$

from which

$$K_z = \frac{q_z \, \Delta t}{\left[\left(\dfrac{\Delta h}{\Delta z} \right)_z + 1 \right] \Delta t}$$

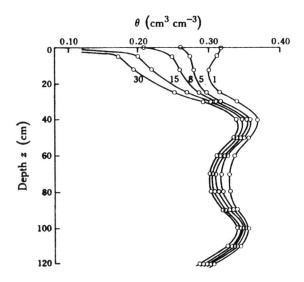

$\theta \; (\mathrm{cm}^3 \; \mathrm{cm}^{-3})$

Figure 11.14 Profiles of volumetric water content, θ, measured on the numbers of days after drainage shown as numerals on the curves. (After Olsson and Rose (1978).)

and so, using Eq. (11.16),

$$K_z = - \frac{E\,\Delta t + \Delta W_z}{\left[\left(\dfrac{\Delta h}{\Delta z} \right)_z + 1 \right] \Delta t} \qquad (11.18)$$

An example of how Eq. (11.18) can be used to determine hydraulic-conductivity characteristics in the field is given in the following example.

Example 11.3

Olsson and Rose (1978) required to know the hydraulic-conductivity characteristics of an orchard soil in order to determine the patterns of uptake of water by the roots of peach trees in the orchard. Whilst the top-soil was a fine sandy loam in texture, change in soil characteristics with depth indicated that hydraulic-conductivity characteristics were likely to vary with depth in the soil profile. To this end they measured how profiles of soil water content and suction head changed with time for a month following irrigation of a bare, uncultivated site, in Victoria, Australia, which wetted the soil to a depth of approximately 1.3 m. The water content was measured gravimetrically in the top 17 cm, and by neutron moderation below that depth to 1.2 m. Lower soil layers exhibited a moderate degree of volume change in response to change in water content. This shrinking and swelling behaviour does not invalidate the use of Eq. (11.18) to determine K_z, provided that h is determined *in situ*, as was the case in these experiments (Olsson and Rose, 1978).

Profiles of the volumetric water content, θ, are shown in Fig. 11.14. Numbers on the profiles show the time in days after evaporation and

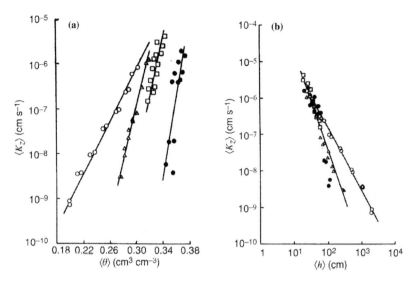

Figure 11.15 (a) Relationships between the hydraulic conductivity, $\langle K_z \rangle$, averaged over the water-content range between successive profile measurements, and the volumetric water content, $\langle \theta \rangle$, in the soil profile of Shepparton fine sandy loam at various depths: 0.125 m (o), 0.275 m (△), 0.425 m (•) and 0.925 m (□). (b) Relationships between the hydraulic conductivity, $\langle K_z \rangle$ and the *in situ* suction head $\langle h \rangle$ for the same depths as in (a). Values of $h > 700$ cm of water were inferred from θ using draining moisture characteristics. (After Olsson and Rose (1978).)

drainage commenced. Successive profiles can be regarded as continuing examples of the pair of profiles labelled with times t_1 and t_2 in Fig. 11.12.

Solution

The amount of evaporation $E \Delta t$ from the bare soil surface was calculated for each of the time intervals between successive profile measurements shown in Fig. 11.14. This was done by locating the depth below the soil surface, z_0, of the zero-flux interface depicted in Fig. 11.13, and then using Eq. (11.15) to determine $E \Delta t$. As the water content declined over the measurement period, z_0 increased from 0.17 m to 0.52 m.

Using this information on evaporation over each observation and calculation period, and the corresponding profiles for θ and h, values of K_z for the average water-content profile between any two successive observation times were calculated. By repeating this process for all the intervals between observations, K_z was calculated as a function of θ and depth z using Eq. (11.18).

The results of these calculations for four depths in the soil profile are given in Fig. 11.15(a) as a function of θ, and in Fig. 11.15(b) as a function of h. In Fig. 11.15 the labels are shown enclosed in angle brackets as $\langle K_z \rangle$, $\langle \theta \rangle$ and $\langle h \rangle$ in order to draw attention to the fact that the values plotted are average values applying over the modest range of water contents between successive profile measurements. Using such time intervals is necessary, since, in order to apply this method of determining hydraulic-conductivity characteristics in practice, the changes in water content between successive profile measurements must be sufficient to exceed substantially the inevitable measurement uncertainty. Units in Fig. 11.15 are those used in the publication.

The straight lines fitted to the relationships between θ and $\langle K_z \rangle$ plotted on a logarithmic scale indicate that there is a rapid power-law type of increase in $\langle K_z \rangle$ with θ. This is consistent with the power-type dependence of K on θ/θ_s which is often used to provide an empirical fit to such data, and is illustrated by Eq. (11.10). The change with depth in the hydraulic-conductivity characteristics shown in Fig. 11.15 reflects changes in a number of soil properties, including mechanical analysis and bulk density, with an increase in clay content in the depth interval 0.3−0.45 m (Olsson and Rose, 1978).

It can be argued that, in non-swelling unsaturated soils, it is the pore-size distribution which most closely controls the value of K, and that pore size is more closely related to h than it is to θ (though the two variables are related by the moisture characteristic). This closer physical relationship may be the reason why less variation in hydraulic-conductivity characteristics with depth is found when they are plotted against $\langle h \rangle$ (Fig. 11.15(b)), rather than against $\langle \theta \rangle$, (Fig. 11.15(a)).

An advantage of determining $K(\theta, z)$ using the method outlined in this example is that it avoids the uncertainty introduced when what one hopes are 'undisturbed' core samples are removed from the soil profile and taken for laboratory determination of K using laboratory methods. Even more uncertainty is removed if both h and θ are measured directly, rather than inferring h from θ via desorption moisture characteristics.

With $K(\theta, z)$ and $\theta(h)$ known, it is possible to solve the fundamental differential equations for flow of water through unsaturated soil, which are given, for example, in Marshall, Holmes and Rose (1996). A convenient computer-based method of carrying out such tasks is to use the efficient numerical Soil Water Infiltration Movement (SWIM) model of Ross (1990).

A method such as that outlined in Example 11.3 provides information on hydraulic conductivity at a particular location. However, the common field experience indicating the dominant importance of spatial variability in infiltration rate, discussed in Section 6.3, is one indication of the commonly found spatial variability in subsurface hydraulic-conductivity characteristics.

When a less-permeable surface soil layer overlies more permeable subsoil (a common feature with sandy subsoils), movement of the wetting front can become unstable, and preferential advance of water in narrow 'fingers' can occur. Another cause of local spatial variability during penetration of water into expanding and shrinking soils is the formation of cracks or macropores, which can be continuous to the soil surface (see Fig. 6.5). Another important source of large pores in soil results from the activity of large soil fauna such as earthworms; such pores are called biopores. Decayed roots of dead vegetation provide channels through which water can move rapidly relative to movement through the bulk of the soil matrix. Whilst such preferred pathways for movement of water can play a major role in determining infiltration characteristics, because of the relatively large size of such channels, these channels are emptied of water as quite low suctions develop.

Applying Eq. (11.11) or Eq. (11.12) to soils assumes that they are reasonably homogeneous porous media, and such equations are not designed to describe movement of water through any channels, biopores, or cracks that may exist. Understanding the effect of such features as fingering and movement of water via preferred pathways is a current area of research. The current general lack of detailed localised field information appears to restrict the application of such understanding as we have. Despite this, progress can be made, even at watershed scales, using broad-scale models. This was illustrated for infiltration in Chapter 6.

Main symbols for Chapter 11

E	Evaporation rate
g	Acceleration due to gravity
h	Suction head ($= -\psi$, and hence is a positive quantity)
H	Hydraulic head
K or $K(\theta)$	Hydraulic conductivity (a function of θ in unsaturated soils)
K_s	Hydraulic conductivity at pore saturation
K_z	Hydraulic conductivity at depth z
p	Water pressure
q	Volumetric water flux per unit cross-sectional area of flow
q_z	Value of q at depth z
s	Distance in a particular direction
T	Time
W	Water storage (equivalent ponded depth)
z	Height above a reference level (such as a water table (Eq. (11.5))); z can also be defined as as distance in a downward positive direction (Eq. (11.6))

θ	Volumetric water content
ρ	Density of water
τ_s	Suction pressure in soil water (Eq. (11.7))
ψ	Pressure head of soil water (commonly used when this pressure is negative or less than atmospheric pressure)

Exercises

11.1 Figure 11.3 shows the spatial variation in the components of the hydraulic head, H, namely the elevation head and the pressure head. These components are shown with the direction of z assumed to be positive in the upward direction from the water table as a datum or reference level.

   Draw a similar diagram showing the two components of H, but using the alternative assumption that z is positive downwards from a reference level taken as the interface between the unsaturated zone and the capillary fringe. Confirm that there is no gradient in H, since a constant value of H is the requirement for equilibrium. (Note that ψ is negative above the water table and positive below it.)

11.2 Write Eq. (11.11) in the form which would be appropriate if the direction of z were to be taken as positive in the upward direction. Then justify to yourself, in a given assumed context, that the flux density q is described as being in the *same* direction whichever sign convention is chosen for z to be positive.

11.3 At a particular time tensiometers record that $h = 1.5$ m at a depth of 0.2 m beneath the soil surface, and $h = 2.0$ m at a depth of 0.4 m. Take the reference height to be the soil surface.
   (a) Calculate the hydraulic head at both depths (0.2 m and 0.4 m) beneath the soil surface.
   (b) What will be the direction of movement of water between these two depths?
   (c) If the average hydraulic conductivity, K, between the two depths is 10^{-7} m s^{-1}, calculate the flux density q in m s^{-1} and mm d^{-1}.

11.4 An empirical form used to relate the hydraulic conductivity, K, to the volumetric water content, θ, is

$$K = K_s (\theta / \theta_s)^k$$

where K_s and θ_s are values at saturation.
   (a) Show that this form of relationship for K suitably describes the conductivity characteristic in Fig. 11.15(a).
   (b) For the depth of 0.125 m in the Shepparton fine sandy loam soil whose hydraulic-conductivity characteristic is shown in

Fig. 11.15(a), $\theta_s = 0.31$. From that figure read off the corresponding value of K_s, and then determine the value of k in the relationship for K.

11.5 Assuming that the water content at field capacity, θ_{fc}, is approximated by the water content corresponding to $\psi = -10\,\text{kPa}$, estimate θ_{fc} for the soil whose moisture characteristic is given in Fig. 11.10.

(Note that $-10\,\text{kPa} = -0.01\,\text{MPa}$. Using this very approximate figure, θ_{fc} can only very approximately be inferred for the range of soils whose moisture characteristics are displayed in Fig. 11.9.)

11.6 An area of bare soil was irrigated and allowed to drain and dry by evaporation from the soil surface. Over the time period between days 5 and 8 following irrigation, the zero-flux plane, defined by the depth at which there is zero gradient in hydraulic head (Fig. 11.13), was located at an average depth of 0.3 m beneath the soil surface.

The average decline in volumetric water content over the effective three-day period was $0.017\,\text{m}^3\,\text{m}^{-3}$. Calculate the total amount of evaporation which took place over the three-day period.

11.7 Use the value of surface evaporation calculated in Exercise 11.6 to calculate the hydraulic conductivity of the soil profile at a depth of 0.9 m using Eq. (11.18), given the following further information applicable to the three-day period from day 5 to day 8 following irrigation.

- The decline in volumetric water content in the soil profile from the surface to the depth of 0.9 m was $0.01\,\text{m}^3\,\text{m}^{-3}$ on average.
- The average suction gradient $(\Delta h/\Delta z)$ at depth 0.9 m over the time period was $4\,\text{mm}^{-1}$.

11.8 With z chosen to be positive in the downward direction, the hydraulic head H is given by Eq. (11.13). Draw several profiles of h similar to, but different from, the shape given on the right-hand side of Fig. 11.13, and give quantitative scales to the axes. Then use Eq. (11.13) to calculate the corresponding profiles of H. From these plots of H versus z, interpret the implications for the direction of flow of water at various levels in the profile.

References and bibliography

Campbell, G. S. (1974). A simple model for determining unsaturated conductivity from moisture retention data. *Soil Sci.* **117**, 311–314.

Hanks, R. J. (1992). *Applied Soil Physics. Soil Water and Temperature Applications*, 2nd edn. New York: Springer-Verlag.

Marshall, T. J., Holmes, J. W., and Rose, C. W. (1996). *Soil Physics*, 3rd edn. Cambridge: Cambridge University Press.

Olsson, K. A., and Rose, C. W. (1978). Hydraulic properties of a red–brown earth determined from *in-situ* measurements. *Aust. J. Soil Res.* **16**, 169–180.

Paltineau, I. C., and Starr, J. L. (1997). Real-time soil water dynamics using multisensor capacitance probes: laboratory calibration. *Soil Sci. Soc. Am. J.* **61**, 1576–1585.

Rose, C. W. (1979). *Agricultural Physics.* London: Pergamon Press.

Ross, P. J. (1990). Efficient numerical methods for infiltration using Richards equation. *Water Resources Res.* **26**, 279–290.

Tomer, M. D. (1999). Comparing observed and simulated water storage during drainage to select hydraulic parameters for volcanic soils. *Aust. J. Soil Res.* **37**, 33–52.

12
Salinity and contaminant transport

12.1 Introduction

Salinity

The term salinity describes the presence of soluble salts in soil or aqueous solution. Whilst the ions Na^+ and Cl^- of sodium chloride are often dominant contributors to salinity, other ions can also provide an important addition to the solute concentration in soils. As the concentration of saline salts in soil increases, plant or tree growth is inhibited, and, at sufficiently high concentrations, death may occur (see the frontispiece to this chapter, showing a saline stream, Quairading, Western Australia). Physical characteristics of soil are also deleteriously affected by salinity: hydraulic conductivity can decline dramatically, and, on drying, soils can become very hard and unsuitable for plant growth.

Especially in those regions of the world where annual rainfall is low and evaporation rates potentially high, areas of naturally saline soil can be found. This is referred to as 'primary salinity', and a salt concentration by weight of only some 0.1% to 0.3% in surface soils can substantially reduce the growth of most plants and trees. It has been estimated that about 1000 million hectares, or 7% of the world's land area, is affected by salt to some extent (Dudal and Parnell, 1986).

However, what has emerged, particularly in recent times, as a major environmental issue in sustainable land management is the large-scale increase in salt-affected land caused by human activity. This is referred to as 'secondary salinity'. World total estimates of secondary salinity are suggested by Dregne, Kassas and Razanov (1991) as some 74 million hectares, which can be subdivided into some 43 million hectares of irrigated land in arid and semi-arid regions, and 31 million hectares of non-irrigated land (or 'dryland').

Secondary salinity has emerged as a major issue in two distinct contexts. Firstly, excessive use or inefficient distribution of water in irrigation for agriculture is one cause of secondary salinity. A second cause is the resultant change in subsurface hydrology due to replacement of deeper-rooted native vegetation by shallow-rooted vegetation such as pastures or agricultural crops. This replacement results in lower year-round evaporation because of the use of a summer fallow period in the case of agricultural crops, and water stress during summer in pastures. The resulting increase in amount of available water can then lead to the saline groundwater rising towards the soil surface and its vegetation. The result is called 'dryland salinity', a term used to distinguish it from 'irrigation-induced salinity'.

As is typical of a broad class of environmental issues, both land clearing and irrigation are carried out with the desirable intent and practical effect of increasing food production, at least in the short term. The issue is that, for such production to be sustainable in the longer term, we must understand the environmental consequences of our well-intentioned activities, and then adjust management practices so that such productive activity can be sustained indefinitely.

Soluble salts have various origins. Salts in the landscape are a product of rock weathering. However, another major source is sea salt, whipped up by wind into the atmosphere as oceanic aerosols, which are eventually brought down to land, often by rainfall, perhaps well inland of the sea; salt with this origin is called 'cyclic salt'. This term implies that salt deposited inland will ultimately return to the sea by river flow; however, this does not always occur. Yet another source of salt in soil is inundation by the sea during past eras, due to changes in sea level, or to subsidence or uplift of land.

Commonly (though not universally), the clay fraction of soils has negatively charged surfaces. Thus the positive ion in salt, Na^+, is attracted to such surfaces, but the negative chlorine ion, Cl^-, is free to move in the soil. Sodium ions constitute only a fraction of all the positive ions in the soil (the 'cations'), that are loosely bound to the soil, but can be exchanged for other ions present in the soil. The percentage of all exchangeable cations in a soil constituted by sodium ions is called the 'exchangeable-sodium percentage', or 'ESP'. When the ESP of a soil becomes sufficiently high, commonly in the range of 6% to 15%, its physical characteristics deteriorate dramatically. Thus the damage caused by salination is not solely due to the effects of soluble salts reducing or stopping plant growth, but also due to deterioration in physical properties of soil, including loss of the ability to form aggregates and a decline in hydraulic conductivity.

Contaminant transport in soil

Apart from naturally occurring saline salts, there is an enormous variety of manufactured substances, which, if improperly used or disposed of, can lead to soil contamination. Sometimes the word 'pollutant' is used in a stronger sense than 'contaminant'. The term pollutant is often used to indicate that there is evidence that the type or concentration of chemical involved is of concern, especially to human life or health, but quite possibly also to other life forms.

The sources of soil contaminants or pollutants are extremely varied, being much wider than the well-known disposal of industrial by-products, or so-called 'wastes', and release of inadequately treated sewage. As is so graphically explained in Rachel Carson's book *Silent Spring* (Carson, 2000), which was first published in 1962, the pressure to increase agricultural productivity by the use of chemicals to control insects or suppress weeds has led to disastrous consequences, which had either not been foreseen or were seriously underestimated. Despite considerable improvement in our knowledge of such consequences, and much progress in control and targeted use, such problems have not disappeared. Human weaknesses, or inadequate instruction in the use of chemicals, the continuous advent of new chemicals, long-lasting effects of chemicals now banned from use and the long residence times of applied, disposed of, or leaked chemicals are just a few of the reasons for the problems requiring current treatment, if this is feasible, and the need for continued surveillance regarding the use and disposal of chemicals.

Even naturally occurring and potentially beneficial chemical species, such as the nitrate ion (NO_3^-), which is a major plant nutrient, can cause environmental problems. Application rates or timing of application in use of nitrogenous fertilizers can be such that much greater concentrations of this ion are formed in the soil than can be taken up by the target crop or pasture. This excess nitrate can be leached down by water flowing through the unsaturated soil to reach the groundwater. Concentrations of nitrate and some herbicides in groundwater have reached levels that make it unfit for human consumption in too many parts of the world. In some situations water-quality problems arise from pollution by sewage.

Such soluble chemicals move through the soil profile with water. Non-soluble chemicals that are bound to the soil can also be pollutants, perhaps through being taken up from the soil by food crops. Such chemicals, which are strongly 'sorbed' or bound to the soil, move chiefly with soil during erosion, either by wind or by water. Transport of sorbed chemicals during erosion by water is discussed in Section 9.4.

The following sections of this chapter focus on understanding some of the physical processes involved in the fate of water-soluble chemicals of either natural or human origin.

12.2 Dryland salinity processes

Dryland salinity can develop where a substantial change in vegetation has occurred, for example when forest or other natural vegetation is replaced by pasture or agricultural cropping. A common result of such a change in land use is that year-round evaporation is reduced by the limited growth period and shallow root systems of the introduced vegetation. The excess soil water resulting from this decrease in evaporation can move down through the profile and add to the groundwater store, leading to a slow rise in the height of the water table if the extra water cannot escape laterally quickly enough.

When this rising water table intersects or comes close to the ground surface, it produces waterlogging, but also (if soluble salts are present), 'salination' of the surface soil due to the accumulation of salt left behind

Figure 12.1 Clearing of deep-rooted perennial vegetation and its replacement by shallow-rooted vegetation (e.g. pastures), from which year-round evaporation is less. This leads to a gradual rise in the groundwater table which first intersects lower elements in the landscape, unless its rise is modified by other geological controls.

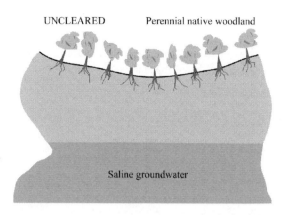

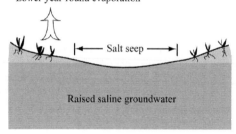

by water evaporating from the soil surface. The processes involved in developing 'salt seeps' in lower landscape elements, which are often the first visible sign of dryland salinity, are sketched in Fig. 12.1.

Reduction in year-round evapotranspiration due to land clearing for agricultural activities or urban settlement is widespread around the world. It is when this type of change in land cover and land use occurs within a context of a number of other climatic, topographic, hydrological and geological factors that dryland salinity can slowly develop. It is this slowness of onset and the lack of early warning signs which contribute to the insidious nature of dryland salinity.

Australia has extensive areas of flat terrain and vast quantities of salt of 'cyclic' or marine-aerosol origin stored in the regolith and ground-water aquifers. This salt has been brought inland over geological time scales by on-shore winds, with salt in aerosol form deposited over large distances inland by dry deposition, or brought to earth in rain-fall. Runoff from arid and semi-arid inland regions commonly does not reach rivers, which could return some of the incoming salt to the sea.

In some countries, such as Australia, some river systems drain to inland salt lakes, never reaching the sea.

Extensive clearing of the natural deep-rooted perennial vegetation from such relatively flat landscapes for replacement by shallow-rooted annual crops is why extensive areas in Ausralia are currently subject to dryland salinity. Knowledge of the processes which have led to the development of this environmental problem, aided by soil mapping, computer groundwater simulation, geological investigation, aerial-photographic interpretation of landform patterns and satellite imagery, allows prediction of the extent of expression of dryland salinity (see the frontispiece to this chapter). The expression of dryland salinity in the Australian context will continue to expand, unless there is a massive and widespread change in farming systems.

In the southern half of Australia, where land clearing has been extensive, rainfall occurs dominantly during the mild winter season, when planting of annual crops takes place. With most growth of crops restricted to the winter and spring, often followed by a summer fallow period, year-round evaporation is less than that which takes place from the original deeper-rooted vegetation which grew all year round, and extracted water to much greater soil depths than do its replacement crops or pastures. As described in detail by Clarke *et al.* (2002) and their extensive references, it is this widespread change in land cover and land use, combined with the features mentioned earlier, of flat terrain with stored salt and saline groundwater, that leads to the changes sketched in Fig. 12.1. Because of the flatness of the terrain in such regions, with land slopes commonly of order 0.1%, lateral flow of groundwater is inadequate to allow lateral dispersal of the extra recharge to the groundwater resulting from the changed land cover. The lower elements of the landscape tend to suffer first from the consequent rise in the saline groundwater (Fig. 12.1). However, the exact location and expression of surface salinity can be significantly modified by subsurface geological and geomorphological features. Whilst the details on such features given in Clarke *et al.* (2002) refer to Australia, the general process description they give is of quite general relevance.

Example 12.1, which follows, is based on data typical of areas in southern Australia subject to dryland salinity.

Example 12.1

In a particular relatively flat semi-arid landscape the groundwater table is extensive and close to horizontal. The average annual rainfall is $450 \, \text{mm} \, \text{y}^{-1}$. When the land was covered by native vegetation (shrubs and low trees) evaporation was $400 \, \text{mm} \, \text{y}^{-1}$, and drainage $50 \, \text{mm} \, \text{y}^{-1}$ to saline groundwater. The depth below the soil surface to the water table was then stable at 20 m.

Following widespread clearing and sowing to wheat in the winter rainfall period of the Mediterranean climate of the region, evaporation fell to 300 mm y^{-1}.

How many years from this change in land use might it take for the water table to reach the soil surface, with its associated problems of salting and waterlogging?

Assumptions. Assume that the entire 100 mm y^{-1} water generated by this change in land use percolates to the groundwater, and that this change is sufficiently extensive for a one-dimensional analysis to be valid. Assume that the average pore-space fraction of the soil profile is 0.4, and give any further assumptions you make in the calculation.

Solution

Denote the rainfall rate by P and the evaporation rate by E. On the assumptions made, and summing over the period of a year, mass conservation of water requires that

$$\Sigma P \, \Delta t = \Sigma E \, \Delta t + U$$

where U is the annual underground drainage of water, expressed as the equivalent ponded depth of water.

With native vegetation on the landscape, $U = 50$ mm y^{-1} led to a stable water-table depth. Thus this 50 mm y^{-1} recharge of the water table was evidently removed by lateral flow. It will be assumed that, when the recharge increases following clearing and wheat cropping, this dispersal of groundwater by lateral flow remains the same. Then, following this change in land use:

$$U = \Sigma(P \, \Delta t - E \, \Delta t) = 450 - 300 = 150 \, \text{mm y}^{-1}$$

Thus the effective value of U which results in raising the water table is $150 - 50 = 100$ mm y^{-1} $= 0.1$ m y^{-1}. Since the porosity is 0.4, only 0.4 m equivalent ponded depth of water is required in order to saturate 1 m depth of soil. Thus the ponded depth of water of 0.1 m resulting from the increase in U will saturate a soil depth of $0.1/0.4 = 0.25$ m of soil.

Let t be the number of years required to saturate the depth of 20 m previously unoccupied by the groundwater. Then

$$0.25t = 20 \, \text{m}$$

or

$$t = 80 \, \text{years}$$

The figures chosen in this example are typical of regions such as the wheat-growing areas in Australia. Since it takes of order 80 years from the commencement of wheat cultivation for dryland salinity problems to emerge, one can imagine the socio-economic difficulties in managing another change in land use such that year-round evaporation is increased, land drainage usually not being economically feasible as a possible solution to the problem in this extensive dryland context.

Land clearing and current agricultural practices set in train changes in the groundwater hydrological regime, which typically take times of order a century to reach a new (and much less benign) equilibrium. (The time scale of 80 years in Example 12.1 is typical in its context.) Such analysis and prediction indicate that, in those broad areas of Australia which are susceptible to dryland salinity, a third of all cleared land is likely to become saline and unproductive if the approach to this new hydrological equilibrium is allowed to proceed unchecked (Clarke *et al.*, 2002).

There has been general acceptance of this daunting prognosis during the last decade or so. This realisation has given greater urgency to attempts to find ways in which agriculture and/or forestry activity can be sustained, whilst at the same time halting or reversing the current rate of development of dryland salinity, which appears inexorable with current agricultural practices.

Since it is the removal of the prior complete cover of deep-rooted tree-based vegetation which has established the hydrological conditions for the development of dryland salinity, a solution from a biophysical point of view to the salinity problem would be to re-establish substantial (if not full) tree cover, if not solely by the original vegetation, then with planted forests. The feasibility of this solution is being investigated, but, for the majority of land owners, the long lead time for an economic return from tree establishment makes this option untenable without substantial financial assistance. Hence there is also considerable investigation of the feasibility and effectiveness of a range of possible 'agroforestry' systems. In an agroforestry system, agricultural crops or pastures are grown on that fraction of the land which is not under forest. Preliminary results show that the greater the fraction devoted to trees the more effective is the control of the rising water table. However, there are many possible options, including the use of deeper-rooted perennial crops such as Lucerne and alfalfa, which use more water than do shallow-rooted annual crops, and so provide better control of the height of the water table.

Whilst the critical depth in the soil above which salinisation will develop depends somewhat on soil type, keeping the water table at least 2 m below the soil surface has emerged as a useful guide. If a saline water table is too close to the soil surface, then the significant upward flow of water to meet the evaporative demand will leave behind in the soil sufficient salt to cause salinity problems.

Thus, in those parts of the world which are experiencing dryland salinity problems, the challenge is to find quite new farming systems (or, in some cases, to return to earlier systems) that are both economically viable and also effective at water-table control. These requirements are likely to be fulfilled by rather different management systems in different climates and landscapes.

In Australia, it appears that, without unduly delaying the process of change in management systems, it is not desirable to await practical demonstration of the range of available options to farmers, since these involve the time scale of tree growth. In this context, computer-based modelling of the consequences of adopting alternative systems is emerging as a timely way in which to aid evaluation of alternative management systems for the range of climates and landscapes involved (Clarke *et al.*, 2002). Such modelling methodologies in current use are illustrated by Pavelic, Narayan and Dillon (1997) and by Stirzaker, Vertessy and Sarre (2002).

The change in land management required to overcome the threat of dryland salinity may often be so great that government social and economic policy actions must play a helpful role in managing the period of transition, at least if serious hardship among those dependent on the land's continued productivity is to be alleviated.

12.3 Irrigation-induced salinity processes

Application of too much irrigation water leads to the water table rising for mass-balance reasons similar to those considered in Example 12.1. A water table too close to the soil surface results in the consequent problems of waterlogging and salinity. Whilst this is the most common cause of irrigation-induced salinity, inadequate leaching or flushing of salt in the soil profile can also restrict plant growth. This can be an issue, either in soils of low permeability, or if the salt concentration in irrigation water is sufficiently high.

In order to ensure that leaching of salt at the bottom of the root zone of vegetation is adequate to prevent a build up of the salt applied in the irrigation water, loss of salt by drainage beneath the root zone must equal the salt input, which commonly comes chiefly with irrigation water, since salt concentrations in rainfall are very small. This 'leaching requirement' is expressed in the simple and much used mass-balance equation for salt, commonly called the 'leaching-fraction model' (United States Salinity Laboratory Staff, 1954).

Let the equivalent ponded depth of irrigation water applied over any given time period be denoted by D_i, and the depth of drainage water at the bottom of the root zone of the irrigated crop be U over the same period. Then, if c_i is the salt concentration of irrigation water, and c_u the salt concentration of water draining beneath the crop, the leaching-fraction model, which assumes that there is no change in storage of salt in the root zone, states that the input and output of salt must be equal. Hence

$$D_i c_i = U c_u$$

and

$$\text{LF} = U/D_i = c_i/c_u \tag{12.1}$$

where LF is called the 'leaching fraction'.

This simple steady-state model assumes that there is no chemical interaction between salt and soil, and can be simply extended to account for any rainfall during the period of analysis. The steady-state condition assumed in the model is effectively quickly achieved in well-structured or permeable soils where rapid leaching can occur. Whilst D_i and c_i are easily determined, c_u requires extraction of a soil sample at the base of the root zone, and a measurement of its salt concentration. Then U can be calculated as the only unknown in Eq. (12.1). The leaching fraction must be sufficient for c_u not to exceed limits set by the sensitivity to salt of the particular crop being irrigated.

In less-permeable soils the steady state implied in Eq. (12.1) might not be achieved for many years, and then a transient mass balance or non-steady solution must be employed (see Marshall, Holmes and Rose (1996), Section 12.2).

Example 12.2

In an irrigation area, average rainfall is $200 \, \text{mm y}^{-1}$, evaporation $800 \, \text{mm y}^{-1}$ and $c_i = 2 \, \text{dS m}^{-1}$. (In irrigation practice, concentrations are commonly expressed in terms of the conveniently measured electrical conductivity of a liquid extract from the soil. The SI unit of electrical conductivity is siemens per metre (S m^{-1}). For historical and practical reasons a unit commonly used for electrical conductivity in irrigation management is the decisiemens per metre, or dS m^{-1}.)

In order for plant growth not to suffer from salinity, it is required that c_u should not exceed some limit, commonly taken as $4 \, \text{dS m}^{-1}$. What should the amount (or depth) of irrigation water, D_i, be in order to satisfy the leaching requirement?

Solution

Including the annual amount of rainfall, $\Sigma P\,\Delta t$, mass conservation of water on an annual basis, assuming that there is no loss by runoff in this arid environment, requires that

$$(\Sigma P\,\Delta t - \Sigma E\,\Delta t) + D_i = U$$

Substituting this expression into Eq. (12.1) gives.

$$D_i = \frac{c_i}{c_i - c_u}(\Sigma P\,\Delta t - \Sigma E\,\Delta t)$$
$$= 1200\,\text{mm y}^{-1}$$

Thus irrigation with this quite salty water ($2\,\text{dS m}^{-1}$) requires a large depth of irrigation water ($1200\,\text{mm y}^{-1}$). Unless an underground drainage system is installed, this large irrigation requirement increases the danger of raising the groundwater, thus exacerbating salinity problems (as will be illustrated in Example 12.3). Construction of a drainage system still leaves the problem of safe disposal of the salty drainage water collected. (As an exercise, show that, if irrigation water of better quality were available, say with $c_i = 1\,\text{dS m}^{-1}$, then an irrigation of only $800\,\text{mm y}^{-1}$ would satisfy the leaching requirement.)

As mentioned earlier, a very common cause for the development of salinity at the soil surface over time is the production of a high saline water table through use of more than the desirable amount of irrigation water. Thus there are two important constraints within which irrigation management must work. Firstly (as illustrated in Example 12.2), enough irrigation water must be applied to keep the salt concentration in the root zone from rising sufficiently high for growth of the irrigated crop to be significantly reduced. This is especially necessary if the water supply used for irrigation has a substantial salt concentration. However, application of more irrigation water than this leaching requirement is likely to raise the water table dangerously close to the soil surface and lead to its salination by the upward flux of water induced by evaporation. This flux of liquid up to the evaporation site brings saline salts with it. These salts then accumulate at the surface as evaporation proceeds. It is found in practice that an evaporative flux of about $1\,\text{mm d}^{-1}$ commonly leads to salinity developing in the surface layers of the soil, though this critical rate depends on climatic conditions at the site and other factors.

Thus it is clear that a degree of skill is needed for sustainable irrigation farming. Limitations in skill constitute one reason why some 43 million hectares of irrigated land is affected by salt throughout the world, largely in the world's semi-arid and arid regions (Dregne, Kassas and Razanov, 1991). Installation of an underground drainage system allows control of the rise of the water table. However, the installation of drainage should not be allowed to encourage wasteful excess use of irrigation water for a number of reasons, one important reason being the likely environmental damage caused by the disposal of large quantities of salty drainage water if the site of irrigation is located far from the sea. Some novel ways of dealing with this problem are being investigated.

12.4 Movement of contaminants in groundwater

In Section 12.2 we saw that the development of dryland salinity involved the slow elevation of the groundwater water table towards the soil surface, combined with the presence of naturally occurring salt in the groundwater or in the previously unsaturated soil profile. Thus the salt which moves with a saline aquifer due to its high solubility, and also any salt stored in the unsaturated soil profile, quickly dissolves as the water table rises. Salt can be stored harmlessly in the unsaturated soil away from the roots of vegetation, because infiltrating water commonly moves in preferred pathways from which salt is flushed, and where roots can proliferate.

However, this section will focus on soluble contaminants and pollutants other than naturally occurring salt. As mentioned in Section 12.1, the origin and nature of such chemicals are many and varied. Soluble contaminants can be spread on unsaturated soil, or be introduced into the soil profile, for example in land disposal sites. All such contaminants can move vertically and laterally throughout the unsaturated soil profile into and through groundwater regions or aquifers. Frequently it has been found that there are preferred pathways for such transport, so that contaminants applied to the earth's surface reach groundwater at depth more quickly than would be expected if the subsurface soil were a more spatially uniform material than is commonly the case.

In somewhat special situations the fluid trapped in the earth is crude oil rather than water. Oil and water don't easily mix with each other, one fluid being termed 'immiscible' in the other. As mentioned previously, some groundwater is salty due to solution of naturally occurring salt. Because of their different densities, fresh groundwater can ride over that which is salty (see Exercise 10.5). However, fresh and salty groundwater can mix together, and so are termed 'miscible'. Thus, when salt water, or water with any other soluble chemical, is displacing fresh water, the process is described as 'miscible displacement'.

The areas of study which contribute knowledge useful to the understanding and management of activities designed to minimize the severity of problems arising from pollution of groundwater include geology, groundwater hydrology and soil science, with mathematical methods as always providing the handmaiden of quantification.

Let us now seek to describe the environmentally important processes which affect the movement of contaminants in groundwater. An ability to describe processes gives clues on how activities might be managed in such a way as to achieve amelioration of undesirable outcomes that have occurred and perhaps their avoidance in future.

If groundwater is to be used for drinking by humans, there are generally agreed World Health Organisation standards that give recommended maximum limits for inorganic, organic, radionuclide and bacteriological contaminants (e.g. Freeze and Cherry (1979)). However, by the time the pollution of an aquifer is discovered, it is usually too late to take remedial measures that will be effective in the short term. This is because of the slow movement typical of groundwater systems. Thus pollution of an aquifer is likely to persist for long time periods, even if the source or sources of pollution are discovered and stopped.

Thus, where there are environmental scientists and engineers charged with protection of groundwater resources, they face a need to be able to identify sources of entry of pollutants into groundwater systems, and to understand the mechanisms of pollutant transport well enough to be able to estimate or predict such movement. A brief introduction to some of the physical concepts involved in this challenging task will now be given. If contaminants are chemically reactive, or groundwater systems are highly heterogeneous in nature, the task is even more difficult, and concepts beyond those introduced here are needed for effective analysis.

Figure 12.2(a) gives the result of a simple experiment in which an immiscible contaminant or solute displaces water in a cylinder due to movement of the piston from left to right in the figure. To the right of the figure is the ratio of contaminant concentration c at the measurement position and c_0, the undiluted concentration of the contaminant. For Fig. 12.2(a) there is a quite abrupt rise in c/c_0 as the contaminant/water interface passes across the measurement position. The change in c/c_0 is rapid, but not completely abrupt, due to 'molecular diffusion', an interpenetration of water and contaminant molecules due to their short-range thermal agitation. Flow that leads to an abrupt increase in contaminant concentration, without even the modifying influence of molecular diffusion, is often referred to as 'piston displacement'. Results of the type shown in Fig. 12.2 are commonly called 'breakthrough curves'.

In Fig. 12.2(b) a miscible contaminant of concentration c_0 is displacing water due to flow induced by the gradient in hydraulic head in the aquifer. The fluid now has to move through tortuous pathways in the porous soil or rock, which is assumed to be homogeneous in nature. As illustrated in Fig. 10.5(a), some tubes of fluid have a longer or more

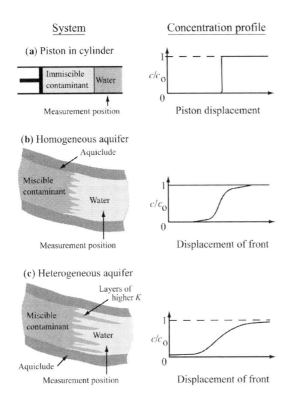

System   Concentration profile

(a) Piston in cylinder

Measurement position   Piston displacement

(b) Homogeneous aquifer

Measurement position   Displacement of front

(c) Heterogeneous aquifer

Measurement position   Displacement of front

Figure 12.2 The concentration at the measurement position of a contaminant, c (expressed as a ratio to c₀, the undiluted contaminant concentration), as displacement of the contaminant/water interface occurs for the following three systems: (a) a piston displacement in a cylinder of an immiscible contaminant, (b) a miscible contaminant in a homogeneous confined aquifer and (c) a miscible contaminant in a heterogeneous confined aquifer with layered beds of higher hydraulic conductivity (e.g. sand or gravel).

tortuous path to take than others, which leads to a spreading out of the contaminant/water interface. This process is described as 'dispersion'. On comparing the right-hand-side figures of Figs. 12.2(a) and (b) it can be seen that, compared with what would be expected in the piston displacement of Fig. 12.2(a), dispersion of the contaminant leads to earlier arrival of some contaminant and later arrival for other components. Dispersion of this type is commonly called 'hydrodynamic dispersion'.

It is not unusual for aquifers to be heterogeneous in their makeup. Figure 12.2(c) illustrates the consequences for contaminant transport of layers of higher hydraulic conductivity in the aquifer. Layers of sand or gravel in a bed of less-permeable material constitute an example of this heterogeneity, which can have a substantial influence on contaminant movement (Fig. 12.2(c)). The effect of such inhomogeneities can be interpreted as a dramatic increase in apparent dispersion, or dispersion on a larger scale than in Fig. 12.2(b). Movement through channels of high hydraulic conductivity relative to other components of the medium is referred to as 'preferential flow'.

In Fig. 12.2(a), the velocity of flow, v, and the volumetric flux density, q, are identical, $v = q$. However, in saturated porous media of porosity ε, as shown in Eq. (10.4), $v = q/\varepsilon$. In a homogeneous porous medium of constant porosity the contaminant will move through space with this velocity v, and Example 12.3 illustrates use of this relationship.

Example 12.3

If it is uninfluenced by substantial salinity gradients, the slope of the piezometric surface gives the rate of decline in hydraulic head as water moves through a section of confined aquifer, as shown in Fig. 10.10. The slope of the piezometric surface is 5 degrees in an aquifer where the flow is dominantly horizontal. The aquifer is assumed to be homogeneous with an average porosity of 0.25. If a soluble pollutant enters the recharge area of the confined aquifer, how long will it take for that pollutant to reach a free-flowing well (or artesian bore) distant $L = 2$ km from the recharge area, if the mean hydraulic conductivity K of the confined aquifer is

(a) $50 \, \text{m/d}^{-1}$ (a possible value for a coarse sand, or a sand and gravel mix) and
(b) $10^{-2} \, \text{m} \, \text{d}^{-1}$ (possible for a clay bed or consolidated sandstone).

Solution
The magnitude (regarded as a positive quantity irrespective of sign) of the hydraulic gradient, and written $|\Delta H / \Delta x|$, is given by $\sin 5° = 8.72 \times 10^{-2}$. (The magnitude is used here because we have no need to follow the direction of motion algebraically.) Then

$$q = K |\Delta H / \Delta x|$$

and

$$v = q/\epsilon \quad \text{(Eq. 10.4)}$$

For (a),

$$v = (50 \times 8.72 \times 10^{-2})/0.25 = 17.4 \, \text{m} \, \text{d}^{-1}$$

Thus the mean transit time is

$$L/v = 2000/17.4$$
$$= 114.7 \, \text{days}$$
$$= 0.314 \, \text{years}$$

For (b),

$$v = (10^{-2} \times 8.72 \times 10^{-2})/0.25 = 3.49 \times 10^{-3} \, \text{m} \, \text{d}^{-1}$$

Thus the mean transit time is

$$L/v = 2000/(3.49 \times 10^{-3})$$
$$= 5.73 \times 10^5 \text{ days}$$
$$= 1571 \text{ years}$$

From the discussion of Fig. 12.2 it follows that serious contamination could commence prior to these estimated mean transit times due to dispersion or preferential-flow effects. This example shows that, because of the enormous range in hydraulic conductivity in naturally occurring granular and consolidated materials shown in Table 10.1, there is a correspondingly great range in transit times for pollutants.

The transport of contaminant by water neglecting dispersion effects is referred to as 'advection'. In some literature this is referred to as 'convection' of the contaminant. However, the term convection has historically been used to describe fluid flow caused by differences in density in the fluid (e.g. due to temperature differences). Thus, although use of the term convection in this context is common, it could lead to confusion.

The mass transport per unit area of contaminant of concentration c (kg m^{-3} of fluid) is given by the product of c and the volumetric flux density of water, q. Denoting the mass flux density of contaminant by qs, then

$$q_s = qc \left(\frac{m^3}{m^2 s} \cdot \frac{kg}{m^3} \right) \quad \text{or } (\text{kg m}^{-2} \text{s}^{-1}) \tag{12.2}$$

or, from Eq. (10.4),

$$q_s = v\varepsilon c \quad (\text{kg m}^{-2}\text{s}^{-1}) \tag{12.3}$$

Hydrodynamic dispersion of the type whose effects are illustrated in Fig. 12.2(b) generates a flux additional to that from advection. This additional flux is found to be proportional to the gradient in the contaminant concentration, or $\Delta c/\Delta x$. Since dispersion in contaminant movement takes place in the direction from where the concentration is high to where it is low, the contaminant is dispersed in the direction of $-\Delta c/\Delta x$, since Δc implies an increase in c. The proportionality constant between the flux density of contaminant and $-\Delta c/\Delta x$ is called the 'dispersion coefficient', D, which is defined so that

$$\text{dispersion mass flux density} = -D\frac{\Delta(c\varepsilon)}{\Delta x}$$

where $c\varepsilon$ is the mass of contaminant per unit volume of space. The dispersion coefficient, D, is commonly taken to include the effect of molecular diffusion caused by random molecular movement. In terms of bulk movement of contaminant in aquifers, molecular diffusion is usually a minor player, except under certain circumstances, such as flow in fissures, or where long time scales are important.

Adding the flux densities of contaminant transport due to advection and dispersion gives

$$q_s = v\varepsilon c - D\frac{\Delta(c\varepsilon)}{\Delta x} \qquad (12.4)$$

Equation (12.4) indicates, as illustrated in Fig. 12.2(b), that displacement with velocity v in the x-direction will be accompanied by a spreading out of the contaminant due to the second dispersive term in the equation. The greater the value of the dispersion coefficient, D, the more noticeable this spreading will be. The coefficient D can be conceptually regarded as quantifying all those processes which lead to a contaminant distribution different from that expected on the basis of piston displacement or advection alone. As shown in Fig. 12.2, these dispersive processes include molecular dispersion, hydrodynamic dispersion and the effects of preferential flow through those parts of an aquifer which are of higher hydraulic conductivity. Laboratory experiments provide evidence that hydrodynamic dispersion effects increase with flow velocity v. An approximation often used to express the two components of D arising from molecular diffusion and hydrodynamic dispersion is

$$D = D_o + \zeta v \qquad (12.5)$$

where D_o is the appropriate molecular-diffusion coefficient for the contaminant diffusing through the saturated pore space of the aquifer, and ζ (zeta) is the 'dispersivity' or 'coefficient of dispersivity'.

In aquifers, fluxes of contaminant commonly change rather slowly with time. However, as seen in Example 12.3, the time scales involved are often lengthy, and over such time scales considerable change can take place, and such change takes place in space as well as with time. Indeed, mass conservation requires there to be a close link between changes in space and with time in contaminant concentration. The reason for this is as follows. Consider the flux of a contaminant into and through a particular volume containing an aquifer. Then any difference between the fluxes into and out of the volume must result in a rate of change in the amount of contaminant stored in the volume.

Even if introduced into a uniform one-dimensional flow, a concentrated or 'point' source of contaminant leads to a flux in three dimensions, since dispersion of the contaminant takes place in all directions. A practical difficulty in predicting contamination of groundwater, especially if the aquifer is heterogeneous, lies in being able to estimate adequately the three-dimensional variation in hydraulic conductivity and dispersion coefficients.

Further mathematical development of these concepts is given in Marshall, Holmes and Rose (1996) and other more advanced texts.

12.5 Contaminant transport in the unsaturated zone

In understanding the mechanisms involved in movement of contaminants through the unsaturated zone it is helpful to distinguish between those contaminants which are soluble in water and those which are not (or are very sparingly soluble). Contaminants that are water-soluble are said to be 'miscible' in water, and, broadly speaking, miscible contaminants closely follow movement of water through the soil, a process referred to as 'miscible displacement'.

There are also contaminants, such as oil, which are almost insoluble in water, and such liquids are said to be 'immiscible' in water. Whilst transport through soil of immiscible liquids is affected by the movement of water, it has characteristic differences in behaviour from miscible displacement of soluble contaminants, referred to as 'solutes'.

Contaminants most commonly originate at or close to the soil surface through a wide variety of sources of natural or human origin (Section 12.1). Examples include accession by rainfall (as in 'acid rain'), erosion and deposition of human sewage or animal excreta, oxidation of metal sulphides and spillage of industrial wastes. Excessive application of agricultural chemicals is also a significant source of contaminants, and the dumping of urban or industrial wastes provides more localised and intensive sources of pollutants.

One concern with such contaminants and pollutants is with their uptake and possible concentration by food crops, pastures and all animal and bird life (i.e. flora and fauna, see Carson (2000)). Another concern is the fact that such contaminants and pollutants can move down into the soil profile through infiltration and subsequent downward movement or 'leaching', thus endangering the water quality of the groundwater, rivers, lakes and oceans. These water-quality concerns are amplified when such water resources are drawn upon for human consumption.

The various mechanisms which affect contaminant transport through soil have commonly been investigated by measuring the inputs into, and losses from, isolated columns of soil. The column may (preferably) be an intact soil core, as in a 'lysimeter' (e.g. Chichester and Smith (1978)), or in special studies in which intact soil cores have been isolated and returned to the laboratory for investigations. Intact soil cores have the advantage of retaining the naturally occurring soil structure and any macropores or biopores that may exist in the profile. Even though such natural features are commonly lost in soil columns formed by repacking soil, studies with such 'disturbed' soil columns have clarified many of the mechanisms of contaminant transport which will now be discussed. Two such processes are convection (or advection) and dispersion, processes already considered in Section 12.4 for saturated soil, as in groundwater.

In unsaturated soil these two processes remain of major importance, and preferential-flow pathways through large pores can assume even greater significance, as will be discussed qualitatively. Brief mention will also be made of two physicochemical processes that can affect contaminant transport, namely ion exchange and ion exclusion.

Convection (or advection) in unsaturated flow through soil

Convective movement of liquids is that described by Darcy's equation, given in Eq. (10.6) for flow in saturated soil, but which is equally applicable to movement of water in unsaturated porous media as discussed in Section 11.4 and given in Eq. (11.9). Darcy's equation is also applicable to liquids other than water (with appropriate interpretation of the hydraulic head).

Consider the fate of a pulse of contaminant of concentration c_0 applied as input to the top of a soil column isolated to form a lysimeter as shown in Fig. 12.3. If the contaminant is immiscible with water, if there are no preferred pathways allowing flow to bypass segments of the soil matrix, and neglecting complications that may occur due to a difference in density between the contaminant and water, then, following rainfall inputs, the concentration of contaminant in the leachate might vary as shown by curve (a) in Fig. 12.4. As with Fig. 12.2(a), this breakthrough curve is referred to as 'piston flow', so named because the contaminant is displaced through the soil as if it were being pushed by a piston. When input of water into the soil column displaces an immiscible contaminant before it, the contaminant will emerge from the base of the lysimeter (Fig. 12.3) following the input of a volume of water equivalent to the entire pore volume of the isolated soil column. Hence, for curve (a) in Fig. 12.4, the immiscible contaminant is shown as emerging following a cumulative leachate volume of one pore volume.

If the immiscible liquid does not break down or react in any way with the soil, then it emerges with concentration c equal to its initial concentration c_0. Alternatively, if a pulse of miscible solute is entered

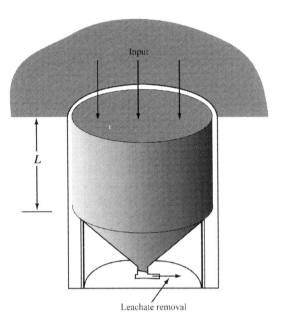

Figure 12.3 The cross-section of a lysimeter in which an intact column of soil has been isolated from its surrounds. Liquid that has exited from the bottom of the soil column as leachate is removed for chemical analysis in the laboratory.

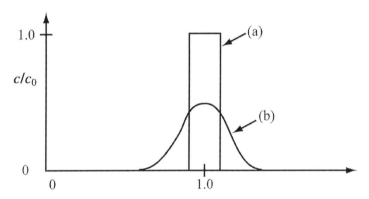

Cumulative pore volumes of leachate

Figure 12.4 The ideal outflow concentration ratio curve (or 'breakthrough curve') for (a) an immiscible liquid and (b) a miscible contaminant that does not react in any way with the soil. The contaminant concentration c is expressed as a fraction of the initial concentration pulse, c_0.

into the top of the soil column of Fig. 12.3, then curve (b) in Fig. 12.4 shows the typical shape of the breakthrough curve with the same ideal assumptions of there being no preferred flow pathways and no reaction with the soil in the column.

The reason for the spreading illustrated in curve (b) of Fig. 12.4 is the process of dispersion, considered in Section 12.4 for groundwater, and discussed further in the following Example 12.4.

Example 12.4

An accidental spill of industrial solvent miscible with water onto the soil surface has provided a pulse of contaminant that can leach into the soil to reach a drainage system installed 1.5 m beneath the soil surface. Assume that leaching occurs through the top 1.5 m of soil with a constant volumetric water content $\theta = 0.25$ m^3 m^{-3} and an average hydraulic gradient in the downward (z) direction of $\Delta H/\Delta z = -0.5$ mm^{-1}. If the hydraulic conductivity of the soil at this water content is 3×10^{-8} m s^{-1}, calculate

(a) the velocity of leaching of the contaminant and
(b) the time taken for the contaminant to move from the soil surface to the drainage system.

Solution

Using Darcy's equation, Eq. (11.9),

$$q = -K(\theta)\,\Delta H/\Delta z$$
$$= 3 \times 10^{-8} \times 0.5 \text{ m s}^{-1}$$
$$= 1.5 \times 10^{-8} \text{ m s}^{-1}$$

From Eq. (10.4), when soil was saturated, the velocity of flow, v, was given by $v = q/\varepsilon$, since the entire pore fraction, ε, was filled with water. In unsaturated soil, the fraction of pore space filled with water is the volumetric water content, θ, so

$$v = q/\theta \qquad (12.6)$$

where θ is less than ε. Hence, in this example,

$$v = 1.5 \times 10^{-8}/0.25$$
$$= 6 \times 10^{-8} \text{ m s}^{-1} \qquad \text{(a)}$$

The time taken for the contaminant to travel a distance L is given by

$$t = L/v \qquad (12.7)$$

Hence, here

$$t = 1.5/(6 \times 10^{-8})$$
$$= 0.25 \times 10^8 \text{ s}$$

or

$$t = 290 \text{ days}$$

Dispersion in unsaturated flow

The spread in the breakthrough curve (b) of Fig. 12.4 was earlier ascribed
to the effect of dispersion which invariably accompanies miscible convective flow or miscible displacement. The degree of mixing that takes
place between a pulse of introduced miscible solute and the body of water
involved in the convective flow is indicated by the degree of spread in
the bell-shaped breakthrough curve of Fig. 12.4, curve (b). This illustrates the general finding that breakthrough curves of the type shown
in Fig. 12.4 reveal information on the processes affecting contaminant
transport in soil.

The mechanisms of hydrodynamic dispersion and molecular diffusion described in Section 12.4 for dispersion accompanying movement
of contaminants in groundwater also apply to flow in unsaturated soil.
However, some of the factors involved only in a minor way in hydrodynamic dispersion in saturated soil become of greater significance in
unsaturated flow; examples of this are the frictional drag close to solid
surfaces and the greater tortuosity of flow restricted to smaller water-
filled pores and films around solid surfaces.

Equations (12.2)–(12.4) given for contaminant transport in groundwater are commonly assumed to apply also to the transport of non-
reactive contaminants in unsaturated flow, provided that the soil porosity
or pore fraction, ε, is replaced by the appropriate value of the volumetric
water content θ. This replacement is necessary since, unless the contaminant is volatile and can move in the gaseous fraction of the pore space,
contaminant can move only through that fraction of the pore space which
is occupied by water (see Fig. 11.5).

Thus the analogue of Eq. (12.4) for contaminant transport in the
downward (z) direction in unsaturated soil is

$$q_s = v\theta c - D\,\Delta(c\theta)/\Delta z \qquad (12.8)$$

where the dispersion coefficient D is now a function of θ, in the same
way as that in which the hydraulic conductivity, K, depends on the water
content in unsaturated soils. This comment also applies to the dispersivity, ζ, in Eq. (12.5), when it is applied to unsaturated soil. It is the
strong dependence on water content of transport and diffusion parameters which adds to the difficulty in predicting movement of contaminants
in unsaturated soils.

As argued in Section 12.4 for groundwater flows, mass conservation
requires a close link between changes in space and with time of contaminant concentration. The same conclusion follows for contaminant flow
in unsaturated soil, and this can be demonstrated as follows. In the following figure, contaminant flux density q_s enters the unit of the shaded

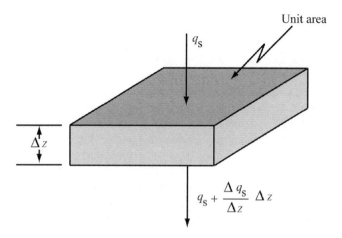

volume of soil of length Δz. With unsteady flow, the flux emerging from the bottom of the soil column of length Δz will be different from q_s by an amount depending on the spatial rate of change of q_s, given by the ratio $\Delta q_s / \Delta z$. Over a distance Δz, the change in flux density is therefore given by $(\Delta q_s / \Delta z) \, \Delta z$. This difference in flux between input and output of contaminant implies a change with time in contaminant concentration, with the rate of this change, $\Delta c / \Delta t$, depending on the difference between the input and output fluxes of contaminant. Mass conservation of contaminant in the soil column of volume $\Delta z \, (\mathrm{m}) \times 1 \, (\mathrm{m}^2) = \Delta z \, (\mathrm{m}^3)$ can be expressed as

rate of increase of contaminant = difference between input and output

flux densities

Hence

$$(\Delta c / \Delta t) \, \Delta z = -(\Delta q_s / \Delta z) \Delta z$$

or

$$(\Delta c / \Delta t) = -(\Delta q_s / \Delta z)$$

where q_s is given by Eq. (12.8).

This equation expressed in terms of differential-calculus notation is referred to as the well-known 'convection–dispersion' equation. This is the simplest form of equation describing the combined convection (or mass flow) and dispersion of a contaminant that is assumed not to react with the soil in any way. Solution of this equation requires analytical or numerical methods beyond the level of this book. (See, for example, Marshall, Holmes and Rose (1996) and Jury, Gardner and Gardner (1991).)

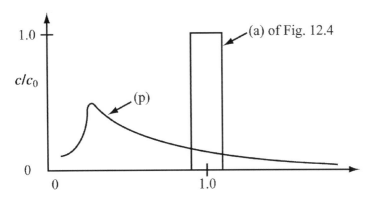

Figure 12.5 The breakthrough curve p illustrates the typical effect of preferential flow through a soil column on contaminant transport. The contaminant concentration c is expressed as a fraction of the initial concentration c_0 of the contaminant pulse.

Preferential-flow effects

Entry of water and contaminants into dominantly unsaturated soil can occur during limited periods of surface saturation, if at all. During transient flow, liquids can enter and move not only through the soil matrix, but also and especially through soil cracks, macropores and biopores. Flows through such relatively large pores can bypass the majority of the soil matrix, and can overwhelm the effects of transport through the soil matrix of the types considered earlier in this section. The type of effect of such preferential flow is illustrated by the breakthrough curve (p) in Fig. 12.5.

Effects of ionic attraction and ion exchange

As discussed in connection with Fig. 1.6, colloidally sized components of soil, namely clays and organic matter, bear an electrical charge on their surfaces, which is electrically balanced by ions in solution of opposite sign.

   If the contaminant is ionic, then ions of opposite sign to that of the charged surfaces (which are commonly negatively charged) will be attracted to these surfaces, whereas charges of the same sign will be repelled. Those ions attracted to the charged colloidal surfaces will be hindered in their transport through the soil, whereas those ions which are repelled can have their transport enhanced. Whist these ionic effects are important in other contexts (as discussed in Section 1.4), they also affect the transport and so the breakthrough curves of the contaminant involved. Such physicochemical effects are considered in Phillips (1993) and in texts such as Jury, Gardner and Gardner (1991).

   In the following Section 12.6, consideration focuses on issues of contaminant transport of particular significance in agricultural contexts.

Figure 12.6 The changing profile characteristics of solute concentration during leaching following a discrete surface application.

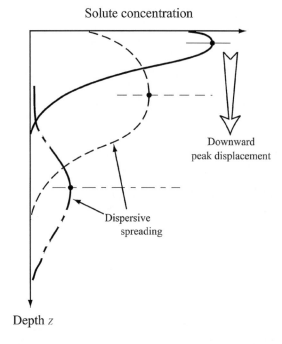

Solute concentration

Downward
peak displacement

Dispersive
spreading

Depth z

12.6 Solute and contaminant transport in agricultural contexts

Understanding physical aspects of movement of solutes in the unsaturated zone is basically similar for all chemicals that are not sorbed to the soil or transformed in some way. However, there is a characteristic source difference between the extensively distributed saline salts considered in Sections 12.2 and 12.3 and the discrete surface application of fertilizer or herbicide which is a common feature of higher-input agriculture. Movement of the nitrate ion (NO_3^-) and of some herbicides and pesticides has received considerable attention in the environmental literature, mainly due to the danger of groundwater contamination discussed in Section 12.4.

In the context of soluble agricultural chemicals applied to the soil surface, interest lies in being able to understand and predict movement of the pulse of solute which results from the discrete application of the chemical of concern. This pulse has the twin characteristics of a maximum, or peak, in concentration and a distribution of solute concentration around that peak due to dispersion or preferential flow. These two characteristics, discussed in Section 12.5, are illustrated in Fig. 12.6, where the displacement of the solute peak can be understood in terms of convective flow using mass-conservation principles.

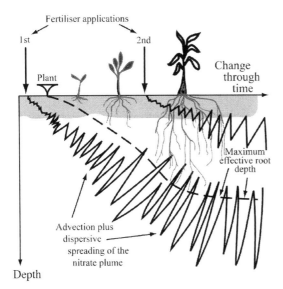

Fertiliser applications

1st

2nd

Plant

Change through time

Maximum effective root depth

Advection plus dispersive spreading of the nitrate plume

Depth

Figure 12.7 The development through time and space of a plume of nitrate concentration following two successive applications of nitrogenous fertilizer, one prior to, and the second following, establishment of an annual crop such as maize.

A simplified analysis following Rose *et al.* (1982), which allows prediction of movement of the solute peak in the agricultural or forestry context of growing vegetation, will now be outlined. To illustrate the analysis, the solute will be taken to be the nitrate ion (NO_3^-). The concentration of nitrate in groundwater should not exceed certain limits if it is to be used for human consumption.

As a crop grows, its maximum effective depth of root penetration increases. Root growth has important implications for transport of the nitrate ion within the soil profile, because of uptake by roots of both water and nitrate, as is illustrated in Fig. 12.7.

Figure 12.7 suggests that much of the nitrate from the pre-planting application of fertilizer will escape below the root zone. Such unused nitrate is able to contribute to groundwater pollution, since at depth in the soil profile nitrate is unlikely to be transformed or taken up. Whereas the timing of application of fertilizer and irrigation water, if used, can be controlled, the timing of rainfall cannot be manipulated. Desirable management objectives could be to minimize loss of water and nitrate to groundwater, whilst ensuring good production by minimising water and nutrient stress on the crop.

The greatest potential for loss of nitrate by leaching below the root zone occurs either in well-aggregated soils or in soils of light texture (e.g. sandy soils). In both of these soil contexts water is able to drain rather rapidly following saturation to an approximately reproducible water content called the field capacity, as discussed in Section 4.5. Expressed as a volumetric fraction, the field capacity will be denoted θ_{fc}.

Three major assumptions will be made in the following simple but practically useful analysis of displacement of a solute peak through the root profile of a crop growing in a fairly uniform soil profile.

(i) Drainage of water through the soil profile to reach the field capacity is completed within the time interval of the calculation, Δt. (Implicitly, Δt is the smallest time interval within which it would be possible to make and carry out any management decision. This allows the detailed dynamics of drainage to be ignored.)

(ii) Any soil–solute interactions that may take place do not significantly affect movement of the solute peak.

(iii) The field capacity, θ_{fc}, is uniform over the soil depth of interest.

(iv) One-dimensional (vertical) movement of water and solute is also assumed.

In the context of a growing crop of root depth r, there are two types of hydrological situations, which have quite different implications for solute displacement. In the first hydrological situation, what is dominant is evapotranspiration at rate E and the associated uptake of water by roots. Because of this dominance of uptake by roots, the vertical bulk movement of water in the profile depth r is minimal (Fig. 12.8(a)). In the second hydrological situation, infiltration at rate I and redistribution of water throughout the soil profile are the dominant processes. If such redistribution is sufficient, it can result in a downward displacement of the peak solute concentration (Fig. 12.8(b)).

As shown in Fig. 12.8(a), the solute peak in the profile is not displaced during an evapotranspiration-dominated period, the reason being that water (and perhaps solute) is taken up by lateral movement to the closest root, which acts as a 'sink'. Roots can take up water preferentially to solute. If so, then uptake of water by roots will be associated with an increase in solute concentration. This is indicated in Fig. 12.8(a) by the concentration profile after uptake being greater than that before it, as shown by the dashed line.

In the second type of dominant hydrological process (Fig. 12.8(b)), infiltration of water is sufficient for the wetting front to move down past the prior position of the peak solute concentration located at depth α beneath the soil surface. The amount of infiltration might not be sufficient for the wetting front to move below depth α, but, if it is, there will be a downward displacement of the position of the solute peak, denoted as $\Delta \alpha$ in Fig. 12.8(b).

Since upward movement of water in a cropping context is minimal, being restricted to near the soil surface, the solute peak does not move upwards, but either remains stationary or moves downwards.

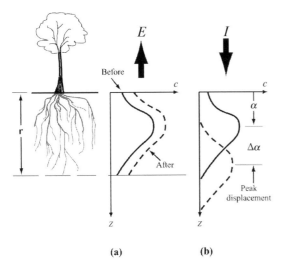

Figure 12.8 (a) The context when evaporation (or evapotranspiration) and uptake of water by root are the dominant processes. The solute concentration can be increased if roots take up water preferentially to the solute. There is no significant change in the peak position of solute concentration. (b) The context where the infiltration of water and the redistribution of water and solute within the soil profile are the dominant processes affecting the position of the solute peak. A displacement $\Delta\alpha$ in the prior depth, α, of the peak occurs if the wetting front moves beyond this prior position of the solute peak.

Equation (12.6) shows that the velocity of movement of water in unsaturated soil, v, is related to the volumetric flux density of water, q (m s^{-1}), by

$$v = q/\varepsilon$$

Thus, if $\theta = \theta_{fc}$, then it follows that

$$v = q/\theta_{fc} \qquad (12.9)$$

The position of the solute peak (α, Fig. 12.8(b)), can be taken as originating at the soil surface where nitrogenous fertilizer, for example, is commonly applied. Denote the equivalent ponded depth of infiltrating water by i. Then $i = I\,\Delta t$, where I is the mean rate of infiltration of water into the soil, either from rainfall or from irrigation, averaged over the time interval Δt, where t represents time. Assuming that the fertilizer applied dissolves and releases nitrate within the time period Δt, then the advective displacement of the solute will be given by the product of the mean velocity v and the assumed duration of movement, Δt. Thus, from

Eq. (12.9),

$$v \, \Delta t = q \, \Delta t / \theta_{fc}$$

In this context, $q = I$, so $q \, \Delta t = I \, \Delta t = i$. Thus the displacement $v \, \Delta t = i/\theta_{fc}$.

Hence the first displacement downwards from the soil surface of the peak concentration, $\Delta \alpha_1$, will be given by

$$\Delta \alpha_1 = i_1 / \theta_{fc} \qquad (12.10)$$

where i_1 is the equivalent ponded depth of the first amount of infiltrating water following application of fertilizer to the soil surface. In this first event, all infiltrating water moves beyond the initial position of the solute located at the soil surface where the fertilizer is applied.

In subsequent infiltration events, however, not all the water infiltrating the soil surface will move below the position of the solute peak, which now lies beneath the soil surface. The way in which Eq. (12.10) can be generalised so that the displacement of the solute peak, $\Delta \alpha_n$, in the general nth infiltration event after application of fertilizer is developed in Box 12.1. The equations developed in this box will later be applied to the field lysimeter data in Example 12.5.

Example 12.5

Experiments on the leaching of ^{15}N-labelled nitrate fertilizer were carried out by Chichester and Smith (1978) in four isolated soil cores or 'lysimeters' of depth 2.44 m in Coshocton, Ohio. (The use of fertilizer labelled with ^{15}N instead of the usual ^{14}N removes uncertainty as to the source of nitrate, which can also come from the soil itself.) Whilst the soil type varied through the profile, measurement showed θ_{fc} to be constant at close to 0.3 m^3 m^{-3}. Rainfall provided inputs, and evaporation was calculated from water content measurements in the lysimeter. Measurements made following the application during May 1972 of a labelled fertilizer to a corn crop are shown in Table 12.1.

The ^{15}N concentration in the leachate collected at the bottom of the lysimeters was measured. The peak in ^{15}N concentration appeared at the lysimeter base in January 1973 for two lysimeters and in March 1973 for the other two.

Using the data in Table 12.1, calculate the expected month of appearance of the peak concentration of fertilizer, assuming that i_n comes before e_n in the calculations. Also use Eqs. (12.14) and (12.15) (developed in Box 12.1) in order to complete the calculations. Equation (12.14) enables calculation of the downward displacement of the solute peak in any general infiltration event in which an equivalent ponded depth of water, i_n, is

Table 12.1 Calculation of the position of the peak in $^{15}NO_3$ concentration following application of fertiliser in May 1972 (data are from Chichester and Smith (1978) (i_n before e_n is assumed) (all lengths are in cm), from Rose, Hogarth and Dayananda (1982))

Year and month	1 i_n	2 $\frac{i_n}{\theta_{fc}}$	3 e_n	4 r	5 $\frac{i_{n-1} - e_{n-1}}{r}$	6[a] $\theta_{n-1} = \theta_{n-2} + 5 \quad \theta_{fc} - \theta_{n-1}$	7 $\theta_{fc} - \theta_{n-1}$	8 (if $r < \alpha_{n-1}$) $\frac{r}{\theta_r}(\theta_{fc} - \theta_{n-1})$	9 (if $\alpha_{n-1} < r$) $\frac{\alpha_{n-1}}{r}(\theta_{fc} - \theta_{n-1})$	10[b] $\Delta\alpha_n = 2 - 8$ or $2 - 9$	11 $\alpha_n = \alpha_{n-1} + \Delta\alpha_n$
1972											
May	6.0	19.9	6.7	20		0.3	0		0	19.9	19.9
June	9.2	30.6	5.9	40	−0.0175	0.2825	0.0175		0	30.6	50.5
July	7.0	23.4	11.8	65	0.0508	0.33→0.3	0		0	23.4	73.9
Aug	7.0	23.4	11.8	65	−0.0740	0.226	0.074	16.0		7.4	81.3
Sep	9.6	32.0	6.9	65	−0.0738	0.152	0.148	32.1		0	81.3
Oct	4.5	15.0	3.1	20	0.1350	0.287	0.013	0.9		14.1	95.4
Nov	14.2	47.4	1.7	20	0.0700	0.36→0.3	0	0		47.4	142.8
Dec	8.8	29.2	1.9	20	0.625	→0.3	0	0		29.2	172.0
1973											
Jan	7.2	24.0	2.7	20	0.345	→0.3	0	0		24.0	196.0
Feb	7.0	23.2	2.5	20	0.225	→0.3	0	0		23.2	219.2
Mar	12.9	43.1	4.3	20	0.225	→0.3	0	0		43.1	262.3

[a] **5** refers to entry in column 5. 0.3 is the maximum value ($= \theta_{fc}$).

[b] **2**, **8** and **9** refer to entries in the respective columns.

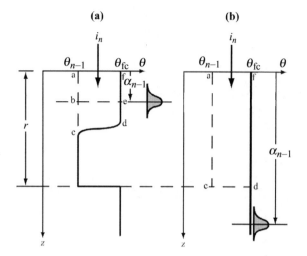

Figure 12.9 Two possible scenarios in which infiltration in the nth event, of amount i_n, can lead to water moving beyond the prior position of the solute peak, described by α_{n-1}. The water content beneath the depth of rooting, r, is assumed constant at the field capacity, θ_{fc}. Following infiltration and drainage, the water content in the root zone is also assumed to be θ_{fc}. In (a) the wetting front, shown as approximately rectangular, does not reach the depth r, as it does in (b). In (b) the possibility that the prior position of the solute peak could be beneath the depth of rooting, r, is indicated by the greater α_{n-1}.

Box 12.1

In order to generalise Eq. (12.10) so that it can describe the displacement, $\Delta\alpha_n$, in the nth infiltration event since application of fertilizer, we need to know the amount of water that has passed *beyond* the prior position of the solute peak. This can be readily calculated if we make simplifying assumptions about water-content profiles and uniformity of θ_{fc} with depth, as illustrated in Fig. 12.9.

The downward displacement of the solute concentration peak is here described in a series of infiltration events, which can be denoted $1, 2, 3, \ldots, n-1, n, n+1, \ldots$. Equations will now be developed for the possible displacement of the solute peak in the general nth infiltration event, for which the prior position of the peak following event $n-1$ is denoted α_{n-1}.

A further assumption now made is that, by the end of an evapotranspiration period (located between successive infiltration events), uptake of water by roots brings the water content in the root zone back to a time-varying but spatially uniform value, denoted θ_{n-1} in Fig. 12.9. This is a reasonable assumption since, in a uniform soil profile, roots tend to extract more water from those levels in the root zone which have the most water. This approximate assumption has some experimental support if evapotranspiration proceeds far enough into a drying cycle (e.g. Rose and Stern (1967)).

Though the wetting front can be rather abrupt in well-drained soils, as shown at cd in Fig. 12.9(a), the exact shape of this profile does not affect the theory developed here.

Below the root zone (of depth r) there is no sink for water, but, if it is free to drain, then the assumption that the water content remains constant at θ_{fc} is made (Figs. 12.9(a) and (b)).

The equivalent ponded depth of water which moves beyond the mean solute position in the first infiltration event after application of fertilizer was denoted i_1 (Eq. (12.10)). In subsequent events following this first downward displacement of the solute peak, the depth of water moving past the solute peak will be less than the depth of infiltrating water. This is because of the need to wet the soil above the depth α_1 from θ_{n-1} to θ_{fc} (Fig. 12.9(a)). Area in Fig. 12.9 represents equivalent ponded depth of water, so that, from Fig. 12.9(a), this amount of water is represented by the area abef, and so is given by $\alpha_{n-1}(\theta_{fc} - \theta_{n-1})$.

Denote by ΔW_n the depth of water which, in the nth infiltration event, moves *beyond* the general position of the solute peak in the prior $(n-1)$th event, namely α_{n-1}. It follows from Fig. 12.9 that a depth of water equal to $\alpha_{n-1}(\theta_{fc} - \theta_{n-1})$ must infiltrate before the wetting front moves below the prior position of the solute peak, located at depth α_{n-1}. It follows that

$$\Delta W_n = i_n - \alpha_{n-1}(\theta_{fc} - \theta_{n-1}) \tag{12.11}$$

where i_n is the amount of infiltration during the (general) nth calculation period (expressed as equivalent ponded depth), and α_{n-1} is the depth of the solute peak in the prior period $(n-1)$.

The expression for ΔW_n in Eq. (12.11) is the generalisation of the expression for i_1 in Eq. (12.10), so

$$\Delta \alpha_n = \Delta W_n / \theta_{fc} \tag{12.12}$$

with ΔW_n given by Eq. (12.11).

Provided that $\alpha_{n-1} < r$, Eqs. (12.11) and (12.12) allow the calculation of $\Delta \alpha_n$. However, if the solute peak is located below the depth of root development (i.e. $\alpha_{n-1} > r$, the alternative possibility illustrated in Fig. 12.9(b)), then it follows for similar reasons to those considered earlier that

$$\Delta W_n = i_n - r(\theta_{fc} - \theta_{n-1}) \tag{12.13}$$

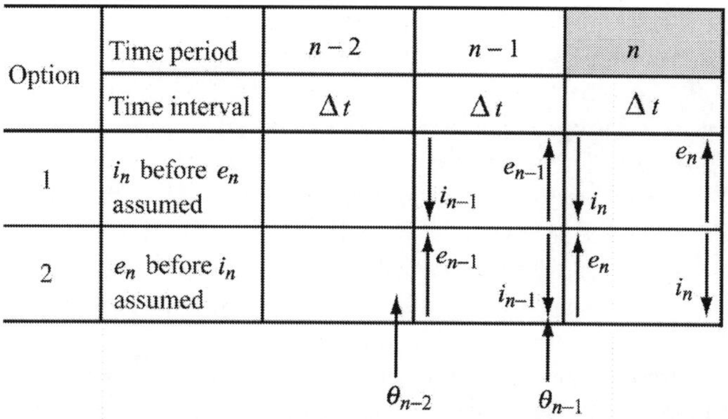

Figure 12.10 Terms in the water-balance or accounting procedure used to follow the water content in the root zone. The displacement of the solute peak is calculated for the nth period.

Option	Time period	$n-2$	$n-1$	n
	Time interval	Δt	Δt	Δt
1	i_n before e_n assumed		e_{n-1} i_{n-1}	e_n i_n
2	e_n before i_n assumed		e_{n-1} i_{n-1}	e_n i_n

θ_{n-2} $\qquad$ θ_{n-1}

Equations (12.11) and (12.13) can be generalised by the following equation:

$$\Delta W_n = \theta_{fc}\, \Delta\alpha_n = i_n - \text{Min}(\alpha_{n-1}, r) \times (\theta_{fc} - \theta_{n-1}) \tag{12.14}$$

where the appropriate coefficient of the term in $\theta_{fc} - \theta_{n-1}$ is whichever is smaller, α_{n-1} or r. In either case, the peak solute displacement $\Delta\alpha_n$ is given by Eq. (12.12).

Also, the general position (or depth) of the solute peak is α_n, which is given by

$$\alpha_n = \alpha_{n-1} + \Delta\alpha_n \tag{12.15}$$

The question now considered is that of how to calculate θ_{n-1} in Eq. (12.14). θ_{n-1} was defined as the water content in the root zone just prior to the nth calculation period. It is the nth period in which the displacement $\Delta\alpha_n$ in the position of the solute peak is calculated using Eq. (12.12). Figure 12.10 shows three general intervals in the time series used in calculating the displacement of the solute peak. The time intervals Δt in Fig. 12.10 do not have to be equal, even though, for computational convenience, this is a common choice. Mass conservation of water in the period denoted $n-1$ requires that θ_{n-1} is related to the water content just prior to the $(n-1)$th event, namely θ_{n-2}, by

$$\theta_{n-1} = \theta_{n-2} + (i_{n-1} - e_{n-1})/r \tag{12.16}$$

where i_{n-1} and e_{n-1} are infiltration and evaporation amounts during period $n-1$. Should θ_{n-1} calculated using Eq. (12.16) exceed θ_{fc}, it is understood that $\theta_{n-1} = \theta_{fc}$, because water in excess of θ_{fc} is assumed to be redistributed and to drain down through the soil profile in less than the duration Δt of the calculation period.

Calculations of the type illustrated in Eq. (12.16) are commonly referred to as a water-balance or accounting procedure, a topic discussed in Section 4.5.

Since it is assumed that the time interval Δt used in the calculation is adequate to allow infiltrating water to be redistributed to the field capacity, calculations can be carried out over time periods of a few days, a week, or a month, depending on availability of data and the rate of extension of the rooting depth r.

The longer the time interval chosen for the calculation, the more significant the issue of the relative timing of infiltration in relation to evaporation events within the period of calculation becomes. The reason for this can be seen by considering the effect of the two contrasting possibilities for the calculated displacement of the solute peak in the nth event shown as options 1 and 2 in Fig. 12.10. For option 1, θ_{n-1} calculated from Eq. (12.16) describes the water content in the root profile just prior to the infiltration event in which the position of the solute peak would be displaced, provided that i_n is greater than $\alpha_{n-1}(\theta_{fc} - \theta_{n-1})$ (Eq. (12.11)).

Alternatively, for option 2, the volumetric water content in the root zone will be reduced below θ_{n-1} by the amount e_n/r before the infiltration event i_n occurs, with the associated possibility of displacement of the solute peak (Fig. 12.10). Consequently, if e_n precedes i_n (option 2 of Fig. 12.10), then θ_{n-1} should be further depleted by e_n/r in the calculation of ΔW_n (and so of α_n). So, with option 2 (e_n before i_n), Eq. (12.14) becomes

$$\theta_{fc} \, \Delta \alpha_n = i_n - \text{Min}(\alpha_{n-1}, r) \times (\theta_{fc} - \theta_{n-1} + e_n/r) \qquad (12.17)$$

If, as can occur in practical applications, there is some uncertainty in the timing of i_n and e_n, calculation of α_n using both Eq. (12.11) (option 1 of Fig. 12.10) and Eq. (12.17) (option 2) allows estimation of the maximum range of uncertainty in the prediction of α_n due to this uncertainty in data (see Example 12.5).

Solute dispersion can have a minor effect on the position of the solute peak, but this can be neglected in the context of this theory for movement of the solute peak in and beneath the root zone of vegetation.

taken in by the soil surface. Equation (12.15) shows how the depth below the soil surface of the solute peak, α_n, is accumulated through successive downward displacements of amount $\Delta \alpha_n$.

Solution

The calculations can be carried out using a spreadsheet of the style shown in Table 12.1, using a uniform monthly time period for calculation.

From the last column in Table 12.1 it can be seen that the calculated value of α_n exceeded the lysimeter depth (244 cm) for the month of March 1973, the same month as that during which the peak concentration was measured to emerge from two of the lysimeters. In Box 12.1 were presented the consequences of adopting either of two alternative possible relative timings of infiltration inputs and evaporation losses within the time period adopted for calculation (one month in Table 12.1). The calculations in Table 12.1 assume that infiltration (i_n) occurs before evapotranspiration (e_n), for which appropriate equations are (12.14) and (12.15).

With the alternative assumption in Box 12.1 of e_n before i_n in the calculations, the predicted month of appearance of the peak concentration at the base of the lysimeter is June 1973. More detailed ^{15}N evidence (Rose, Chichester and Phillips, 1983) indicated the likelihood that the decomposed silt stone and permanently fissured clay shale bedrock in the lower part of the lysimeters could have allowed some preferential movement, which would have accelerated the emergence of the solute peak.

How to interpret the effects of dispersion on solute transport in unsaturated media is shown, for example, in Marshall, Holmes and Rose (1996).

Main symbols for Chapter 12

c Concentration of contaminant at a measurement position

c_i Salt concentration of input irrigation water

c_o Undiluted concentration of a contaminant

c_u Salt concentration of water draining beneath the root zone of a crop

D Dispersion coefficient of a contaminant

D_i Equivalent ponded depth of applied irrigation water

D_o Dispersion coefficient of a contaminant in saturated pore space

e Equivalent ponded depth of evaporation over a given period

H Hydraulic head

i Depth of irrigation water applied

I Rate of application of irrigation water

K Hydraulic conductivity of a porous medium

L Distance of contaminant transport

P Rainfall rate

q Volumetric flux of water per unit cross-sectional area of flow

q_s Mass flux density of contaminant (Eq. (12.2))

r Depth of rooting

t Time

U Equivalent ponded depth of drainage water beyond the depth
of the root zone

v Mean velocity of flow of water through a porous medium

x Distance

z Distance of downward displacement

α Displacement of peak in solute concentration

$\Delta \alpha$ Change in α

ΔW_n Equivalent ponded depth of water that, in the nth infiltration
event, moves beyond the position of the solute peak in the prior
$(n - 1)$th event.

ε Porosity

ζ Dispersivity, also termed coefficient of dispersivity (Eq. (12.5))

θ Volumetric water content

θ_{fc} Volumetric water content at field capacity

Σ Summation symbol

Exercises

12.1 (a) What is meant by saying that a soil is saline?

(b) With the aid of sketches, discuss the origins of dryland salinity.

(c) Describe how salinity problems can arise due to the use of
either an excess amount of irrigation water or too little.

(d) In a semi-arid region, where annual rainfall is $400 \, \text{mm} \, \text{y}^{-1}$,
native vegetation was cleared in order to grow wheat. Prior to
clearing, the long-term depth beneath the soil surface of the
water table was 25 m. The sum of evaporation and transpiration
(evaporation from plants) is often called evapotranspiration.
For the land with native vegetation, the evapotranspiration was
$360 \, \text{mm} \, \text{y}^{-1}$ on average. Thus $40 \, \text{mm} \, \text{y}^{-1}$ drained to groundwater (assuming that there was no runoff). Since the ground-
water depth was constant, evidently this amount of drainage
was dispersed, probably by lateral groundwater flow.

When winter-grown wheat was introduced into the cleared land,
the evapotranspiration fell to $310 \, \text{mm} \, \text{y}^{-1}$ on average. Still assuming that there was no runoff, and that the amount of rainfall is
unaltered, there is excess water to drain to the water table.

In how many years' time (following clearing, and with continued annual cropping) might the water table be expected to reach
the soil surface, leading to problems of waterlogging and salinity?

Table 12.2 *Data on weekly water inputs (i$_n$) and evaporative losses (e$_n$), given as equivalent ponded depths of water in cm, following application of fertilizer on 25 April 1972*

Week/date in 1972	i_n (cm)	e_n (cm)
25–30 April	0.5	0.2
Week 1 of May	4.3	0.2
Week 2 of May	4.3	0.2
Week 3 of May	4.3	0.2
Week 4 of May	4.3	0.2

Assume that the porosity of the soil is 0.4, and state any further assumptions you make in your calculation.

12.2 Saffigna, Keeney and Tanner (1977) observed the times of emergence of concentration peaks of chloride and nitrate ions which resulted from application of fertilizer to a potato crop grown in a 1.5-m-deep field lysimeter in Wisconsin, USA. (Both these ions are negatively charged, so the movement of these two ions in the (slightly) negatively charged soil would be expected to be identical. There was no source of chloride in the soil other than from the application of fertilizer, giving it an advantage over nitrate as a tracer of solute movement.)

A field lysimeter allows solute emerging from the bottom of the isolated soil column to be collected and its concentration determined. The soil was a uniform loamy sand, with a field capacity of approximately 0.1 m^3 m^{-3} (a typically low value for a sandy soil).

The experimental plots were irrigated every 5 days to replace evaporative loss. Thus it is appropriate to carry out calculations with a weekly time interval, and to assume that evaporation precedes infiltration (Eq. (12.17)), using the data in Table 12.2.

Table 12.2 provides data for the first application of fertilizer, on 25 April 1972, when root development of the crop was very limited, and an effective root depth of 10 cm will be assumed. Also assume that the initial value of the volumetric water content in the soil profile is at the field-capacity value.

The solute concentration measured at the base of the lysimeter exhibited a broad peak (the peak had been broadened by dispersion), which emerged during the fourth week of May 1972. Comment on how your calculations agree with this observation.

Note. Calculations can be set out in a manner generally similar to those in Table 12.1 if you wish. However, there will be differences due to the use of Eq. (12.17) instead of Eq. (12.14).

12.3 In extracting gold from crushed ore, leaching is carried out using cyanide solution (as sodium cyanate, NaCN). The solution can be sprayed onto the ore, which is placed in a liner designed to prevent loss of the dangerous cyanide solution. Cyanide decays or is degraded by several different chemical and biological processes, and this rate of decay with time, t, can be described by the exponential decay function

$$c/c_0 = \exp(-kt)$$

where c is the cyanide concentration at time t, c_0 is the initial concentration of cyanide (at $t = 0$) and k is a degradation coefficient.

Assuming that $k = 0.002\,\mathrm{h}^{-1}$ under the particular conditions of the heap leach pad, calculate the time required for the initial cyanide concentration to decay to (a) 50% and (b) 1% of the initial concentration.

(Data from Dr Ian Phillips in this exercise are acknowledged with thanks).

References and bibliography

Carson, R. (2000). *Silent Spring*. London: Penguin Books.

Chichester, F. W., and Smith, S. J. (1978). Disposition of 15N-labelled fertilizer nitrate applied during corn culture in field lysimeters. *J. Environ. Qual.* **7**, 227–233.

Clarke, C. J., George, R. J., Bell, R. W., and Hatton, T. J. (2002). Dryland salinity in south-western Australia: its origins, remedies, and future research directions. *Aust. J. Soil Res.* **40**, 93–113.

Dregne, H., Kassas, M., and Razanov, B. (1991). A new assessment of the world status of desertification. *Desertification Control Bull.* (United Nations Environment Programme) **20**, 6–18.

Dudall, R., and Purnell, M. F. (1986). Land resources: salt affected soils. *Reclamation Revegetation Res.* **5**, 1–10.

Freeze, R. A., and Cherry, J. A. (1979). *Groundwater*. Englewood Cliffs, New Jersey: Prentice-Hall International, Inc.

Hanks, R. J. (1992). *Applied Soil Physics. Soil Water and Temperature Applications*, 2nd edn. New York: Springer-Verlag.

Jury, W. A., Gardner, W. R., and Gardner, W. H. (1991). *Soil Physics*, 5th edn. New York: John Wiley and Sons, Inc.

Marshall, T. J., Holmes, J. W., and Rose, C. W. (1996). *Soil Physics*, 3rd edn. Cambridge: Cambridge University Press.

Pavelic, P., Narayan, K. A., and Dillon, P. J. (1997). Groundwater flow modelling to assist dryland salinity management on a coastal plain of southern Australia. *Aust. J. Soil Res.* **35**, 669–686.

Phillips, I. R. (1993). Solutes and their transport through soils. In *Environmental Soil Science*, eds. I. F. Fergus and K. J. Coughlan. Queensland Branch, Australia: Australian Society of Soil Science Inc., pp. 133–159.

Rose, C. W., Chichester, F. W., Williams, J. R., and Ritchie, J. T. (1982). A contribution to simplified models of field solute transport. *J. Environ. Qual.* **11**, 146–150.

Rose, C. W., Chichester, F. W., and Phillips, I. (1983). 15N-labelled nitrate transport in a permanently fissured shale substratum. *J. Environ. Qual.* **12**, 249–252.

Rose, C. W., Hogarth, W. L., and Dayananda, P. W. A. (1982). Movement of peak solute concentration position by leaching in a non-sorbing soil. *Aust. J. Soil Res.* **20**, 23–36.

Rose, C. W., and Stern, W. R. (1967). Determination of withdrawal of water from soil by crop roots as a function of depth and time. *Aust. J. Soil Res.* **5**, 11–19.

Saffigna, P. G., Keeney, D. R., and Tanner, C. B. (1977). Nitrogen, chloride and water balance with irrigated Russet Burbank potatoes in central Wisconsin. *Agron. J.* **69**, 251–257.

Stirzaker, R., Vertessy, R., and Sarre, A. (eds.) (2002). *Trees, Water and Salt: An Australian Guide to Using Trees for Healthy Catchments and Productive Farms*. Canberres, ACT: Joint Venture Agroforestry Program Publications, Rural Industries Research and Development Corporation.

United States Salinity Laboratory Staff (1954). *Diagnosis and Improvement of Saline and Alkali Soils*. Agriculture Handbook No. 60. Washington: United States Department of Agriculture.

Appendix

Table A.1 *A list of conversion factors additional to those in Table 1.2*

Quantity	Unit	Conversion to SI unit
Length	mile	$\times\ 1609 = $ m
Area	km^2	$\times\ 10^6 = m^2$
Volume	cubic yard	$\times\ 0.765 = m^3$
Mass	tonne (metric ton)	$\times\ 10^3 = $ kg
	ton (2240 lb)	$\times\ 1016 = $ kg
Density	$g\,cm^{-3}$	$\times\ 10^3 = kg\,m^{-3}$
Time	hour	$\times\ 3600 = $ s
Velocity	miles per hour (m.p.h.)	$\times\ 0.447 = m\,s^{-1}$
Force	dyne (c.g.s. unit)	$\times\ 10^{-5} = $ N
Pressure	atmosphere	$\times\ 1.013 \times 10^5 = $ Pa
	mbar (millibar)	$\times\ 100 = $ Pa
	mm water	$\times\ 9.81 = $ Pa
	p.s.i. (pounds per square inch)	$\times\ 6894 = $ Pa
Heat energy	calorie (cal)	$\times\ 4.18 = $ J
Power	horse power (hp)	$\times\ 746 = $ W
Temperature	degrees Fahrenheit ($^\circ$F)	$(^\circ F - 32) \times 5/9 = {}^\circ C$
	degrees Kelvin (K)	$K - 273 = {}^\circ C$

Table A.2 *Some properties in SI units*

Property	Value
Density of water (at 4 °C)	1000 kg m^{-3}
(at 20 °C)	998 kg m^{-3}
Density of air (at 20 °C and 100 kPa pressure)	1.20 kg m^{-3}
Specific heat of air	1.01 kJ kg^{-1} K^{-1}
Latent heat of vaporisation of water at 20 °C	2454 kJ kg^{-1}
Universal gas constant	8.31 J mol^{-1} K^{-1}
Stefan–Boltzman constant	5.67 × 10^{-8} W m^{-2} K^{-4}

Note. For more detailed information on conversion factors, quantities and properties, consult references such as Kaye, G. W. C., and Laby, T. H. (1995). *Tables of Physical and Chemical Constants*, 16th edn. Harlow: Longmans.

Answers to exercises

Chapter 1

Exercises on units and conversions

1.1 $5 \times 10^5 \, \text{m}^2$;   50 ha; $5.38 \times 10^6 \, \text{ft}^2$.

1.2 $8.29 \, \text{ft}^3 \, \text{min}^{-1}$; $0.00391 \, \text{m}^3 \, \text{s}^{-1}$; $14\,080 \, \text{l} \, \text{h}^{-1}$.

1.3 $1.6 \times 10^4 \, \text{m}^3$; $5.65 \times 10^5 \, \text{ft}^3$.

1.4 $1.47 \times 10^6 \, \text{N} \, \text{m}^{-2}$ (or Pa); $30\,700 \, \text{lb wt.} \, \text{ft}^{-2}$; $213 \, \text{lb wt.} \, \text{in}^{-2}$ (or p.s.i).

1.5 If the 'weight' (mass really) is 75 kg, and the area of each foot is $180 \, \text{cm}^2$, then the pressure is $20\,420 \, \text{Pa}$, or $20.4 \, \text{kPa}$.

1.6 The dimension of both sides of the equation is T.

1.7 For mean flow, the percentage is 1.4%. For minimum flow, the percentage is 219%.

1.8 $90 \, \text{kg} \, \text{km}^{-3}$.

1.9 The error is 0.9 acre (about 1 acre).

1.10 The percentage error is 14.7%; the possible error is $2690 \, \text{m}^3$.

Exercises on thermal radiation

1.11 See the text.

1.12 The radiant emittance ratio is 1.83×10^5.

1.13 (a) 5794 K;   (b) 10.1 μm.

Chapter 2

2.1 (a) See Fig. 2.11.   (b) See the text.

2.2 See the text. The physical dimension of all terms is $\text{ML}^{-1}\text{T}^{-2}$.

2.3 (d) Substituting the values given, and solving the resulting equation, gives $\alpha = 34°$ approximately.

2.4 (i) Cohesive strength $C = 6.51 \, \text{lb wt.} \, \text{in}^{-2}$.
   (ii) $C = 44\,880 \, \text{Pa} = 44.9 \, \text{kPa}$.
   (iii) $\phi = 20.8°$.
   (iv) ϕ would be the same.

2.5 $\phi = 27°$; $C = 250 \, \text{Pa}$.

Chapter 3

3.1 $gV(\sigma - \rho)$.

3.2 The viscosity coefficient is 0.0144 Pa s.

3.3 (b) $Q = 2.945 \times 10^{-9} \, \text{m}^3 \, \text{s}^{-1}$.   (c) $n = 943$.

3.4 (a) See the text.   (b) The fluid velocity is $11.1 \, \text{m s}^{-1}$.

3.5 (b) The Reynolds number is approximately 4, so flow will be laminar.

3.6 (a) $C_{\text{shallow}} = 198 \, \text{m s}^{-1}$.

   (b) The period is 0.468 d.

   (c) $C_{\text{shallow}} = 5.4 \, \text{m s}^{-1}$.

3.7 (b) $C_{\text{deep}} = 0.260 \, \text{m s}^{-1}$. The depth of a shallow-water wave is 6 mm.

3.8 (b) $\theta_r = 20.7°$; $\theta_i - \theta_r = 9.3°$.

Chapter 4

4.1 The equivalent ponded depth is 94.8 mm.

4.2 20.7 mm.

4.3 $127 \, \text{mm h}^{-1}$.

4.4 Use $d\theta$ as the ponded depth of water, D, in depth d.

4.5 (a) See Section 4.5.

   (b) See Table 4.3.

Chapter 5

5.1 (b) 1396 Pa.

5.2

	Higher temperature	Lower temperature
Relative humidity	0.4 or 40%	0.3 or 30%
Saturation deficit	$59 - 24 = 35 \, \text{g m}^{-3}$	$19 - 6 = 13 \, \text{g m}^{-3}$
Dew point temperature	$26 \, °\text{C}$	$3 \, °\text{C}$
Vapour density	$24 \, \text{g m}^{-3}$	$6 \, \text{g m}^{-3}$

5.3 (a) $L_{\text{oe}} = \varepsilon \sigma T_a^4$.   (b) $75.3 \, \text{W m}^{-2}$.

   For (c) and (d), see the following plot.

Table 4.3 Solution to Exercise 4.5(b) (the water-storage capacity (WSC) is 5 in)

Column No.	1	2	3	4	5	6	7	8
Quantity	$\Sigma P \Delta t$	Δe	$\Sigma E \Delta t$	$\Sigma (P-E)\Delta t$	M_{t-1}	$M_t(1)$	$S+U$	M_t
	(in)	(mm Hg)	(in)	(in)	(in)	(in)	(if +ve)	(in)[c]
Comment				C1 − C3[a]	5 in or C8t[b]	C5(t − 1) + C4t	C6 − 5 in	
Month								
Initial (Dec)					5.00			
Jan	7.30	0.49	5.9	1.40	5.00	6.40	1.40	5.00
Feb	7.60	0.51	6.1	1.50	5.00	6.50	1.50	5.00
Mar	8.90	0.60	7.0	1.90	5.00	6.90	1.90	5.00
Apr	8.70	0.59	6.9	1.80	5.00	6.80	1.80	5.00
May	8.10	0.54	6.4	1.70	5.00	6.70	1.70	5.00
Jun	5.10	0.53	6.3	−1.20	3.80	3.80	0.00	3.80
Jul	5.80	0.53	6.3	−0.500	3.30	3.30	0.00	3.30
Aug	6.80	0.53	6.3	0.500	3.80	3.80	0.00	3.80
Sep	7.70	0.54	6.4	1.30	5.00	5.10	0.100	5.00
Oct	12.6	0.51	6.1	6.50	5.00	11.5	6.50	5.00
Nov	13.5	0.40	5.0	8.50	5.00	13.5	8.50	5.00
Dec	13.2	0.38	4.8	8.40	5.00	13,4	8.40	5.00

[a] C refers to column number.

[b] C8t implies that, from January onwards, the value of M_{t-1} is obtained by transfer from column 8 in the same month (t).

[c] If $S+U > 0$, $M_t = 5.0$ in; if $S+U$ is shown as zero, $M_t = C6$.

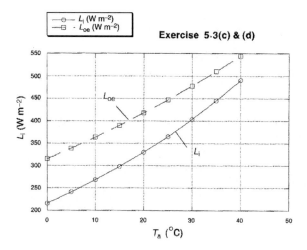

5.4 (a) 68 SU arrives at the earth's surface.

(b) The overall albedo is 0.168, assuming that solar radiation reflected from the earth's surface suffers no loss in its transmission through the earth's atmosphere.

5.5 (b) $E = 1.20 \times 10^{-4}\,\mathrm{kg\,m^{-2}\,s^{-1}}$.

(c) (i) $\Sigma E = 4.05\,\mathrm{mm}$;   (ii) 101.3 tonne.

5.6 (a) $R_n = 365\,\mathrm{W\,m^{-2}}$.

(b) $E = 1.225 \times 10^{-4}\,\mathrm{kg\,m^{-2}\,s^{-1}}$.

The ponded depth evaporated in 9 h is 3.97 mm.

5.7 $E = 2.03\,\mathrm{mm\,d^{-1}}$.

5.8 (a) See Section 5.5.

(b)

Table 5.2 *Hourly calculation of evapotranspiration rate using data from Slatyer and McIlroy (1961), but assuming that $D_0 = D$*

Date	2 April 1961			17 April 1961		
Time (h)	07–08	08–09	09–10	08–09	09–10	10–11
E (mm water d^{-1})	0.6	2.7	4.8	1.7	3.1	3.4
E (lysimeter) (mm d^{-1})	2.1	5.5	7.9	3.7	4.0	4.6
Date	1 July 1961			25 September 1961		
Time (h)	12–13	13–14	14–15	11–12	12–13	13–14
E (mm water d^{-1})	2.6	2.2	1.6	4.6	6.0	6.8
E (lysimeter) (mm d^{-1})	3.7	3.7	4.3	7.6	7.0	10.1

Note. As expected, because of the assumption that $D_0 = D$, the calculated value of E is consistently less than the evapotranspiration measured by the lysimeter.

5.9 See Section 5.6.

5.10 (a) 0.352 m.   (b) 6.72 m.

5.1 Chapter 6

6.1 $I = 3 + 880/\Sigma I \, \text{mm} \, \text{h}^{-1}$.

  I is shown as a function of ΣI in the accompanying Fig. 6.21.

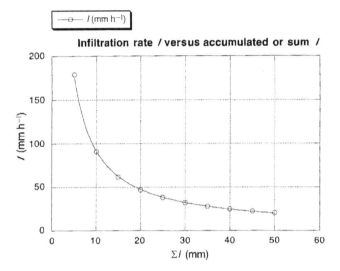

Figure 6.21 For Exercise 6.1.

6.2

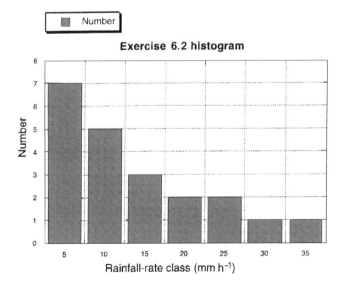

Figure 6.22 For Exercise 6.2.

Figure 6.23 For Exercise 6.2,
fitted with $k = 0.0676$.

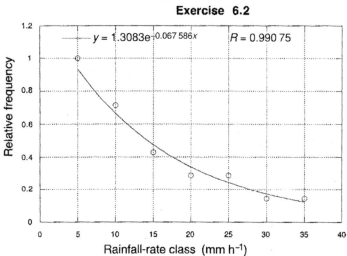

6.3

Figure 6.24 For Exercise 6.3.

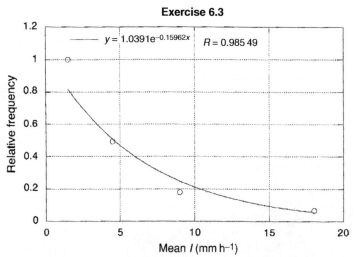

From the curve fit to Fig. 6.24, $I_m = 1/k = 6.3\,\mathrm{mm\,h^{-1}}$.

6.4

Figure 6.25 For Exercise 6.4.

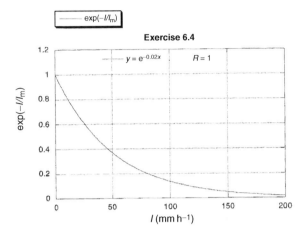

6.5 $\Sigma Q = 42.7$ mm.

6.2 Chapter 7

7.1 $\bar{I} = I_m[1 - \exp(-P/I_m)]$
$= 177.8[1 - \exp(-P/177.8)]$
Neglecting hydraulic lag, Q_{calc} is calculated from $Q_{calc} = P - \bar{I}$ (as in Table 7.1). The calculation for this part of the exercise is given in Table 7.3.

Table 7.3 *Calculation of Q_{calc} (neglecting hydraulic lag) for the data of Exercise 7.1*

i	P_i	$\bar{I}$	Q_{calc}	Q_{meas}
(min)	(mm h^{-1})	(mm h^{-1})	(mm h^{-1})	(mm h^{-1})
70	57.90	49.42	8.48	3.20
71	93.00	72.42	20.6	4.00
72	104.7	79.13	25.6	5.80
73	117.5	85.98	31.5	13.6
74	104.7	79.13	25.6	27.5
75	93.00	72.42	20.6	27.3
76	81.20	65.19	16.0	32.9
77	104.7	79.13	25.6	33.7
78	81.20	65.19	16.0	18.0
79	104.7	79.13	25.6	19.6
80	117.5	85.98	31.5	18.0
81	69.50	57.53	12.0	17.2

Recognising hydraulic lag, $\bar{I}$ is calculated with Q_i calculated from

$$Q_i = 0.78Q_{i-1} + 0.22R_i$$

The calculation is given in Table 7.4.

Table 7.4 *Calculation of Q_{calc} (w. lag) for the data of Exercise 7.1*

i (min)	P_i (mm h^{-1})	$\bar{I}$ (mm h^{-1})	R_i (mm h^{-1})	Q_{calc} (w. lag) (mm h^{-1})	Q_{meas} (mm h^{-1})
70	57.90	49.42	8.48	1.87	3.20
71	93.00	72.42	20.6	5.99	4.00
72	104.7	79.13	25.6	10.3	5.80
73	117.5	85.98	31.5	15.0	13.6
74	104.7	79.13	25.6	17.3	27.5
75	93.00	72.42	20.6	18.0	27.3
76	81.20	65.19	16.0	17.6	32.9
77	104.7	79.13	25.6	19.4	33.7
78	81.20	65.19	16.0	18.6	18.0
79	104.7	79.13	25.6	20.1	19.6
80	117.5	85.98	31.5	22.6	18.0
81	69.50	57.53	12.0	20.7	17.2

Plot Q_{calc} and Q_{calc} (w. lag) against Q_{meas}, and you will see a distinct improvement in calculated or predicted values if the effect of hydraulic lag is recognised.

7.2 The sum of P_i is 1607 mm h^{-1}, so (sum of P_i) × 0.6 h = 964 mm.

The sum of Q_{meas} is 1230 mm h^{-1}, so (sum of Q_{meas}) × 0.6 h = 738 mm.

Hence $\Sigma Q = 947$ mm (which may be compared with the measured value of 738 mm).

7.3 Students will need to locate the relevant hydraulic-engineering guidelines used in their country for flood forecasting, and compare the methodology with the basic structure given in Sections 7.4 and 7.5.

7.3 Chapter 8

8.1 to 8.3   See the text.

8.4 $v = 1.6$ mm s^{-1}.

8.5 Equation (8.10), that $d_i = v_i c_i$, applies at any location where the concentration of sediment size class i remains at c_i during the short period of sedimentation. Mass conservation of sediment in a volume such as that given in Fig. 8.4 implies that the mass of sediment of size class i entering and leaving it must be the same. Research has shown that spatial variation in sediment concentration during erosion occurs only slowly, in fact much more slowly than linearly with time. Thus lateral displacement of sediment during the short period of sedimentation will have no effect on the rate of deposition of sediment onto the soil surface.

8.6 $d = 135$ tonne ha^{-1} min^{-1}.

8.7 $d = 8$ kg m^{-2} s^{-1}. $q_s(\text{IN}) = 1.3 \times 10^{-2}$ kg m^{-1} s^{-1}, which is two orders of magnitude smaller than the rate of deposition, d.

8.8 (a) $c_t = 104.5$ kg m^{-3}.   (b) $q_s = 1.79$ kg m^{-1} s^{-1}.
  (c) The mass deposited is 516 tonne.

8.9 (a) Manning's equation, Eq. (7.7), gives

$$V = (S^{0.5}/n)D^{2/3}$$

Hence

$$V = (\sin 7^\circ)^{0.5} \times (2 \times 10^{-2})^{2/3}/0.3$$
$$= 0.857 \, \text{m s}^{-1}.$$

Thus, from Eq. (8.14),

$$c_t = \frac{0.1}{0.2} \times \frac{2000}{1000} \times 1000 \times \sin 7^\circ \times 0.857$$

(b)
$$= 104.5 \, \text{kg m}^{-3}$$

$$q_s = qc_t$$
$$= DVc_t$$
$$= 2 \times 10^{-2} \times 0.857 \times 104.5$$
$$= 1.79 \, \text{kg m}^{-1} \text{s}^{-1}.$$

(c)

$$\text{Mass deposited} = \text{unit sediment flux} \times \text{perimeter} \times \text{time}$$
$$= 1.79 \, \frac{\text{kg}}{\text{m s}} \times 120 \, \text{m} \times 40 \, \text{min} \times \frac{60 \, \text{s}}{\text{min}}$$
$$= 5.16 \times 10^5 \, \text{kg}$$
$$= 516 \, \text{tonne}$$

8.10 Denote the average total volumetric flux of water collected by f_A (m^3 min^{-1}). Then the average value of the volumetric runoff rate per unit plot area, Q, is given by

$$Q = 3.97 \times 10^{-5} f_A \, \text{m s}^{-1}$$

Each value of Q in the following table persisted for 5 min. Denote the sum of the seven values of Q in the table by ΣQ (m s^{-1}).

Then the total volume of water collected during the entire runoff event of duration 35 min is $4.0\,m^3$.

Period	f_A ($m^3\,min^{-1}$)	Q ($m\,s^{-1}$)	$Q^{1.4}$
1	0.050	1.98×10^{-6}	1.04×10^{-8}
2	0.090	3.57×10^{-6}	2.37×10^{-8}
3	0.15	5.96×10^{-6}	4.84×10^{-8}
4	0.21	8.34×10^{-6}	7.75×10^{-8}
5	0.16	6.35×10^{-6}	5.30×10^{-8}
6	0.10	3.97×10^{-6}	2.74×10^{-8}
7	0.040	1.59×10^{-6}	7.61×10^{-9}
Sum		3.18×10^{-5}	2.48×10^{-7}

Thus $Q_e = 19.4\,mm\,h^{-1}$.

(*Note.* From its defining equation, check that an appropriate unit for Q_e is the same as for Q, namely $m\,s^{-1}$ in SI units.)

8.11 $\beta = 0.9$. If $F = 0.2$, $\beta = 0.76$.

Chapter 9

9.1 The volumetric discharge is $2572\,ft^3\,s^{-1}$.

9.2 The volumetric discharge is $4560\,m^3\,s^{-1} = 9.66 \times 10^6\,ft^3\,min^{-1}$.

9.3 For $G = 500\,l\,s^{-1}$, $G_s = 0.992\,kg\,m^{-1}\,s^{-1}$.
  For $G = 3000\,l\,s^{-1}$, $G_s = 1.85\,kg\,m^{-1}\,s^{-1}$.

9.4 and 9.5 Answers are given in the text.

Chapter 10

10.1 (a) The area is $4.7\,m^2$.

  (b) Because the hydraulic gradient through the sand bed is unaltered.

10.2 The lateral rate of loss of water to either side of the conveying channel is $0.082\,m^2\,d^{-1}$ (a total of $0.164\,m^2\,d^{-1}$).

10.3 (b) The value is as in Exercise 10.2.

10.4 $K = 3.27 \times 10^{-6}\,m\,s^{-1} = 11.8\,mm\,h^{-1}$.

10.5 $h_2/h_1 = 40$.

10.6 (a) $q = -K_s(\Delta H_1/z_s)$.

  (b) Also $q = -K_c(\Delta H_2/z_c)$, so
    $z_c/K_c = z_s\,\Delta H_2/(K_s\,\Delta H_1)$

(c) Determining K_s from an extracted column of lake sediment,
with dimensions measured, would allow the calculation of K_c,
and hence the loss of water from the lake by deep drainage.

10.7 The basin water volume is 0.172×10^{12} acre foot.

10.4 Chapter 11

11.1 The figure follows:

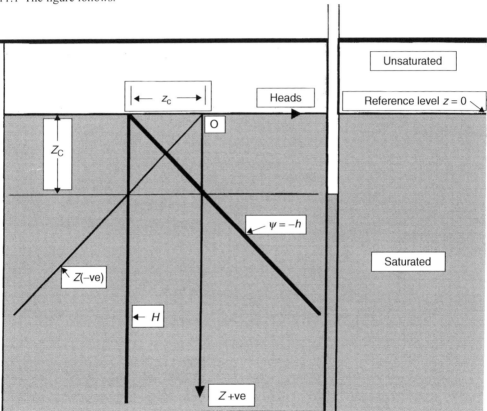

11.2

$$q = -K(\theta)\frac{\Delta\psi}{\Delta z} - K(\theta) \quad (z \text{ positive upwards})$$

11.3 (a) At $z = 0.2$ m, $H = -1.7$ m.
At $z = 0.4$ m, $H = -2.4$ m.
(b) Downwards.
(c) $q = 3.5 \times 10^{-7}$ m s^{-1}
$= 30.2$ mm d^{-1}

Table 12.3 *Example 12.2: calculation of displacement of solute concentration peak using the data of Table 12.2 (all lengths are in cm)*

	1	2	3	4[a]	5	6	7	8	9	10
								$\theta_{fc} \Delta\alpha_n$	$\Delta\alpha_n$	
Time period	i_n	e_n	r	$\dfrac{i_{n-1} - e_{n-1}}{r}$	$\theta_{n-1} = \theta_{n-2} + \mathbf{4}$	$\theta_{fc} - \theta_{n-1} + \dfrac{e_n}{r}$	$\mathrm{Min}(\alpha_{n-1}, r)$	$\mathbf{1} - (\mathbf{7} \times \mathbf{6})$	$\mathbf{8}/0.1$	$\alpha_n = \alpha_{n-1} + \Delta\alpha_n$
25–30 April	0.5	0.2	10	$\dfrac{(0.5-0.2)}{10} = 0.03$	0.1	0.02	0	0.5	5	5
Week 1, May	4.3	0.2	10	0.41	0.13 → 0.1	0.02	5	4.2	42	47
Week 2, May	4.3	0.2	10	0.41	→ 0.1	0.02	10	4.1	41	88
Week 3, May	4.3	0.2	10	0.41	→ 0.1	0.02	10	4.1	41	129
Week 4, May	4.3	0.2	10	0.41	→ 0.1	0.02	10	4.1	41	170

[a] **4** refers to column number 4.

11.4 (a) From the given relationship,

$$\ln K = \ln K_s + k \ln \theta - k \ln \theta_s$$
$$= k \ln \theta + \text{constant}$$

Hence a plot of $\ln \langle K_z \rangle$ against $\ln \langle \theta \rangle$ would be expected to yield a straight-line relationship such as those shown in Fig. 11.15(a).

11.4 (b) $k = 7.66$.

11.5 From Fig. 11.10, $\theta_{\text{fc}} = 0.5$ (non-dimensional).

11.6 The total amount of evaporation is 5.1 mm.

11.7 $K = 3.27 \times 10^{-4}\,\text{m}\,\text{d}^{-1} = 3.78 \times 10^{-7}\,\text{cm}\,\text{s}^{-1}$.

11.8 Whilst there is no particular solution to this exploratory exercise, it is suggested that, for simplicity, equal scales be used for the downward-positive axis using the soil surface as datum and the (positive) suction head, h, plotted as in Fig. 11.13.

11.5 Chapter 12

12.1 (a) See the text. (b) See Fig. 12.1 and the associated text.
(c) See Section 12.3. (d) The time required is 1000 y.

12.2 See Table 12.3.

12.3 (a) 15 d.   (b) 96 d.

Index

absolute temperature 19
actual mean flow velocity of water in soil 331
adsorbed cations 11
aeolian processes 7, 16
agroclimatic assessment 142
agroforestry 387
agronomic production 142
alluviation 7, 16, 259
angle of repose 59, 61
    angle of internal shearing resistance or
        friction angle 62
aquifer
    aquitard and aquiclude 340
    artesian well 347
    confined aquifer 340, 341
    freshwater lens 188
    salt-water intrusion 347
    unconfined aquifer 326
Archimedes' principle 88
arid zone 180, 208
    banded vegetation 208
artesian basins 341, 342
atmosphere 6
    atmospheric humidity 157
    atmospheric pressure 87, 158
Atterburg limits 76
    liquid limit 77
    plastic limit 76
available water content 145

beach erosion 119
Bernoulli's equation 103–107, 115, 328,
    334
biopores 202, 205
biosphere 6
black body 24, 167
Bowen ratio 171–172, 183
breakthrough curve 403
Brownian motion 12
Bryce Canyon 4

capillarity and surface tension 72–75, 85,
    357
    capillary fringe 327, 352–353, 356
    capillary pull 7
carbon dioxide
    a greenhouse gas 18, 26, 27
    absorbed in photosynthesis 3, 154
    concentration in soil 350
    concentration in the atmosphere 18
carnivores 156
catchment 293 (*also see* watershed)
    catchment health 291
cations 11
    attracted to humus 13
    calcium cation 12
    cation exchange 12
    hydrated ions 12
chemical transport (*see* nutrient and other chemical
    transport)
Chezy 237
    Chezy's C
clay
    characteristics of 11–13
    dispersion or deflocculation 13
    double layer 12
    formation 9–13
    kaolinitic clays 11
    layer-lattice clays 11
    secondary minerals 9
    suspension 12
coastline–ocean interactions 116–120
cohesive soils 68–69
cohesive strength of soil 62, 68, 73
colloids 11
    effect of electrolyte concentration 13
    flocculated state 13
    short-range or van der Waals attractive
        forces 12
compaction of soil 59
consistency of soil 76–79

contaminant
    concentration 399
    contaminants in groundwater 326
    soil-related contaminants 307–311
contaminant displacement and transport 382–383
    advection (or convection) 395
    breakthrough curves 392, 399, 403
    convection and dispersion equation 402
    dispersion coefficient 395, 401
    dispersion during contaminant transport 393, 395, 399,
        401–403
    dispersivity (or coefficient of dispersivity) 396, 401
    hydrodynamic dispersion 393, 401
    in agricultural contexts 404–414
    in the unsaturated zone 397–403
    ion-attraction effects 403
    ion-exchange effects 403
    miscible displacement 391, 392, 393, 397
    molecular diffusion 401
    piston flow 393, 398
    preferential flow 393, 403
contour hedgerow or buffer strip (see soil-conservation
        methods)
convection 22, 169, 183
    convective turbulence 180
    forced convection 183, 184
    free or natural convection 170
conversion factors for units 35–37, 419
Coulomb's failure law 68, 74
crop-yield variation 142
cyclic salt 382
cycling of chemical elements 6

Dalton 161
    Dalton-type evaporation equation 162, 175–177
Darcy's law 332–340
    Darcy flux or unit discharge 335, 347
    in the unsaturated zone 364
decomposers 155
depositability 269
deposition rate 263, 268
dew-point temperature 161
drainage basin (of a river) (see watershed)
drainage flux 145, 369
Dupuit–Forschheimer assumption 338, 344

eddies 22, 168, 182
    eddy diffusivity in air 170
elevation head 329, 330, 354
emissivity 167

energy and fluid flow 99–107
    energy conservation 101
    frictional forces 100
    heat energy 100
    kinetic energy 101–103
    potential energy 99–103
    work done 99
energy-conservation principle 163, 164
energy budget 20–23
energy exchange at the earth's surface 18–29, 153–185
    night-time energy exchange 23
enrichment ratio 308
environmental concerns
    agricultural pollutants 404–414
    associated with liquid flow 85–86
    beach erosion 119
    carbon dioxide, sequestration by soils 261
    enhanced greenhouse effect 27–29
    flooding and flood prediction 246, 248
    groundwater quality 326, 338, 391–397
    involving soils 16–18, 79–80
    land degradation 16–18, 79–80
    over-clearing of forests 292, 306
    over-pumping of groundwater 325, 341
    ozone hole 26
    pollutants 86, 307–311, 326, 351, 391–403
    river health 316–319
    salinity 351, 380–382
    soil erosion by water 259–284, 292
    tsunami 110
    waste dumping 110
    water-stress effects on plant growth 146, 364
equipotential surfaces 338
    flow net 339
equivalent ponded depth of water 136
error and data limitations 40–42
    in estimating water storage in the soil profile 130
    rules of error calculation 41
essential elements for plant growth 154
evaporation (also see evapotranspiration)
    definitions 21
    evaporation flux density 169, 170
    evaporation pan 161
    from an open water surface 20, 84, 110, 161–162
    from the soil 128, 369, 370–372, 374
    potential evaporation rate 179
evapotranspiration 21, 128, 139, 146, 153–182
    comparison of measurement methods 174
    measurement using energy conservation 162–175
    measurement using standard meteorological data 175–182

excess rainfall rate 209–212, 219
exchangeable-sodium percentage (or ESP) 382
exfiltration 227, 248
exponential function 217

field capacity 144, 362, 405, 410
fingers and fingering in water flow 376
flood prediction 246, 248, 257
flow concentration and rilling 282–283
flow of fluids
    around submerged solids 107–110
    boundary-layer concept 93
    coefficient of dynamic viscosity 93
    drag coefficient 108
    eddies in turbulent flow 94, 168
    eddy shear stress 95
    form drag 107
    kinematic eddy viscosity or eddy diffusivity 95
    kinematic viscosity 95
    laminar flow 91, 169
    Reynolds number 95–97, 108
    skin friction 107
    Stokes range 108
    turbulent flow 91
    viscosity effects
fluid density 86, 91
fluid pressure and buoyancy 86–90
    buoyancy force 88
food chain 154
force
    lift force 107
    drag force 107
fractional (or partial)-area runoff concept 212, 221

geographical-information-system (GIS) methodologies 300,
    307
geostatistics and geostatistical techniques 207
glaciation
    as a process 16–18
    glacial period 8
grassed waterway 241
gravimetric soil water content 57
gravitational systems of units 35
Green–Ampt infiltration equation 204, 369
greenhouse effect 18, 26, 27
    enhanced greenhouse effect 27–29
groundwater
    aquifer 324
    as a major source of streamflow 325
    as a source of potable water 325

at equilibrium 327–330
confined aquifer 328
groundwater flow 324–342
groundwater table (*see also* water table) 325
groundwater zone 127, 324
hydrogeological structures 340
in natural subsurface formations 340–342
over-pumping 325
quality 325
unconfined aquifer 328
GUEST erosion-prediction methodology 283–284

head
    elevation (or gravitational) head 330, 354, 367
    hydraulic head 330, 354, 367, 371
    pressure head 330, 353, 354, 358–359, 367
    suction head 355, 356, 367
heat-energy 156
    advected flux of heat energy 165
    sensible heat 164
heat flux
    heat-flux plate 168
    into the atmosphere 164, 170, 182–185
    into the ground 164, 168, 185–189
    non-radiative sensible heat flux into the atmosphere
        182–185
Hortonian flow 208
humidity
    absolute humidity 158
    saturation and saturation vapour pressure 159
    specific humidity 158
humid zone 209, 296, 297
humification 13
    humus 13, 51
hydraulic
    conductivity 333, 365, 372–375
    gradient 332
    head 329, 330, 354
hydrogen-ion concentration (*see* pH)
hydrograph 214, 294, 296
hydrological cycle 197–199
hydrological (or hydraulic) lag 229, 239, 247, 249–254, 295
hydrological year 139
hydrology at watershed and catchment scales 292–300
hydrosphere 6, 19
hyetograph 214, 294

ideal-gas law 158
infiltration 52, 126, 197–199, 248–249, 407
    amount 410

at a range of scales 199
effect of surface sealing 17
effects of soil cracks and worm holes 202
into small areas 199–207, 368–369
infiltration-excess overland flow 208
infiltration rate 141, 199
actual and potential rate 210
apparent infiltration rate 215
case studies at the plot scale 213–216, 248
field determination 248
flux-controlled 200
general model of spatially variable infiltration 216–220
measurement by rainfall simulation 202, 203
spatial mean infiltration rate 219
spatial variability 207–222
infiltrometer ring (or infiltrometer) 200–201, 207, 368
internal-drainage method of hydraulic-conductivity
measurement 372
ion exchange 14
Isaac Newton (*see* Newton)

James Joule 99
Joseph Fourier 27, 28

land degradation 16–18
landslides 62
rapid mass movement 64
latent heat of vaporisation of water 20, 160, 164
leaching 14, 15, 326, 389, 397, 404
leachate 399
leaching requirement 388
liquid behaviour 84, 110
environmental significance 85–86
liquids in motion (or hydrodynamics) 90–98
lithosphere 2, 6–23
buckling process 2
earth's mantle 2
lithospheric or tectonic plates 2, 3
molten magma 3
transverse fracture zones 2
volcanoes 2
long-shore sediment transport 118
lysimeter 141, 180, 397, 399, 408

macropores 202, 205
Manning 237
Manning's equation 238
Manning's *n* 238, 240
mass conservation of water 130, 242, 369–370
mass movement of soil 50

mathematical modelling 216–218, 246, 249
parameter optimisation 251
mechanical analysis of soil 52–53
environmental implications 79–80
soil texture 53
mechanics 48–51
momentum 48–49
statics 48
meteorological data 175
micrometeorology 169–171
measurement of heat and water flux into the atmosphere
170–171
miscible displacement (*see* contaminant displacement and
transport)
MODFLOW 340
moisture characteristics (*see* soil moisture
characteristics)
mole (written mol) 158
molecular diffusion 392
momentum flux 170

natural erosion 5, 17
net radiation flux density 165
net radiometer 166
Newton, Isaac 31
first law of motion 31, 48
second law of motion 32, 48
third law 48–49
shear stress in fluids 93
nitrogen
conversion into different forms 309
displacement of the nitrate ion in soil 405–414
mineralisation 309
nitrate concentration 383, 405
nitrate concentration and human health 326
nitrate ion 6, 351, 404
nitrification and de-nitrification 309
normal flow 236
nutrient and other chemical transport
enrichment during erosion 308
in watersheds 307–311
transport of soluble chemicals 309–311
transport of sorbed chemicals 307–309
nutrient cycling 10, 155, 309

oceans and waves 110–120
mass-transporting currents 111
tides 111
organic matter in soil 51
overconsolidation of clay soils 58

overland flow 226–254
  dynamics 242–248
  resistance 234–235, 242
  routing parameter 251
oxygen
  an air component 3
  cycle 155
  released in photosynthesis 5
ozone 26
  hole in ozone layer 26

pan evaporation rate 140
peak lag time 214
perenial vegetation 384
permanent wilting point 363
permeameter
  falling-head type 345, 346
pF 361, 362
pH 14
photosynthesis 3, 5, 20, 26, 154
photosynthetically active radiation 26
piezometer 328, 340
  piezometric surface 340, 341
piezometer tube 328, 340, 347
plane of failure 62
planet earth 1–2
plant-available water store 129, 364
plant growth and water accounting 142–144,
  146
Poiseuille's equation 205, 365
pollutants 86, 110, 128, 156, 310, 317, 382
ponding of water 203, 205
pore space 54, 331
  porosity 331
pore water pressure 75
precipitation and its measurement 20, 197–199
preferential flow 207, 351, 393, 403
pressure
  dynamic pressure 99–103
  overburden pressure 54
  pressure head 329, 330, 353, 354
  static pressure 105
Priestley–Taylor equation 179
principle of conservation of energy 101
principle of mass conservation (or continuity)
  104
psychrometer (or hygrometer) 158
  dew-point hygrometer 161
  wet-and-dry-bulb hygrometer 158
psychrometric constant 159

radiant emittance 24
radiation
  absorption 26
  diffuse radiation 20
  distinction between short- and long-
    wave 24–26
  global radiation 20
  global short-wave radiation 164
  incoming long-wave radiation 167
  infrared radiation 19
  long-wave radiation 19–28, 164
  net long-wave radiation 167
  outgoing emitted terrestrial radiation 167
  Planck's law 24, 44
  reflection coefficient (or albedo) 164
  short-wave radiation 19–26, 164
  solar radiation 22–26
  spectral radiant emittance 24
  terrestrial radiation 25
  Wien's law 44
  window of radiation exchange 26
raindrop impact on soil 261
rainfall amount or depth 131, 133
rainfall rate or intensity 131
  rain gauge 131
rainfall threshold prior to runoff 228
relative humidity 192
renewable energy 28
respiration 185
Reynolds number 95–97, 109
Richardson number 184, 185
ring infiltrometer (see infiltrometer)
riparian zone or strip 304, 310
rivers 311–319
  agricultural development effects 318
  alluvial rivers 311
  bedload 314
  dams, effects on 312
  dissolved load 315
  environmental flows 317
  eutrophication 309
  flooding 312
  form resistance to flow 313
  grain resistance to flow 313
  higher-order streams 314
  hydraulic radius of flow 312
  in-stream biota 311
  over-allocation of river water 318
  rotational flows 315
  saltation load or wash load 315

sand or sediment slugs in 318
   sediment transport in 316
   suspended load 315
   turbidity 318
river basin 292
   alluvial deposits 311
   riparian zone 318
   river management 317
river flow or discharge 137, 294
   base flow 294
river health 316–319
   human health risks 317
   in-stream and riparian communities 316
rock cycle 2–5
   crustal uplifting 3
   geology 2
   igneous rocks 3, 5, 8
   lithification 3
   mass wasting 3
   metamorphic rock 4
   radiogenic heat 4
   rock weathering 3, 6, 9
   sedimentary rock 3, 5, 8
   weathering products 6
root water uptake 129, 142, 144, 358, 405, 406, 407
root zone 145, 405, 406
runoff 134
   effective runoff rate per unit area 279
   partial-area concept of runoff generation 221
   rate measurement by tipping-bucket device 134
   rate per unit area 242
   runoff-model components 228
   runoff modelling 228–254
   routing 229

saline stream 380
salinity 15, 18, 351, 380–382
   dryland salinity 381, 383–388
   irrigation-induced salinity 381, 388–391
   primary salinity 381
   salt seeps 384
   secondary salinity 381–382
saturated zone (in soil profile) 127, 209
saturation deficit 161, 162
saturation overland flow 209, 254
saturation ratio 361
saturation vapour pressure (see humidity)
scale effects on model use 310
scale-model design 98

SCS curve-number method of watershed-runoff prediction 256, 297–300
secondary minerals 10–11
sediment 59
   deposition characteristics 266–270
   transport in watersheds 300–307
sediment concentration 264
   event-average value 278
sediment flux 264
sedimentation and sediment deposition 262, 267
settling-velocity characteristics 263, 268
   average settling velocity 268
snow accumulation 132, 139
soil
   aggregation 15, 17, 52, 206
   average density 55
   bulk density 56
   factors of soil formation 7–8
   formation processes 8–16
   fractions 52
   heat flux in 168
   horizons 7, 14, 15
   matrix 206
   morphology 7
   pedological organization 206
   pedology 7, 15
   physical characteristics 51–59
   porosity 56
   profile 7, 9
   profile development and soil horizons 7, 14–15
   rock weathering and clay formation 9–13
   temperature 185–189
   the interface of earth environments 5–8
   type 7
   voids ratio 56
soil-conservation methods 303–306
   buffer strips or contour hedgerows 303–304, 305, 310
   minimum tillage 306
   protective surface cover 306
   riparian strip 303–304, 310
   terraces 303–304
soil creep 65
soil-erodibility factor 277, 280, 281
soil erosion (see also water erosion and deposition)
   and deposition studies using isotopic tracers 307
   and surface transport 15–16
   natural erosion 5, 17
   rate 270–273
   theory 273–278

soil moisture characteristics 359–364
  hysteresis in characteristics 359, 361
  pressure-membrane apparatus 361
  suction plate 360
soil strength 46–51, 59–71, 83
  Coulomb's failure law 68, 74
  effects of water content 72–76
  elastic limit 76
  shear strength 66, 67
  tillage effects 47
  unconfined compressive strength
soil water
  in equilibrium 356–364
  measurement of suction or pressure head 358–359
  movement in the unsaturated zone 350–376
  profile storage 135–137
  suction head 355, 356
  suction or negative pressure 73, 126, 352–353, 356,
    357
  units of head or pressure 362–364
solarimeter 26
  Campbell–Stokes recorder 26
solute transport in agricultural contexts 404–414
  movement of the solute peak concentration 405–414
solutes and solute transport 397
specific gravity 89
specific heat of air 159
spectral radiant emittance 24
static coefficient of friction 61
Stefan–Boltzman law 43
Stevenson screen 158, 175–177
strain 64
  tensile strain 64
stress 50, 63
  compressive stress 64
  effective stress 74–75
  normal stress 54
  shear stress 64, 66, 240
  tensile stress 63
Stokes' law 267
stomata 157
stream power 241, 261
  effective excess stream power 274
  excess stream power 273
  threshold stream power 273
stream tube 104
suction plate 360
surface
  detention 138
  runoff 145

  seal 222
  transport processes 15–16
sun 19, 20
  solar energy 154
suspended load 267
sustainable production 144

temperature
  absolute temperature 19
  diurnal variation in soil surface 188
  gradient 186
  phase delay in soil 188
  profile in soil 187
tensiometer 358, 359
terminal velocity of settling 266
thermal
  capacity 187
  conduction 22, 186
  conductivity 187
  diffusivity (or thermometric conductivity) 188
thermal radiation 19–29
  light 19
  long- and short-wave radiation 24–26
  part of the electromagnetic spectrum 19
  visible waveband 19
thermodynamic laws 153
time of concentration of water flow 294
tipping-bucket rainfall-rate recorder 131
translocation 9
transmission zone in soil 200
transpiration 20, 23, 157, 163
transport-limit sediment concentration (in soil erosion)
    273
  theoretical expression for sediment concentration 275–276
tsunami 110
turbulence
  convective 180, 183
  mechanical 180, 183

ultraviolet radiation 26
units and conversions 29
universal gas constant 158
unsaturated zone 127, 327
US Weather Bureau Class-A evaporation pan 140

vadose zone 327
valency 11, 12
vapour density of water 158
vapour-phase movement of water in soil 186
vapour pressure of water 157–161

vector and scalar quantities 32–34
vegetation
    nutrient uptake 9–10
    stomata 21
    toxic effects 14
    water-deficit effects 171
vegetation-based ecosystems 153–157
viscometer 92
volume flux of water 234, 331
    unit flux 234–235, 242
volumetric soil water content 56, 135

water
    content 56, 192
    content measurement 359
    dissociation of 14
    earth/atmosphere exchange systems 18–23
    movement through the unsaturated zone 350–376
    polar nature of 12
    quality 85, 338
    quality and eutrophication 310
    quality for human use 392
    storage capacity (WSC) 145
    vapour 26, 84, 157
    water budget, water balance, or water accounting 129, 130,
        137–149, 412
water, erosion and deposition by 259–284
    averaging soil loss over an erosion event 278–282
    entrainment 263
    flow-driven erosion 263
    gully erosion 284
    inter-rill areas 282
    off-site effects 261
    on-site effects 260
    rainfall-driven erosion 282
    re-entrainment 263, 273

    saltation and saltation load 263, 267
    use of caesium-137 in evaluation of net erosion/deposition
        307
watershed (see also catchment) 125–149, 245, 293
    definition 292
    hydrology 259
watersheds and rivers 289–319
watertable (or groundwater table) 326, 327, 354, 384
waveband 24
wavelength 24, 26, 112
waves (see also oceans and waves)
    deep-water or short waves 114
    deep-water wave velocity 115
    longitudinal travelling wave 111
    phase of a wave 112
    shallow-water waves 116
    shallow-water wave velocity 116
    stationary or standing wave 111
    transverse waves 110
    travelling or progressive wave 111
    wave amplitude 112
    wave frequency 112
    wave height 112
    wave period 112
    wave refraction 117
    wave velocity, celerity, or phase velocity 113
weathering and clay formation 9–13
weight and gravitational units 34–35, 50
WEPP erosion methodology 284, 301, 307
wetted perimeter of river 313
wetting front 200–201, 368, 407, 410
wilting-point water content 144

Young's modulus 64

zero-flux interface 371, 374

An Introduction to the Environmental Physics of Soil Water and Watersheds

BY CALVIN W. ROSE

ACKNOWLEDGEMENTS

Cover image: Furrow irrigation of newly planted sugar cane crop near Clare (Queensland, Australia.) © CSIRO Land and Water.

Chapter 1 frontispiece: Meteosat image of the world from space. (Provided by a suspended image service provided by the University of Nottingham.)

Fig. 1.2: Natural erosion at Bryce Canyon, Utah, USA. (Photo by the Author.)

Fig. 1.3: Soils have developed and continue to change at the interface of major earth environments. (After McTainsh and Boughton (1993) in Land Degradation Processes in Australia, Longman Cheshire.)

Fig.1.8: Components of the daytime heat-energy exchange at the earth's surface. (From Rose (1979) in Agricultural Physics, reproduced with permission from Pergamon Press.)

Fig 1.9: Components of the night-time heat exchange at the earth's surface. (After Rose (1979) in Agricultural Physics, reproduced with permission from Pergamon Press.)

Chapter 2 frontispiece: Landslide in Nepal. (Reproduced with permission from Australian Associated Press Ltd.)

Fig 2.2: Soil aggregates or crumbs, with water within and between adjacent aggregates. (After Rose (1979) in Agricultural Physics, reproduced with permission from Pergamon Press.)

Fig. 2.20: Effects of suction and degree of saturation and stress on the effective stress of a beach sand drying from saturation. (From Marshall, Holmes and Rose (1996) in Soil Physics, 3rd Ed, p238; reproduced with permission from Cambridge University Press.)

Fig. 3.12: The relationship between the drag coefficient for a sphere and the Reynolds number. (After Monteith and Unsworth (1990) in Principles of Environmental Physics, 2nd Ed, Edward Arnold.)

Photograph on page 62: Children standing near the edge of a landslide in East Timor. (Reproduced with permission by Lisette Wilson and Kevin Austin, taken when they were with the United Nations Transitional Administration in East Timor. Lisette is currently with the World Wide Fund for Nature, South Pacific Program.)

Fig. 4.4: A runoff plot defined by boundaries across which there is no flow except at its lower end. (After Rose (1993) Chapter 14 in Hydrology and Water Management in the Humid Tropics, eds Bonnell, Hufschmidt and Gladwell, p321; reproduced with permission from Cambridge University Press.

Fig. 4.8: The relationship between the ratio evapotranspiration/loss and the dried mass per hectare of a growing tropical legume pasture crop. (After Rose et al. (1972) in Agricultural Meteorology (now Agriculture and Forest Management) Vol 10, p167; reproduced with permission from Elsevier.)

Fig. 4.9: A comparison of estimated and observed changes in availability of soil water under a wheat-fallow sequence. (After Fitzpatrick and Nix (1969) in Agricultural Meteorology (now Agriculture and Forest Management) Vol 6, p317; reproduced with permission from Elsevier.)

Fig. 4.10: Relationships between the yield of grain sorghum and a computed water-stress index for five grain sorghum varieties. (After Nix and Fitzpatrick (1969) in Agricultural Meteorology (now Agriculture and Forest Management) Vol 6, p317; reproduced with permission from Elsevier.)

Chapter 5 frontispiece: Coastline, Wellington Point, Queensland, Australia. (Photo by the Author.)

Fig. 5.4: The sensor head of a net radiometer. (After Szeicz (1975) in Vegetation and the Atmosphere, Vol 1, ed Monteith; reproduced with permission from Academic Press.)

Fig. 5.7: The history of variation on an almost cloud-free day of the components of the energy-budget equation. (After Rose et al. (1972) in Agricultural Meteorology (now Agriculture and Forest Management) Vol 9, p392; reproduced with permission from Elsevier.)

Fig. 5.16: The relationship between temperature and vapour density and relative humidity at sea level at atmospheric pressure. (After Campbell (1988) in An Introduction to Environmental Biophysics, 2nd Ed, p23; reproduced with permission from Springer-Verlag.)

Fig. 6.12: Illustrating the response of the average field infiltration rate to the rainfall rate. (After Rose (1985) in Advances in Soil Science, Vol 2, p10; reproduced with permission from Springer-Verlag.)

Fig 6.14: The apparent infiltration rate as a function of the rainfall rate during a thunderstorm. (After Yu et al. (1997) in the Transactions of the American Society of Agricultural Engineers, Vol 40 (5) p1297; reproduced with permission from the American Society of Agricultural Engineers.)

Fig 6.17: The relationship between the spatial mean infiltration rate and the rainfall intensity. (After Yu et al. (1997) in the Transactions of the American Society of Agricultural Engineers, Vol 40 (5) p1299; reproduced with permission from the American Society of Agricultural Engineers.)

Chapter 9 (1st) frontispiece: Watershed with pineapple production, south-east Queensland, Australia. (Photo by the Author.)

Chapter 9 (2nd) frontispiece (also Fig 9.16): Ringarooma River, Tasmania, Australia. (Reproduced with permission by Dr Rebecca Bartley of CSIRO Land and Water, Australia.)

Fig. 9.9: The relationship between the amount of prior 10-d rainfall and the initial infiltration amount. (From Yu et al. (2000) in the Soil Science of America Journal Vol 64; reproduced with permission from the Soil Science Society of America.)

Fig. 10.10: Schematic cross-section of an unconfined aquifer. (After Bouwer (1978) in Groundwater Hydrology, Fig. 1.2; reproduced with permission from The McGraw-Hill Companies.)

Fig. 11.7: A suction plate with ancillary suction-control equipment. (After Rose (1979) in Agricultural Physics, reproduced with permission from Pergamon Press.)

Fig. 11.9: Desorption moisture characteristics for a range of soil types in the USA. (After Hanks (1980) in Applied Soil Physics, 2nd Ed, p42; reproduced with permission from Springer-Verlag.)

Fig. 11.14: Profiles of the volumetric water content, measured on the number of days after drainage. (After Olsson and Rose (1978) in the Australian Journal of Soil Research Vol 16, p174; reproduced with permission from CSIRO Publishing.)

Fig. 11.15: a) Relationships between the hydraulic conductivity and the volumetric water content. b) Relationships between the hydraulic conductivity and the in situ suction head. (After Olsson and Rose (1978) in the Australian Journal of Soil Research, Vol 16, p175; reproduced with permission from CSIRO Publishing.)

Chapter 12 frontispiece: A saline stream, Quairadong, Western Australia. (Copyright CSIRO Land and Water, Australia.)

The author would also like to acknowledge the expert assistance of Mr. Walter Mack and Mr. Gerry Loiacono in preparing much of the artwork for the book.

Made in the USA
Las Vegas, NV
28 January 2022

42488335R00251